D1191971

Even the casual observer of nature soon realizes that there are palpable differences in the breadth and diversity of resources used by species, even quite closely related ones. Species also show disparate propensities to occupy habitat types, some restricted to a very few, while others are to be found in almost any habitat within their geographic ranges. The variation in the breadth of resource use (ecological versatility) and in habitat use (ubiquity) has important implications for understanding ecological diversity. This book is the first to draw back from particular disciplinary foci, such as host-plant use in phytophagous insects, bilateral mutualisms or competitive coevolution, to develop a broader perspective of versatility and ubiquity. This is done by addressing three main questions (1) how do ecologists study versatility and ubiquity, and what do we know from these studies? (2) how well does existing theory account for observations, and what are the common threads between disciplines? and (3) what is the relationship between versatility and ubiquity? The analyses are undertaken from an ecological rather than evolutionary perspective. The outcomes of the review indicate some promise of unification and systematicization. However, there are exceedingly demanding challenges that ecologists must face in their quest for a more thorough understanding of ecological versatility.

Ecological Versatility and Community Ecology

Cambridge Studies in Ecology presents balanced, comprehensive, up-to-date, and critical reviews of selected topics within ecology, both botanical and zoological. The Series is aimed at advanced final-year undergraduates, graduate students, researchers, and university teachers, as well as ecologists in industry and government research.

It encompasses a wide range of approaches and spatial, temporal, and taxonomic scales in ecology, experimental, behavioural and evolutionary studies. The emphasis throughout is on ecology related to the real world of plants and animals in the field rather than on purely theoretical abstractions and mathematical models. Some books in the Series attempt to challenge existing ecological paradigms and present new concepts, empirical or theoretical models, and testable hypotheses. Others attempt to explore new approaches and present syntheses on topics of considerable importance ecologically which cut across the conventional but artificial boundaries within the science of ecology.

CAMBRIDGE STUDIES IN ECOLOGY

ALSO IN THE SERIES

H.G. Gauch, Jr	*Multivariate Analysis in Community Ecology*
R. H. Peters	*The Ecological Implications of Body Size*
C. S. Reynolds	*The Ecology of Freshwater Phytoplankton*
K. A. Kershaw	*Physiological Ecology of Lichens*
R. P. McIntosh	*The Background of Ecology: Concepts and Theory*
A. J. Beattie	*The Evolutionary Ecology of Ant–Plant Mutualisms*
F. I. Woodward	*Climate and Plant Distribution*
J. J. Burdon	*Diseases and Plant Population Biology*
J. I. Sprent	*The Ecology of the Nitrogen Cycle*
N. G. Hairston, Sr	*Community Ecology and Salamander Guilds*
H. Stolp	*Microbial Ecology: Organisms, Habitats and Activities*
R. N. Owen-Smith	*Megaherbivores: The Influence of Large Body Size on Ecology*
J. A. Wiens	*The Ecology of Bird Communities*
N. G. Hairston, Sr	*Ecological Experiments*
R. Hengeveld	*Dynamic Biogeography*
C. Little	*The Terrestrial Invasion: An Ecophysiological Approach to the Origins of Land Animals*
P. Adam	*Saltmarsh Ecology*
M. F. Allen	*The Ecology of Mycorrhizae*
D. J. Von Willert *et al.*	*Life Strategies of Succulents in Deserts*
J. A. Matthews	*The Ecology of Recently-deglaciated Terrain*
E. A. Johnson	*Fire and Vegetation Dyanmics*
D. H. Wise	*Spiders in Ecological Webs*
J. S. Findley	*Bats: A Community Pespective*
G. P. Malanson	*Riparian Landscapes*
S. R. Carpenter & J. F. Kitchell (Eds.)	*The Trophic Cascade in Lakes*
R. J. Whelan	*The Ecology of Fire*

Ecological Versatility and Community Ecology

RALPH C. MAC NALLY
The Department of Ecology and Evolutionary Biology
Monash University, Melbourne, Australia

CAMBRIDGE
UNIVERSITY PRESS

Published by the Press Syndicate of the University of Cambridge
The Pitt Building, Trumpington Street, Cambridge CB2 1RP
40 West 20th Street, New York, NY 10011–4211, USA
10 Stamford Road, Oakleigh, Melbourne 3166, Australia

First published 1995

Printed in Great Britain at the University Press, Cambridge

A catalogue record for this book is available from the British Library

Library of Congress cataloguing in publication data

MacNally, Ralph C.
Ecological versatility and community ecology / Ralph C. MacNally.
p. cm. – (Cambridge studies in ecology)
Includes bibliographical references (p.) and index.
ISBN 0 521 40553 X (hardback)
1. Ecology 2. Biotic communities. 3. Habitat (Ecology)
I. Title. II. Series.
QH541.M225 1995
574.5–dc20 94-41872 CIP

ISBN 0 521 40553 X hardback

SE

For Erica, Aleck and
especially Marty

Contents

Preface

When Professor John Birks kindly offered me the opportunity to submit an outline for a book for the *Cambridge Studies in Ecology* series, I decided that it would be worthwhile to analyze comprehensively ecological specialization and generalization in natural communities (commonly referred to as *niche breadth*, or *niche width*). Of course, there has been no shortage of review articles on particular groups of organisms, especially insects, looking at this question (e.g., Fox and Morrow 1981, Schemske 1983, Berenbaum 1990, Jaenike 1990, Andow 1991). Nor has there been any lack of theoretical attention (e.g., MacArthur and Levins 1967, Van Valen and Grant 1970, Roughgarden 1972, Slatkin and Lande 1976, Keast 1977, Siegismund *et al.* 1990). Futuyma and Moreno (1988) provided an excellent short review of this topic from an evolutionary perspective. However, it seems that a more extensive treatment of reasons for specialized or generalized resource use and its relationship to community dynamics would be an appropriate subject for a book in this series. I did not realize at the time that this seemingly well circumscribed topic would so thoroughly ramify throughout community ecology. However, a retrospective reading of Futuyma and Moreno's (1988) article had (correctly) said as much in the very first paragraph.

An important dichotomy is developed in Chapters 1 and 2 based on specialization–generalization at the local scale, and the capacity of species to occupy few or many different types of habitats. Fox and Morrow (1981) drew attention to the difference between the degree of specialization or generalization at the scale of the local population, and the degree when integrated over the entire range of a species (i.e., all populations of a species). They noted that some species could be 'local' specialists, relying on one species of host plant, for example, in one type of habitat, yet switch hosts in different habitats. Thus, while maintaining a similar degree of specialization of local resource use, the species might still be perceived as a generalized one over the variety of habitats it occupies. Cody (1974) was another who recognized that niche width might be partitioned into local and global components. He believed that some species of food specialists (e.g., parulid warblers) occupy many types of habitats, while species restricted in habitat use often were food generalists (e.g., emberizid finches). This distinction between local and global specialization and generalization is a crucial one, and dictates the content of this book. Almost all of this volume is concerned with specialization and generalization at the local scale. I ask questions like: how is it framed? how is it studied? what do we know? which ecological processes affect it? and how can we model it? Specialization and generalization in the use of habitats are considered more briefly, being the main focus of Chapter 8. But, generally speaking, the book addresses local specialization and generalization.

Many people have contributed in one way or another to getting this book written, but four were particularly helpful and generous with their time and conceptual and editorial criticisms. My wife, Dr Jane Doolan, deftly wielded her editorial pencil on parts of the manuscript. The project happened to coincide with the arrival of our children, Erica and Aleck, so that Jane also bore the brunt of attending to their (ongoing) constant demands and wishes during this period. For these Herculean tasks, I thank her very fondly.

The main conceptual reviewers of the manuscript (apart from the *Studies* editors) were Dr Peter Fairweather, formerly of Macquarie University but now with the CSIRO, and Dr P. S. 'Sam' Lake, of Monash University. I was indeed fortunate to be able to call upon two such fine and versatile scholars and ecologists for their advice and comments. Their marine littoral and limnological backgrounds helped, no doubt, to broaden the subject-matter from my exclusive experience in temperate, terrestrial systems.

The fourth substantial contributor was Dr Craig Blundell, formerly of BHP Research in Melbourne. Although he is a geophysicist, his broad interests in science and philosophy spurred me on continually. I like to think that I contributed modestly to his doctoral studies spanning the same period in which this book was written. His mathematical acumen and knowledge were a great boon during the development of the modelling algorithms discussed in Chapters 6 and 7 and Appendix B. We shared many tortuous paths before reaching the eventual solutions.

I am also grateful for comments on some chapters by Professor Peter Petraitis of the University of Pennsylvania (Chapters 2 and 3) and Dr Barbara Downes of Melbourne University (Chapters 1 to 4), both of whom prompted important clarifications or additions. I should add the usual caveat that all of the opinions in the book ultimately are my responsibility and that none should necessarily reflect on the scientific credibility of any other persons named here.

I also thank Gerry Quinn, Barry Traill, Linc McIntosh, Niall Richardson, Dugal Wallace, Angela Bowles, Tim Monks, Ian Hoyle and 'both' of my families for their assistance or encouragement. The support of Professor J. W. Warren and the Department of Ecology and Evolutionary Biology at Monash University was invaluable. I also thank the Australian Research Council for some support during the latter phases of writing.

And last, but hardly least, I thank the editorial and production staff of Cambridge University Press. Professor Birks provided critical and editorial advice for which I am most grateful; I hope he is pleased with the outcome. I am especially indebted to Professor John Wiens, whose criticisms, suggestions and thoughtful comments on the manuscript were priceless. He kindly devoted time to the task when he was on sabbatical leave at the University of British Columbia, which clearly indicates a high degree of altruism on his part. I also wish to acknowledge Dr Alan Crowden for maintaining a subtle correspondence of coercion across the world.

RCM
August 1994, Melbourne

1 · *An introduction to ecological versatility*

All scientific disciplines tenaciously seek a unified view of their subject. For example, the quest in particle physics and cosmology is to integrate the disparate physical forces in nature into a single, united framework, with as little or no dependence on empirically derived values for the model parameters (Green *et al.* 1987, Rees 1987). Although the objective is not quite so grand in community ecology, it is just as, or in some respects, more challenging because we have to deal not only with the contemporary dynamics of communities but also with the evolution of organisms. We seek a comprehensive picture of how and why resources are distributed among individuals and populations in the ways that they are, and the effects on population and community dynamics that stem from these distributions (J. Roughgarden cited in Lewin 1986, Hall and Raffaelli 1993). The immense diversity of living organisms and the wide range of physical and climatic variation on Earth, when coupled with organic evolution, have provided a seemingly endless supply of novel circumstances and outcomes. This variety has impeded the progress of the science of community ecology, which many judge to have been almost excruciatingly slow (Oksanen 1991a). Nevertheless, a rich and dynamic variety of new ideas aiming to unify community ecology continues to emerge, such as the 'macroecology' (Brown and Maurer 1989) and 'metapopulation' (Gilpin and Hanski 1991) concepts of relatively recent vintage.

In any event, one of the most obvious features of ecological communities is that species display manifestly different levels of ecological specialization (McNaughton and Wolf 1970, Futuyma and Moreno 1988). But why should specialists and generalists exist at all? Why don't all species show the same level of expertise in using resources? Generalists often are thought to hold advantages over specialists in having access to greater amounts of resources, which permits potentially higher densities and increased opportunities to satisfy or optimize nutritional requirements, if the resources are foods (McNaughton and Wolf 1970, Emlen

1973, Westoby 1978, Bernays and Graham 1988). No single food resource will be nutritionally optimal for a consumer, so that 'extreme' specialists have few options in balancing use between resources to achieve an optimal nutrition. Thus, a specialist may have to rely on resources that are either predictable or stable in availability and that entirely satisfy the specialist's requirements. On the other hand, specialists may be able to use resources more efficiently than generalists. In some cases, specializations leading to a use of resources (especially foods) with limited variation in 'quality' may have a distinct and demonstrable selective advantage (e.g., Stockoff 1993).

It is worth keeping in mind that the concepts of 'specialist' and 'generalist' are idealizations and that both are extreme cases in a continuum of patterns of resource use (Fox and Morrow 1981). In addition, few organisms show a rigid level of specialization or generalization – most seem to exhibit some degree of plasticity as circumstances change (Glasser 1982, Greenberg 1990a).

Most explanations for the patterns of specialization and generalization in natural communities can be regarded as belonging to one of two main classes. The first are interaction-based explanations that attempt to link the effects of interactions between organisms to patterns of resource use. Herbivory, parasitism, predation, omnivory, symbiosis (including mutualism) and interspecific competition comprise this set. Competition between species has been regarded as a major structuring force in natural communities, particularly by theorists (e.g., MacArthur 1969, 1970, 1972, May 1973, 1981). The other interactions have attracted varying degrees of theoretical attention and, like competition, are reviewed here in the context of their effects on patterns of resource use.

The second class of explanations consists of mechanisms within populations that affect the span of resources used. The degree of phenotypic differentiation within a population affects our perception of the diversity of resources used by that population. One obvious example is the ontogenetic, ecological differences in species displaying complex life cycles (Wilbur 1980). For example, in anuran amphibians and fishes, juvenile stages may be exclusively herbivorous while adults are exclusively carnivorous (Wilbur and Collins 1973, Werner and Gilliam 1984). Gender-based differences in resource use also are common (e.g., Selander 1966, Freeman *et al.* 1976, Cox 1981, Hedrick 1993). Like ontogenetic differentiation and gender-specific patterns, polymorphism, polyphenism (Moran 1992), and ecological plasticity each can influence the degree of diversity of resource use within populations. Each of these mechanisms is considered in this book.

The first class of explanations for patterns of resource use that are based on interspecific interactions has an extensive theoretical basis. This is especially true for interspecific competition (Law and Watkinson 1989). Much of this work uses community *stability* to derive the expected patterns of resource use in natural populations. For example, in competition theory, the stability criterion is equivalent to asking either: (1) which patterns of resource use allow competitor populations to coexist at stable densities (*asymptotic stability*)? or (2) which patterns of resource use make sets of coevolved species invulnerable to invasion by immigrant populations (*invasibility*)?

The coevolutionary method had its heyday in the 1960s and 1970s, but there has been an increasing tide against it over the past decade or so (e.g., Levins 1979, Speith 1979, Hastings 1988, Getz and Kaitala 1989). Workers now are more reluctant to accept that coevolution between populations is the most profitable way of thinking about how interspecific interactions might affect resource use. As Bock (1987) noted: 'assemblages are dominated numerically by widespread species, whose past histories and present population dynamics cannot have anything specific to do with communities as we have tended to delimit them'. This means that the numerically dominant elements of many local assemblages display patterns of resource use that are characteristic of the species at large, and that are hardly affected by the local milieu (see also Gleason 1926, Brown 1984).

But if the stability approach is no longer as fashionable as it once was, what alternatives are there? Connell and Sousa (1983) considered 'persistence within stochastically defined bounds' to be more applicable than stability based on a survey of published material on natural systems. A related concept is that of *permanence* (Jansen 1987), which involves looking for community dynamics in which the population densities of all species remain positive and finite. Under the permanence criterion, the densities of populations in natural communities should be 'bounded away from zero', even though these dynamics may be chaotic or cyclical (i.e., non-equilibrial, Law and Blackford 1992). In other words, as populations become rare, there should be mechanisms encouraging their recovery (e.g., density-dependent switching by predators). Another method was presented in an interesting series of papers by J. W. Glasser and his colleagues, who turned the problem around by effectively asking: what are the dynamic consequences and implications of specifying patterns of resource use first (Glasser 1982, 1983, 1984, Glasser and Price 1982)? Rather than deriving patterns of resource use by looking for stable configurations, what happens if specialists, generalists, and facultative strategists interact with one another in model communities? Does this approach offer insights that coevolutionary theory has missed? The vagaries of natural environments (climate, resource fluctuations), community composition (particular sets of competitors, mutualists and natural enemies; migratory fluxes) and a host of other factors make the Glasser approach a more reasonable one than a reliance on the coevolution of syntopic populations. Modelling the dynamics of alternative strategies of resource use and the interactions between rival strategies requires much more work and refinement, and for this reason, forms a substantial part of this book.

Niche breadth (or niche *width*) is the usual term for describing the variety of resources used by species. There has been some confusion about the precise meaning of specialization and generalization, as evidenced by the landmark review of Fox and Morrow (1981). They noted that some species display extremely restricted use of resources at any one place (*local specialists*) while others are able to use most resources (*local generalists*). However, by substituting similar though distinct resources from place to place, and thereby retaining local specialization, some species nevertheless can occupy many types of habitat (e.g., Nitao *et al.* 1991). Some local specialists are so specialized that they can subsist on only one resource, leading to habitat specificity if that resource is restricted to one form of habitat. Local generalists may or may not show the capacity to occupy many different habitats. These observations lead one to distinguish between specialization and generalization in a local sense, that is, in populations within single habitats, and specialization and generalization in the ability of *the populations of a species* to occupy a variety of habitats.

The issues associated with the extent of the geographic distributions of species – *range* – have been covered in great detail elsewhere because range forms an interface between many disciplines, including biogeography (e.g., Cain 1944, Darlington 1957, Brown and Gibson 1983, Myers and Giller 1988, Hengeveld 1990), evolution and genetics (e.g., Simpson 1965, Endler 1977), animal behaviour (e.g., habitat selection, Rosenzweig 1991, and territoriality, Cody 1978, Murray 1981), landscape ecology (Forman and Godron 1986), and areography (Rapoport

1982). However, range and specialization and generalization of habitat use need not be directly related, depending upon the geographic distribution of habitats. Thus, some species might occur over vast areas but only in specific habitats (e.g., rocky outcrops), while others might be geographically localized, but within that small area, occupy all available habitat types (see Hesse *et al.* 1951).

So, there really are three levels to consider in terms of ecological specialization and generalization: (1) the diversity of resource use at the local scale; (2) the diversity of habitats occupied; and (3) the extent of geographic range. The latter has been covered at length in comparatively recent reviews (e.g., Brown and Gibson 1983, Hengeveld 1990), so there seems to be no need to study the determinants of range in much detail here. The niche, so to speak, of the current book is resource use at the local scale, plus a consideration of the relationships between local- and habitat-based specialization and generalization. To further clarify the distinction between these scales, a short overview of some of the issues involved is presented next. Species are viewed as consisting of sets of populations communicating (by migration and/or dispersal) to a greater or lesser extent depending upon their vagility and the barriers to interchange (Hesse *et al.* 1951, Briggs 1974, Opdam 1991). Populations usually are thought of as sets of potentially interbreeding or interacting conspecific individuals.

The local scale

The use of a wide range of resources implies a high niche breadth (generalization), while using few resources, or a reliance on a single resource, connotes a low breadth (specialization). Populations can be regarded as being located along a conceptual continuum from specialized through to generalized. It is clumsy to have to refer continually to the specialized–generalized continuum, so I adopt the simpler term *ecological versatility*. Ecological versatility can be defined formally as *the degree to which organisms can fully exploit the available resources in their local environment.*

I use the term *exploit* to mean the acquisition, handling and/or processing, and conversion of resources into gains in fitness, where the latter comprises all aspects related to fitness, including maintenance, reproduction, and survival. Resource *utilization* is used in this book to refer to just the acquisition of resources, with no explicit reference to how the resource use is converted into a gain in fitness. There has been no formal distinction between these terms in the literature, so these definitions of exploitation and utilization are specific to this book.

Ecological versatility explicitly depends upon both the fitness returns from utilizing resources and the relative availability of resources. How fully a set of resources is exploited is gauged by how closely the fitness gained from exploiting a resource corresponds to its relative availability. An idealized generalist matches the fitness gains precisely with the relative availabilities of a suite of resources. It is worthwhile expanding on the fitness and resource availability aspects to explain their significance in greater detail.

Why is fitness so important, which amounts to why focus on the exploitation of resources rather than on their utilization? By way of illustration, consider food resources. It turns out that the nutritional aspects of food use and the fitness gains associated with their use need not be related in a straightforward way (Real 1975). This has led to a contention between models of foraging strategies in large herbivores (e.g., ungulates) and carnivores because of the much wider range of nutritional values of plants compared with animal prey (Owen-Smith 1988, du Toit and Owen-Smith 1989, Verlinden and Wiley 1989). When one tries to predict the optimal diet for a predator, one can largely ignore the nutritional content and relative digestibility of alternative prey types, which are more or less the same for all prey types. The solution depends much more on the distributions and handling times of alternative prey types (Pyke 1984, Stephens and Krebs 1986). On the other hand, herbivores use food types that may differ markedly from one-another in nutrient content and the concentrations and kinds of chemical deterrents and inhibitors (digestibility). Thus, although acquisition rate is thought to dominate the returns for most predators, digestion rate probably controls diet selection in herbivores (Belovsky 1986, Verlinden and Wiley 1989). The actual fitness returns from the components of the diets of predators therefore depend upon things like how alternative prey are distributed in space and how long it takes to handle them. Either of these factors may be affected by differential exposure to natural enemies in using alternative prey, which can modify the fitness returns. In large herbivores, fitness gains often are more closely associated with the sheer bulk of fodder that has to be digested than with finding it, leading to complicated expressions linking together nutritional content, food bulk, digestion rates, and gut volume (see Belovsky 1986).

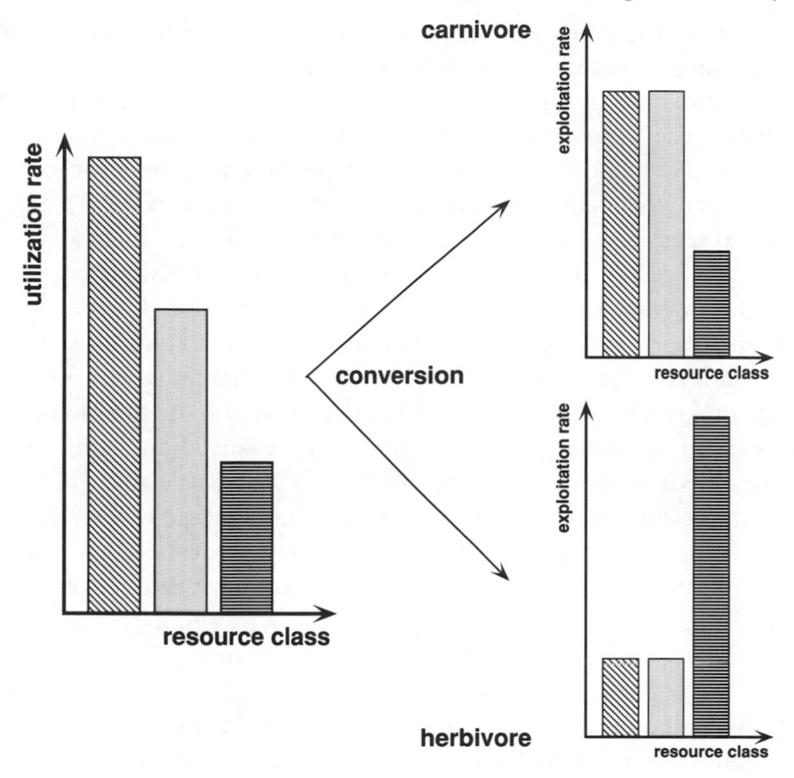

Fig. 1.1 Hypothetical example illustrating the difference between utilization and exploitation rates. The relative value of three dietary constituents remains much the same after conversion in a carnivore but, due to nutrient content or digestibility perhaps, the respective values for a herbivore differ markedly from the relative utilization rates.

To make this contrast more concrete, consider a predator and a herbivore each having diets consisting of three components, in the proportions 3 : 2 : 1 (i.e., utilization rates; Fig. 1.1). The fitness returns to the predator may be much the same set of ratios, 2.5 : 2.5 : 1, say, but the distribution for the herbivore may be much more biased because of nutritional imbalances and digestibilities, 1 : 1 : 4, say. In this example, the predator depends more evenly on the fitness returns from all three resources than does the herbivore, and can be considered to be more generalized (ignoring relative availabilities for this argument). This clearly is a contrived example between very different organisms and related to a single part of the resource spectrum (i.e., food), but the principle is generally valid.

The relative availability of alternative resources is the other key issue in determining versatility. How does availability influence versatility? Let several resources provide more-or-less similar returns in fitness per unit utilized. Then, a generalized consumer appears to be indiscriminant because it exploits resources in the proportions in which the resources occur in its environment (Petraitis 1979, Fox and Morrow 1981). In contrast, a specialist exploits resources differentially, so that some resources are used in higher proportions than their availability, and hence, other resources are underutilized (hence a low versatility).

Although ecological versatility may appear to be synonymous with niche breadth or niche width, there are important distinctions. The most crucial of these is that versatility explicitly depends upon the correspondence between resource exploitation and resource availability. Estimates of the breadth of resource use that do not refer to the relative availability of the resources are not of much use for the reasons discussed by Petraitis (1979) and others. In particular, differences in breadth may merely reflect biased resource availabilities rather than indicating the selectivity of, or suitability for, consumers. Versatility also hinges on resource exploitation, which involves the derived fitness increments from using resources rather than the relative use of resources themselves. Niche breadth is based on utilization, which can cause problems for deciding the units on which to gauge breadth (e.g., numbers or volumes of prey? Case 1984). Niche breadth also suffers from being a general term encompassing not only the diversity of resources used at the local scale, but also other aspects of the ecological flexibility of a species, including the variety of habitats used (e.g., McNaughton and Wolf 1970). Fox and Morrow (1981) have drawn attention to the problems associated with such a broad definition. In summary then, ecological versatility, strictly set at the local level of the population, is a more constrained, better-defined concept than its counterpart niche breadth.

Habitat use

The preceding definitions refer to the strictest meaning of the term ecological versatility – the degree of correspondence between the exploitation and availability of resources at the local or population scale. But the capacity of the populations of a species to occupy many distinct habitats also is an inherently important ecological characteristic (Bock 1987). Indeed, the versatility of habitat use historically has attracted more interest than local versatility, perhaps because of the evolutionary and biogeographic implications (e.g., Jackson 1974).

It is important to re-emphasize that versatility in habitat use is a quite-distinct concept from that of range. Hesse *et al.* (1951: Chapter 8) established this distinction well. They used the term *ubiquity* to denote the diversity of habitat types that the populations of a species occupy. Thus, a *ubiquitous* species has populations occupying many distinct types of habitat while a *restricted* species is confined to a limited number of habitat types. Note that some workers seem to be unaware of this priority for the term ubiquity (e.g., Burgman 1989, Rahel 1990), so one must be wary about the precise meaning of ubiquity in this book. For the extent of geographic range, Hesse *et al.* (1951) used the term *cosmopolitanism*. Thus, a *cosmopolitan* or widespread species has a large geographic range, while a *localized* species occurs in only a small geographic area. 'Large' and 'small' geographic ranges are somewhat arbitrary terms of course, but Hesse *et al.* (1951) used the area of geographic extent compared with the vagility of members of a species as a heuristic to clarify the point. I follow the terminological conventions of Hesse *et al.* (1951) throughout this book.

Several authors have recognized that ubiquity amounts to the evenness of densities of populations of a species among alternative habitat types (e.g., McNaughton and Wolf 1970, Rice *et al.* 1980). A completely ubiquitous species would exhibit the same density irrespective of habitat type. This means that the species responds weakly to large-scale habitat heterogeneity. The degree of ubiquity is sensitive to the changes in the relative densities of populations in different habitats. Even though a species has populations in every habitat type that is sampled, it can be regarded as effectively restricted if the density in one habitat is much greater than in any of the others. This idea of the evenness of the spread among habitats is central to the concept of ubiquity.

These points can be demonstrated by looking at some patterns of densities among various woodland and forest habitats of year-round resident species of birds in central Victoria, Australia (Fig. 1.2). The densities of the grey shrike-thrush (*Colluricincla harmonica*), for example, were not significantly different among the five habitats, thereby fulfilling the definition of a ubiquitous species. The eastern rosella (*Platycercus eximius*), a parrot, occupied all habitat types, but in significantly higher densities in the more open riparian (dominated by river red gum, *Eucalyptus camaldulensis*, and grey box, *E. microcarpa*) and lowland

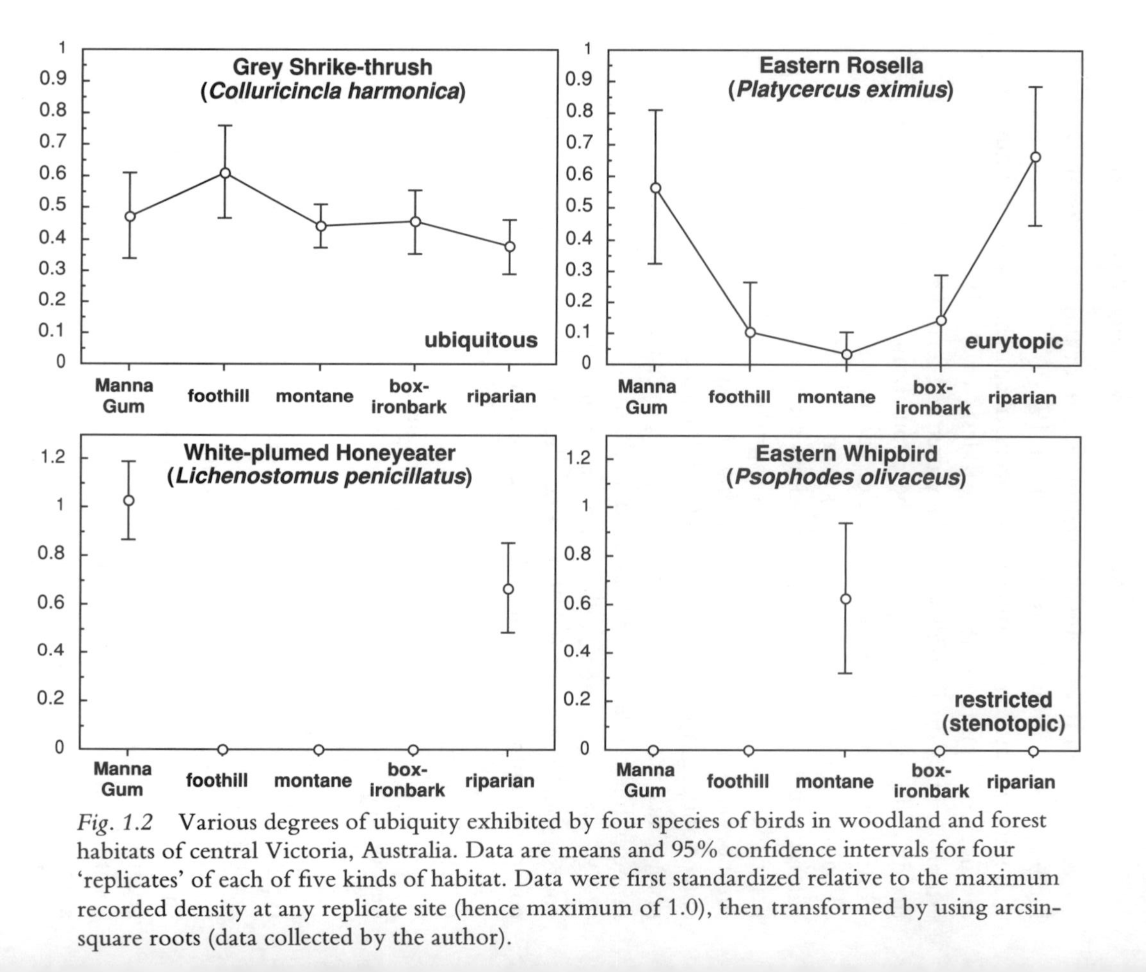

Fig. 1.2 Various degrees of ubiquity exhibited by four species of birds in woodland and forest habitats of central Victoria, Australia. Data are means and 95% confidence intervals for four 'replicates' of each of five kinds of habitat. Data were first standardized relative to the maximum recorded density at any replicate site (hence maximum of 1.0), then transformed by using arcsin-square roots (data collected by the author).

Gippsland manna gum (*E. pryoriana*) woodlands. Thus, the eastern rosella was not ubiquitous, but rather, eurytopic (from the Greek roots, *eurus* – broad or wide and *topos* – a place). White-plumed honeyeaters (*Lichenostomus penicillatus*) eschewed all habitats except the two just mentioned, while eastern whipbirds (*Psophodes olivaceus*) were recorded in montane forests alone – a restricted (or stenotopic, *stenos* – narrow) species (Fig. 1.2). I hasten to add that unlike geographic range, ubiquity assessments such as the ones just described are sample estimates, being predicated on the habitats that one includes in a survey (see Chapter 8).

Three questions on versatility

Having reached the end of this short discussion on local versatility and ubiquity, it seems appropriate to list the main questions addressed in the book. The principal objective of this book is to answer these questions, which, hopefully, will contribute towards that more unified picture for community ecology. However, if the questions cannot be answered yet, then the book should at least provide an overview and report of the current status of our knowledge of factors affecting resource use in natural communities. The three questions are:

(1) how much is known about the patterns of versatility in natural communities, and how consistent are studies in terms of the sorts of quantities that are monitored to summarize and explain variation in ecological versatility?
(2) how well does existing theory account for observations, what are the sources of contention between rival schools of thought, and what sorts of insights can be derived from modelling alternative strategies of resource use (i.e., different versatilities)?
(3) what determines the ubiquities of species?

What is the significance of ecological versatility?

It may be useful to provide a definitive statement on the usefulness of reviewing ecological versatility in depth. In other words, what is the significance of ecological versatility, and why is it worth considering in such great detail (see also the introduction in Sherry 1990: 337)? There is little doubt that ecological versatility is a nexus between many diverse disciplines in natural history, and as such, ramifies throughout much of ecological and evolutionary thought (Futuyma and Moreno 1988).

Many community ecologists believe that the similarity of resource use in one way or another limits local diversity (see Cornell and Lawton 1992). Resource availability is finite owing to the limited availability of energy and nutrients in natural systems and the loss of useable energy at each stage of the trophic hierarchy (i.e., losses due to entropy). It seems logically inescapable that the ways in which syntopic populations use resources and how that use translates into population densities must have important implications for local diversity. There appear to be density thresholds below which populations are bound to become extinct locally (e.g., Pimm *et al.* 1988, Wissel and Stöcker 1991, Possingham *et al.* 1992, Tracy and George 1992), which implies that communities should not consist solely of very large numbers of very rare species. Thus, local populations must have access to sufficient resources to maintain moderately high densities, which should limit the number of coexisting species (Hubbell and Foster 1986, Cornell and Lawton 1992; ignoring, for the moment, metapopulation and landscape-scale dynamics and related ideas, see Gilpin and Hanski 1991, Opdam 1991, Dunning *et al.* 1992, Tilman 1994). If this is so, then an important objective in community ecology must be to explain or understand the distributions of the densities and diversities of resources used by each syntopic population (McNaughton and Wolf 1970, Brown and Maurer 1989). Ecological versatility also has important repercussions for evolutionary biology, such as in the ongoing debate on the relationships between specialization and evolutionary radiation, and the ultimate fate of highly specialized taxa (e.g., Mayr 1976, Moran 1988, Wiegmann *et al.* 1993). Evolutionary biologists frequently have linked diversity to ecological specialization in taxonomic lineages (e.g., Price 1980, Mitter *et al.* 1988, Thompson 1988b, Dial and Marzluff 1989). Versatility also is central to behavioural research too, both in terms of optimization theories (e.g., Emlen 1968, Pyke 1984, Gleeson and Wilson 1986) and resource selection (e.g., Westoby 1978, Demment and Van Soest 1985, Belovsky 1986, Verlinden and Wiley 1989, Abrams 1990). These few examples show the range of applications that the concept of ecological versatility has in ecological and evolutionary thinking, and therein lies its importance and worth as a review topic.

Structure of the book

The heart of the book consists of seven chapters (Chapters 2 to 8). In short, these chapters involve semantics and definitions (Chapter 2), a survey of the ways in which ecological versatility in natural populations have been studied and recorded (Chapter 3), literature reviews of the factors that influence ecological versatility within populations (interspecific interactions except for competition, Chapter 4; differentiation within populations, Chapter 5), models of versatility within populations and how versatility may relate to population dynamics (Chapter 6), the influence of interspecific competition on patterns of versatility (Chapter 7), and a review and discussion of the nature of the versatility of habitat use, or ubiquity (Chapter 8). Chapters 2 to 7 deal mostly with patterns evident at the 'local' scale of the population, or of two or more syntopic populations of different species. In the last chapter (9), the main points arising from previous chapters are reconsidered briefly, and an attempt is made to answer the three questions just posed on the basis of the literature reviewed here. A more complete outline is presented now.

A detailed discussion of semantics, such as meanings of terms like 'versatility', 'resources', 'utilization' and 'exploitation', forms the first part of Chapter 2. This can be done most simply by addressing these questions: (1) what is ecological versatility? (2) what are resources? (3) what are the problems in quantifying ecological versatility? and (4) how is versatility measured? By responding to each question in term, the semantic framework for this book is developed. The chapter also includes an expansion on some of the distinctions between geographic range and the diversity of habitats used by populations of species.

A survey of published material on ecological versatility in natural populations is analysed in Chapter 3. The survey includes an assessment of the degree to which studies have considered the critical issues raised in Chapter 2. For most points, this analysis is not primarily intended to be critical *per se*, but it does demonstrate the lack of procedural consistency evident in field studies (see Peters 1988, 1991). With respect to some aspects, such as the importance of measuring resource availability to gauge versatility, there can be few excuses because these issues have been widely discussed for many years (e.g., Hurlbert 1978, Petraitis 1979).

Reviews of the factors influencing ecological versatility in populations are provided in Chapters 4 and 5. An overview of the effects of interspecific interactions on versatility is provided in Chapter 4, in which herbivory, parasitism, predation, omnivory and mutualism are considered. Versatility theory has been dominated by interspecific competition, but for structural reasons within this book, the discussion of this topic is postponed until Chapter 7. Chapter 5 is devoted to the effects that differentiation within populations have on versatility, which include phenotypic and ecological plasticity, polyphenism, polymorphism, and stage-structure in populations (i.e., ontogenetic divisions, gender, etc.).

There has been much attention paid to modelling resource use in populations, especially by using mathematical analysis. In the past, the main approaches referred to above have been used to try to deduce the patterns that might be expected to characterize ecological communities (i.e., asymptotic stability, invasibility). The use of these criteria is questionable (see Chapter 7) so that in Chapter 6, an alternative tack is employed. Population dynamics and temporal changes in the patterns of resource use in populations displaying different exploitation strategies are modelled, in a way not too dissimilar to that developed by J. W. Glasser (Glasser 1982, Glasser and Price 1982). These models show how these patterns depend upon specific patterns of variability in resource availability. This alternative picture does not depend upon mathematical equilibria or related criteria and shows some of the novel dynamics that might occur in natural systems.

Chapter 7 begins with an overview of thought on the impact of interspecific competition on patterns of resource use, the formal approaches having been pioneered by R. H. MacArthur and his colleagues (e.g., MacArthur and Levins 1964, 1967, MacArthur 1970, 1972, May and MacArthur 1972). These methods rely on stability and invasibility criteria to derive the expected patterns of resource use in sets of competing populations. Logistic models of population dynamics underlie much of the theory in this area. The models usually are phenomenological, which means that they do not closely couple population dynamics either with mechanisms of competition or with explicit variation in resource availability (León and Tumpson 1975, Ives and May 1985). The assumptions of logistic-based methods are so restrictive that other approaches need to be explored. Therefore, the modelling basis developed in Chapter 6 is used to consider the relationships between alternative exploitation strategies (i.e., different patterns of local versatility) and to investigate a number of important phenomena. The latter include an analysis of the factors promoting 'competitiveness', invasibility, permanence, the impact of different forms of resource variation, and the coexistence problem (i.e., how many populations can a given number of resources support?).

The last of the main chapters, Chapter 8, is concerned with ubiquity. One of the principal questions raised in the chapter is: does ecological versatility promote ubiquity? Does the ability to use a wide range of (local) resources lead to the capacity to persist in a wide range of habitat types? There has been surprisingly little attention paid to this rather obvious question (although see Cody 1974).

In summary then, this book aims to provide an overview of the diverse elements that can be related to disparities of resource use within natural populations. This involves definitions and the analysis of existing work to assess the consistency of methods used among workers for, without a common grounding, inferences on the determinants of patterns of versatility are difficult or impossible to draw. It is also extremely important to recognize the diversity of processes and factors that determine local versatility. In particular, there seems to be a tendency to concentrate on certain processes (especially interspecific competition, hence the length of Chapter 7) or phenomena (the diversity of hosts used by herbivorous insects) that probably colour thinking on this topic to a greater extent than is warranted. All of the main ecological interactions have the capacity to affect significantly patterns of resource use. In addition, populations are more-or-less heterogeneous collections of individuals, which can have significant repercussions on the realized versatility of populations and also on population dynamics. The modelling presented in Chapters 6 and 7 indicates how influential alternative exploitation strategies may be to population and community dynamics. Whether local versatility affects (or effects) ubiquity also is considered in the penultimate chapter.

A directory

Like most books, this one can be read in a variety of ways depending upon the information in which the reader is most interested. For those concerned with patterns of resource use at the scale of 'local' populations and the mechanisms that might dictate those patterns, a sequence of Chapters 2, 3–5, 7 (part) and 9 might suffice. A focus on large-scale patterns (e.g., biogeography) would include Chapters 2, 3, 8 and 9. Those interested in modelling probably should concentrate on Chapters 2, 3, 6, 7, and 9, with less attention on Chapters 4 and 5. I have endeavoured to make most of the chapters self-contained, but Chapters 2 and 9 would seem to be necessary reading because they provide the semantic and terminological setting and the conclusions respectively. The main exception to this compartmentalization is Chapter 7. This chapter entails both a discussion of theories of interspecific competition and its possible effects on resource use, and an examination of some of the conclusions derived from that theoretical work by using the modelling structure that is developed in Chapter 6.

Scope

I concede here that this book is heavily weighted towards processes, thought and examples drawn from the literature on animals, especially terrestrial ones. Although some attention is paid to plant and microbial communities, overall the book is necessarily biased to keep it to a manageable size. Plant and microbial ecologists have already considered aspects of the constraints on resource use and their work should be consulted for a more complete picture on ecological versatility. For example, patterns and processes in microbial and fungal communities have been reviewed by Jennings and Lee (1975), Rayner and Todd (1979), Fenchel (1987), Smith and Douglas (1987), Carroll and Wicklow (1992), and Werner (1992). There is a wealth of literature on resource use in plant communities, including Kuijt (1969), McLean and Ivimey-Cook (1973), Grime (1979), Newman (1982), Tilman (1982, 1988), Calder and Bernhardt (1983), and Silvertown (1987).

2 · *Defining and measuring versatility*

An ecologist's judgement of the ecological versatility of a species is very much a filtered one. The most important filters are the ecological characteristics of the taxa under study, and the taxonomically inspired biases of investigators. For example, one reads of specialized and generalized insectivores (e.g., Freeman 1979), but surely this can only mean a relative degree of versatility nested within a coarser degree of food specialization or generalization? The biases of the investigator, engendered by the characteristics of the taxon that she or he studies, often lead to this type of loose terminology.

However, this is by no means the only source of inexactness. Ecological scale – the spatial and temporal scales at which organisms operate – also alters the perspectives of how specialized or generalized species are (Allen and Starr 1982, Dayton and Tegner 1984, Wiens *et al.* 1987, Orians and Wittenberger 1991, Levin 1992). This works in two ways, one organismal and one investigator-based. Many workers have noted that different organisms show different levels of discrimination, and hence, dissimilar responsiveness to spatial variation in resource availabilities (e.g., Addicott *et al.* 1987, Pahl-Wostl 1993). This often leads to an altered perception of the ecological versatility of the population or species over different spatial scales. The often-cited review of Fox and Morrow (1981) showed how the level of perceived specialization of a species depends upon how many populations are considered. Populations of a species may show high resource specificity locally, but shift to, and specialize on, alternative resources elsewhere (e.g., Nitao *et al.* 1991). Therefore, the species as a whole actually displays greater generalization at a large scale than do its local populations.

Investigator biases originate from difficulties in circumscribing and integrating resources at different scales. The ongoing problem of distinguishing between microhabitat, macrohabitat, and habitat is one example of this difficulty – no two definitions are quite the same (e.g., Morris 1987a, Wiens *et al.* 1987), yet perceived versatility clearly depends upon how one partitions spatial scales. Thus, ecologists impose their own idiosyncratic schemes on nature, which may or may not correspond well with scales of variation to which organisms are responding (Levin 1992).

It is worth noting that scale is influential not only in terms of space, but also in time. For example, organisms often may respond much faster to changes in resource availability than studies are designed to detect. This can lead to resource use being deemed to be less specific than it actually is. Such erroneous perceptions must be more severe in studies in which availabilities of resources are not monitored, for information on resource use then is of a low grade.

As with any specialized topic, ecological versatility necessarily involves a semantic framework within which the key concepts are defined and related to each other. This is the purpose of the current chapter. Most readers probably will have an aversion to a chapter devoted almost entirely to semantics and definitions, and understandably so, but there is undoubtedly a need for as precise a set of definitions as possible for the key concepts upon which the remainder of this book is based.

It also is crucial to make a clear statement of the limitations and operational difficulties that bedevil the application of the conceptual framework to field studies. Rosenzweig (1991) has written of the processes by which scientific ideas are developed and tested. One of the interesting points he made was that there is a stage at which mechanistic theories have to be 'unpacked', where unpacking is the construction of specific tests that can be used to test general theories. This unpacking phase can be likened to the transition from the definition of terms to the erection of an operational study. Much of the art (or resourcefulness) in science is the capacity to make operational experimental or observational analogues for abstract concepts. To do so, however, requires a good understanding of the pitfalls and difficulties that are encountered when one attempts to unpack the concepts. Therefore, much of Chapter 2 involves a discussion of the operational problems that have to be contended with when field measurements of ecological versatility are attempted.

Although most of the chapter is concerned with ecological versatility at a 'local' scale, the penultimate section is designed to reinforce the distinction made in Chapter 1 between ecological versatility and the versatility of habitat use, or ubiquity. This section may seem somewhat out of place here, but I think that it is important to collect together in the one place as many of the semantics as possible. The distinction also helps to place ecological versatility in perspective, and reduce confusion in later chapters between it and ubiquity.

2.1 What is ecological versatility?

To begin this chapter, I restate from Chapter 1 what is meant by the term 'ecological versatility'. It is the degree to which organisms can fully exploit resources in their local environment. *Exploitation* involves gaining access, processing (or occupying in the case of resources such as nest holes), and translating resources into increases in fitness. On the other hand, *utilization* refers to just the acquisition of resources, with no regard as to how that acquisition relates to fitness gains.

Ecological versatility is a function of both resource exploitation and relative availability of resources. How fully a set of resources is exploited is gauged by how closely fitnesses gained from exploiting resources match relative availabilities of resources. An idealized generalist would match fitness gains precisely with relative availabilities of a set of resources. It is worthwhile expanding on the fitness and resource availability aspects to explain their significance in greater detail.

The meaning of fitness

It clearly is crucial to define precisely the meaning of the term *fitness* because of its importance here. Different definitions for fitness abound in the literature (see Endler 1986: 33–49 for a detailed discussion). For example, Cooper (1984) presented a case for using the expected time to extinction (ETE) of an evolutionary unit as a measure of fitness of that unit, where a unit may be a genotype, subpopulation of a larger population, or a species as such. He argued that ETE is a notion closely affiliated with the Darwinian meaning of fitness, namely, the *probability of extantness* (Cooper 1984: 620). Thus, a fit population unit would have a relatively long ETE compared with potential rivals, ensuring a greater likelihood of being observed at some future time. Others consider the capacity of populations to maintain densities indefinitely as a measure of fitness (e.g., Holt and Gaines 1992). This definition places fitness at the level of the population, and, in most respects, identifies fitness with what most ecologists normally regard as the *persistence* of a population (e.g., Harrison 1979, Pimm 1984, Schoener and Spiller 1987).

It is becoming increasingly clear that one cannot treat populations as consisting of identical individuals, even in clonal organisms (Jaenike *et al.* 1980, Werner and Gilliam 1984, Jones *et al.* 1986, Milligan 1986, Weider and Hebert 1987, Łomnicki 1988, Barnard *et al.* 1990). Much of the understanding of patterns of versatility requires detailed knowledge of individual behaviour (e.g., Werner and Sherry 1987, Whitfield 1990, Holbrook and Schmitt 1992) and vulnerability (e.g., Jones *et al.* 1986). Thus, the term fitness will be used here for the set of quantities pertinent to individuals within populations rather than to population characteristics. *Fitness* in the current context involves many facets related not only to the acquisition of mating opportunities and the production of offspring, but also to the capacity of individuals to survive, grow, and place themselves in a (better) position to reproduce and survive. Resources are often directed at improving the chances of individuals to breed and survive. For example, energy stores accumulated during nonbreeding phases are used to permit or improve performance in breeding events (e.g., anuran amphibians, Mac Nally 1981, 1987). The distinction between the actual reproductive output and the capacity of individuals to survive and/or accrue sufficient resources to reach a status at which reproduction is possible, or more likely to be worthwhile, is accommodated in models developed in later chapters.

2.2 What are resources?

To date, I have used the term resource without stating precisely what is meant by this expression. As with fitness, there are many different meanings for 'resource' in the literature, some of which are specific to particular applications, such as consumer-resource theory (Haigh and Maynard Smith 1972, Tilman 1982, 1986, Abrams 1988a, 1988b). Plant and plankton ecologists often use the terms resource and nutrient interchangeably (e.g., Bryant *et al.* 1989, Carson and Pickett 1990, Grover 1990, Schimel *et al.* 1991). The following sections involve general comments on semantics associated with the term 'resource'. Resources also are distinguished from factors having deleterious effects on organisms, or at least on the capacities of organisms to exploit resources. Such elements commonly are called limiting factors (Haigh and Maynard Smith 1972, Schoener 1976, Tilman 1982) or malentities (Browning 1963, Andrewartha and Birch 1984).

Resources – the fitness basis and exclusive access

An *environmental component* refers to anything that impinges on an individual organism (Andrewartha and Birch 1984). Components include food, refuges, nesting sites, environmental gradients or patches such as salinity, light or trace elements, space and time, predators, conspecific individuals, and so on. Components can be classified according to their effects on an individual organism. Some components will, if exploited by an organism, help the organism to survive and reproduce. Utilization of these components increases the fitness of the organism. On the other hand, many pose serious threats to the organism if they cannot be tolerated, ameliorated, or avoided.

Any environmental component to which an organism can gain exclusive access for some period of time, and that by its use *increases* the organism's fitness, is a (*true*) *resource*. This definition includes a diverse range of components, such as food, inorganic nutrients, and shelter from biotic or abiotic threats to the organism's well-being.

For an environmental component to be a true resource, exploitation implies exclusive access to that feature for some period of time (see also Abrams 1988a). Why exclusive? Consider several nectarivorous insects feeding simultaneously on one flower. What is the actual resource in this case, the feature giving rise to increased fitness? The flower itself is not the resource from this perspective – rather, the nectar volume consumed by individual insects provides the increase in fitness. Floral morphology may act as an important limitation on the ability of some insects to utilize the nectar secreted by the flower (Harder 1985), but the nectar is *the* resource. To some extent, this might be viewed as too much reductionism. Perhaps the 'true' resource in this case is the sugar content of the nectar. However, the sugars cannot be acquired independently of the remainder of the nectar, so that the nectar is the logical resource in this example. Exclusiveness is a useful criterion for distinguishing resources when analysing patterns of resource utilization.

In many existing classifications of resources, there has been a tendency to assign resources to one of three major categories: food, space, and time (Schoener 1974a). Many hold space to be true resource (e.g., Hastings 1980, Paine 1984, Yodzis 1986). Wiens (1989a: 321) considered space itself ('living space') to be a true resource for some organisms, notably sessile ones. However, more commonly he believed that 'space may affect the distribution of factors that are resources than be a resource itself'. The second part of Wiens' statement seems to me to be more consistent and correct, and that neither space nor time is a true resource. Nevertheless, temporal and spatial variations in true resources do occur and apparently allow segregation of resource use by syntopic populations in some cases (Schoener 1974a).

In summary, then, it always is helpful to look for those components contributing directly to increased fitness, and to which organisms can gain exclusive access, to identify resources. Usually, this approach clarifies an analysis.

Critical and extraneous resources

In practice, gauging versatility depends upon the number of resource categories that are used by an investigator. Ecologically meaningful values can be generated only by excluding or reducing the influence of resource categories that artificially inflate or depress perceived versatilities (see Mac Nally 1994a for an example based on the related topic of foraging versatility). Many workers have recognized the importance of the appropriate delineation of resources to estimating niche measures (e.g., Petraitis 1979, Case 1984, Wiens 1989a). Real (1975) noted that superabundant resources probably do not exert much influence on the way in which individuals of populations manipulate their patterns of resource utilization (see also Tilman 1982, Wiens 1989a). He assigned such resources zero values in his analyses, indicating that they were in most respects irrelevant to patterns of resource use. Thus, any individual can utilize as much or as little of these superabundant resources as they need to without being forced into competitive situations or optimization decisions.

As an illustration, Real (1975) pictured a xeric habitat in which magnesium salts were very abundant in all foods. In this example, the use of magnesium salts as a resource would add 'noise' to a versatility measurement because organisms can use them freely and without limitation. On the other hand, the use of water would exert a strong influence on utilization patterns and hence, should affect versatility. Thus, water would be a *critical* resource state for assessing versatility, but magnesium salts would be *extraneous*. Identification of the nature of resource categories to this extent clearly is needed to characterize better ecological versatility. Of course, whether a resource is superabundant or not, and hence whether it is critical or extraneous, very much depends upon the current ratio of supply and demand, which led Wiens (1989a: 321) to view resources in terms of their *potential* use by organisms. Therefore, extraneous resources probably should be regarded as those resources that are consistently superabundant over time spans much longer than generation times.

Complementary and substitutable resources

León and Tumpson (1975) distinguished between ecological resources on the basis of their substitutability and complementarity for consumers. They defined *perfectly substitutable* resources to be ones satisfying the same requisite needs. They envisaged that complex nutrients contained in different plant species for herbivores or various prey items for carnivores could be freely interchanged. The net result of perfect substitutability is that the total amount of all substitutable resources limits the per capita growth rate of the population.

Perfectly complementary resources must be gathered together by a consumer because they fulfil distinct needs, and are commonly encountered in nutritional and growth studies. The resource acquired (and assimilated) at the slowest rate effectively limits the per capita growth rate. There is evidence for complementary resources in natural systems. For example, Pennings *et al.* (1993) demonstrated substantial improvements in mass-gain in mixed algal diets than in single-species diets in the sea hare, *Dolabella auricularia*, which they attributed to dietary complementarity.

A third case considered by León and Tumpson (1975) was *imperfect substitutability*, in which subsistence is possible on any one of the imperfectly substitutable resources, but an appropriate mix of these resources yields higher population growth rates.

These ideas can be embodied by using indifference curves (León and Tumpson 1975). All combinations along the curve yield identical per capita growth rates. Thus, in Fig. 2.1 a, curves are shown for a pair of per capita growth rates for two perfectly substitutable resources. Linearity of the curves indicates that the same per capita growth rate occurs for any combination of values along the lines. The farther the line is from the origin the greater is the per capita growth rate. Thus, depending upon conditions, the same per capita growth rate can be sustained by exploiting, say, mostly resource *A* or mostly resource *B*. Indifference curves for perfectly complementary resources are rectangular in shape (Fig. 2.1 b) because of the limiting nature described above. The rationale for the rectangular curves is that if only a certain amount of one particular resource can be acquired, then the per capita growth rate cannot be increased by consuming other complementary resources, irrespective of their availabilities. The shapes of indifference curves for imperfect substitutes (Fig. 2.1 c) are convex to the origin; the reasons for this are discussed by León and Tumpson (1975).

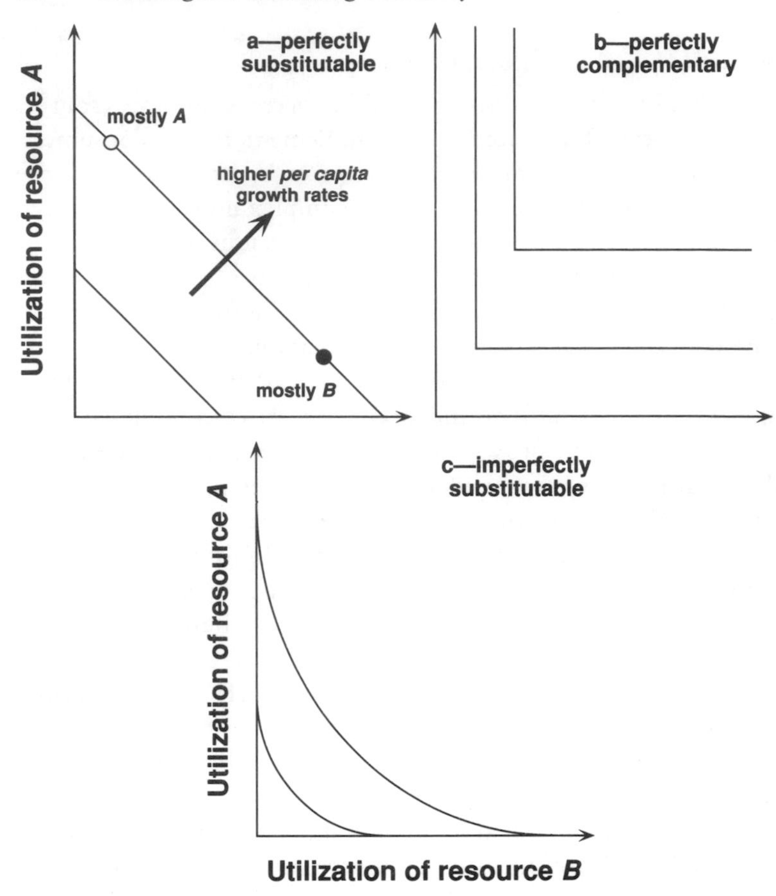

Fig. 2.1 Indifference curves for types of resource recognized by León and Tumpson (1975): (a) perfectly substitutable; (b) perfectly complementary; and (c) imperfectly substitutable.

Why have I given this amount of attention to these ideas of complementary and substitutable resources? One of the main reasons is that ecological versatility is closely related to the capacity of individuals to exploit fully an array of resources. Consider a pair of phytophagous insects occupying a habitat in which there are, say, ten species of shrubs. Suppose one of the insects (species '*A*') can exploit all ten species of plants in a substitutable way, while the other species ('*B*') cannot. Species *A* therefore is more ecologically generalized than species *B*, perhaps due to morphology, perhaps due to physiology or biochemical processing capabilities. The fitness gains associated with exploiting the entire set of shrubs at least potentially will be more closely related to the availability of those shrub species in species *A* than in species *B*.

Constraints

Many environmental components reduce fitness. Therefore, avoiding or ameliorating such components prevents a loss of fitness. There are many diverse components falling into this category. Limitations caused by physical phenomena such as desiccation, toxic substances, salinity, oxygen tensions and the like have been recognized as critical factors in distributions of species and population dynamics for many years (e.g., Hesse *et al.* 1951, Colwell and Futuyma 1971). Predators, parasites, and interspecific competitors also have deleterious effects on the fitness of organisms (Andrewartha and Birch 1984). Westoby (1978) distinguished between true resources (nutrients in his application), and components that modify the effective nutritional value of foods, or 'negative nutrients'. To avoid this clumsy terminology, negative influences on the fitness of organisms are referred to here as *constraints*.

Constraints are similar to effects such as *limiting factors* (Abrams 1988a) and *malentities* (Andrewartha and Birch 1984: Chapter 6). Limiting factors are populations (e.g., predators, parasites, intraspecific and interspecific competitors) or abiotic or biotic substances whose effects on the target population increase as the size of the target population increases, and eventually act to limit the growth rate of the target population. According to Andrewartha and Birch (1984: 120), malentities act directly on organisms, causing reduced fitness, and the effects of malentities do not increase with greater population densities. The term constraint will be taken here to mean something slightly different to a limiting factor or malentity. Constraints are: (1) not couched in density-dependent terms; (2) related specifically to fitnesses of individuals rather than of populations; and (3) functions of the instantaneous 'amount' of the constraint to which an individual is exposed.

2.3 What are the problems in quantifying ecological versatility?

The operational use of concepts such as those described in §2.2 almost always is easier said than done. Cut and dried statements like 'compare relative fitness gained from each resource with the relative availability of that resource' are (relatively) well-defined objectives, but the practicalities of doing so are rather less clear. Why is there such a difficulty in translating simple objectives to operational procedures? The purpose of the current section is to describe a series of problems that arise when one sets out to measure the ecological versatility of species in the field. Some problems are acutely related to matching or identifying the correct spatial and temporal scales at which to conduct the measurements. Other difficulties involve the correctness or otherwise of the array of resource categories used, especially the combination of possibly heterogeneous resources into arbitrarily defined operational categories, and the sheer difficulties involved in actually measuring resource use and availability.

Spatial scale – ecologists and organisms

The spatial resolution of resources (and constraints) is not an absolute in community ecology because of the importance of the biology of the organisms involved and the scale of resolution (Kotliar and Wiens 1990, Rahel 1990, Holling 1992, Levin 1992, Milne *et al.* 1992, Tilman 1994). That is, the perceptual abilities and mobilities of organisms determine the relative scales at which spatial variability appears significant to those organisms, and at which an organism can and does respond to that perceived variability (Wiens *et al.* 1987, Kotliar and Wiens 1990, McLaughlin and Roughgarden 1992).

Spatial heterogeneity in biological and physical structure (and hence of resources and constraints too) at the scale over which populations are active is a continuum (see Addicott *et al.* 1987: 341, Pahl-Wostl 1993), but many schemes have been proposed that attempt to impose or recognize significant 'breaks' along that continuum. In optimal foraging theory, for instance, there are two spatial scales of interest, the *patch* and the *habitat*, each of which is contingent on the organism in question. The difference between them is that a habitat is regarded as an 'infinite' patch in which the behaviour of the foraging organism does not alter the potential rate of gain of food nor the time for which that rate is maintained (Stephens and Krebs 1986). By definition, patches are depleted by the activities of foragers.

A more general scheme is that put forward by Morris (1987a). He recognized three main levels: (1) *habitat* – characterized by a distinct suite of chemical and physical properties; (2) *macrohabitat* – a distinguishable area within which an individual undertakes all of its biological functions in a typical activity cycle (e.g., home range); and (3) *microhabitat* – a patch whose qualities can be discriminated such that an individual will choose to spend an amount of time within that patch commensurate with the suitability of the patch (food, shelter, etc.).

A habitat is defined in Morris' (1987a) scheme independently of the biology of the study organisms, whereas both macrohabitat and microhabitat explicitly depend upon the characteristics of organisms. Habitats themselves are *not* true resources in the sense adopted here – they properly should be viewed as integrals of all true resources exploited and constraints contended with by individuals within the habitats. Fitness effects within a habitat type therefore are summations of the exploitation of resources that actually lead to increased fecundity, survivorship, growth, and so forth, modulated by the impact of the constraints encountered within that habitat. Macrohabitats are more uniform in character than habitats, but still more heterogeneous than microhabitats (Morris 1987a).

Microhabitats are more or less homogeneous, and have been viewed as patches 'containing' true resources, and the densities of those resources are the means by which microhabitats are exploited differentially (e.g., MacArthur and Pianka 1966, Emlen 1973, Holbrook and Schmitt 1988). Others think of microhabitats as providers of shelter or refuge in their own right, which thereby actually increase fitness (Dobkin 1985, McEvoy 1986). In the first case, the microhabitat would be similar to a habitat in being a battery of resources; in the latter view, the microhabitat is a true resource. This dichotomous terminology is so deeply entrenched in the literature that no single usage now seems possible.

Other workers eschew these relatively artificial distinctions, aiming for more general pictures of the relationships between spatial heterogeneity and the responses of organisms. For example, Addicott *et al.* (1987) presented a scheme explicitly recognizing that scales of interpretation are characteristic of each organism. The means by which they did this was the concept of the *ecological neighbourhood*. This involves determining the appropriate spatial and temporal scales for each organism that are functions of the ecological processes involved, the degree of environmental heterogeneity *as it appears* to the organisms in question, and the effective radius of activity or influence of the organisms over those time scales.

Although undoubtedly a reasonable development, the ecological neighbourhood methodology need not lead inexorably to a better understanding in all circumstances. For example, Wiens *et al.* (1987) suggested that hypotheses concerning causal processes (e.g., interspecific competition, speciation) should be tested only at the scale at which the process is likely to be influential. Hence, they argued that there is little point in searching for the expression of adaptations expected under the operation of a process at an inappropriate (spatial) scale. A similar argument presumably can be made for temporal scales (see Connell 1980). However, tests for the effects of a particular process need not reveal expected patterns even at the 'correct' scale and under the actual influence of that process, particularly if the process is viewed in isolation. For example, Maurer (1985) suggested that gene flow is so high in many continental species of birds (Avise 1983) that the spatial scale of variation of phenotypic characteristics is more coarse than the scale of variability in habitat structure. Even if interspecific competition, for example, were the most potent force in shaping ecological characteristics, high gene flow probably smooths out any phenotypic changes arising from interspecific competition across quite different assemblages (Maurer 1985; see also James *et al.* 1984 and Endler 1986: Chapter 4). Thus, a more holistic picture of bird assemblages over contiguous habitats would not necessarily lead to an expectation of ecological and morphological differentiation even at the same scale at which that process was actually operating (Connell and Sousa 1983, Maurer 1985, Addicott *et al.* 1987).

But why is ecological scale a significant problem for evaluating versatility? The perceived components of versatility depend upon both the ecologist who undertakes the study and the methods she or he adopts. Thus, the apparent versatility of the same population of individuals with respect to a specific component of ecological versatility could differ substantially depending upon how the study is circumscribed (Bock 1987, Morris 1987a).

Specifying versatility of spatial occurrence is a classic case of the difficulty of demarcation with respect to scale. Notwithstanding the differences in viewpoints between ecologists, the scale at which versatilities are gauged by an observer often will not be in accordance with the scale at which individual organisms perceive and respond (Brown 1981, Addicott *et al.* 1987, Norbury and Sanson 1992: 2). Hence, versatility estimates often may be poor or inadequate reflections of the ecological capabilities of organisms (Maurer 1985). The ecologist's and the organisms' yardsticks often will be quite different (Levin 1992: 1959, Milne *et al.* 1992).

The difficulties associated with ecological scale mean that there is little prospect of success in trying to generalize definitions for resources to apply to all cases and all situations. Thus, the best that can be achieved is clear, justified statements of the bases used for estimating versatility. This means that all conclusions drawn for such a study are conditional on the resource categories that are defined by a worker. In this sense, ecological versatility is a relative notion that is not easily generalized to all scales (see Wiens *et al.* 1987).

Resource ordering and aliasing

Two points, related to enumerating how many resource categories there are, and how 'similar' the resources are (in some sense), need to be considered as difficulties in measuring ecological versatility. The first point is the problem of lumping and splitting resource categories. Errors in estimating versatility will arise by selecting resource categories in either a too-coarse (lumped) or too-fine (split) fashion than is ecologically meaningful. This phenomenon is reminiscent of *aliasing* in signal processing (Platt and Denman 1975), which is the generation of artefacts by using inappropriate sampling rates. Aliasing is rife in many areas of ecological thought, for example, in food web theory (see Paine 1988, Martinez 1991, Pahl-Wostl 1993) and in predator–prey studies (see Greene and Jaksić 1983 for a discussion of the effect of aliasing). I take some liberties in generalizing the term 'aliasing' in the context of versatility to mean the generation of estimation errors by ecologically erroneous lumping or splitting (see also Mac Nally 1994a). Resource aliasing clearly will have serious effects on our impressions of versatility, as is shown by the simulated effects of arbitrary lumping and splitting on breadth estimates by Petraitis (1979) and others.

The second point concerns the effect that resource ordering has on perceived versatility. By resource ordering, I mean the position of resource categories on a conceptual gradient so that one resource type can be said to be more similar to a specific alternative resource type than to another. If resource categories in a study were prey taxa for example, then versatility of resource exploitation depends upon how different the prey taxa are in terms of their biologies, for instance, catchability. Although there are objective ordination and statistical techniques for assessing similarities in habitat structure (Mac Nally 1990a; see also Chapter 8), objective means generally do not exist for ordering resources. Some ways of addressing this problem have been suggested (e.g., Gray 1979, Gray and King 1986), but these are based on the discriminatory abilities of the consumers of the resources so that the categorizations are not objective in this sense. Nevertheless, a measurement of versatility with respect to an ordered set of resource categories has a higher information content than an analogous measurement with respect to a purely nominal array of categories (McNaughton and Wolf 1970: 133). The problem of ordering is overcome to some extent by using conformance statistics, the basis for which are discussed in §2.4, although superior versatility measures might be constructed if resources could be meaningfully ordered.

The practicalities of measuring utilization and availability

Many workers, particularly theorists, tend to overlook practical issues involved in determining the mix of resources used by organisms and the actual availability of those resources. These issues largely are ignored under the banner of producing general models that may be applicable to many systems. It may be useful to discuss briefly some aspects associated with field measurements of utilization and availability so that these problems can be borne in mind.

Unlike many areas of community ecology (see Chapter 3), there has been no shortage of interest in determining the degree of correspondence between utilization and availability of food in generalist terrestrial herbivores (Westoby 1974, Belovsky 1986). Therefore, this short discussion is largely based on these animals.

In many instances, realized diet may be difficult to determine because of the means by which dietary information has to be collected. For example, Norbury and Sanson (1992) discussed reasons why many common methods of measuring utilization may yield inaccurate views of utilization. If food use is determined by calculating the difference between the amounts available prior to consumption and the residue following grazing or browsing, then factors unrelated to consumption may impair the accuracy of estimated consumption (e.g., differential regeneration of plants, or differential susceptibility to trampling). Sampling the contents of the mouth, oesophagus, and stomach, or faeces, each has a set of problems, largely related to different vulnerabilities of plants to mastication and digestion.

The estimation of the apparent availability of alternative plants also can be fraught with problems. The scale over which the measurement is made is crucial, as is relative conspicuousness (Feeny 1976). Norbury and Sanson (1992) conjectured that small-sized herbivores may be better able to distinguish between food types on a finer scale than large herbivores, thereby producing better nutritional selections. Thus, availabilities may be perceived differently by organisms on as simple a basis as body size.

Holmes and Schultz (1988), Wiens (1989a) and many others have emphasized the difference between the actual density of a resource (its 'abundance'), which is independent of any user of that resource, and 'availability', which is the perceived density of a resource with respect to a specific user. Clearly, the resource 'availability' that human workers measure probably will be very different from the perceived availability as it relates to the study organisms, and availabilities will differ between even those organisms that are syntopic with one another.

Many workers recently have advocated that dietary selectivity (hence versatility) should be determined under field conditions because laboratory experiments can yield misleading results due to several causes (e.g., mix of alternative plant types, climate, other species, detection, accessibility, hunger status, handling time (the latter also affects field studies, see Fairweather and Underwood 1983)) (Soberón 1986, Crist and MacMahon 1992, Norbury and Sanson 1992). The sheer logistics of gathering data on relative availability also may deter accurate measurement (Belovsky 1986). Thus, ecologists almost always deal with approximations to the availability and use of resources, and the accuracy of these measurements must vary from organism to organism and from researcher to researcher.

Overview

Resources comprise the reference frame by which we gauge ecological versatility. This means that it is important to delineate meaningfully and limit the resource categories that are used to estimate the ecological versatility of a population. But this process always has been a serious bugbear because of the ingenuity living organisms show in finding novel ways to exploit ecological opportunities. Therefore, unfortunately, the notion of a resource is a complex one in which there is no shortage of existing definitions. I have used the conditions that to be a resource for an organism, an environmental component (*sensu* Andrewartha and Birch 1984) must be one to which an organism gains exclusive access (for some appropriate period of time) and through that access, the organism increases its fitness. (Conspecific mates are not regarded as resources although they fulfil these conditions. Ecological versatility properly refers to environmental components subject to nonsexual selection, see Endler 1986: 8.) This definition serves only to describe in a general way the qualities that resources should have. I believe that there is *no* completely general way of delineating resources that can accommodate all organisms under all conditions, and under all spatial and temporal scales, which has been emphasized by some workers (e.g., Addicott *et al.* 1987). However, some restrictions seem relatively easy to apply. First, resources that are consistently superabundant with respect to the demands of consumers cannot influence optimization choices in either proximate or evolutionary terms, and so should not be included in resource sets. And second, it is important to avoid aliasing resource categories as far as is possible. For example, Greene and Jaksić (1983) have argued, correctly in my mind, that aliased size-classes of prey populations (i.e., allocation of prey into size-class bins irrespective of taxonomy) should *not* be regarded as resource categories; rather, the resources should be viewed as the individual species of prey. Lumping prey species by using size or higher taxonomic units seems fraught with danger because of the arbitrary way in which partitions can be selected. In general, the difficulties involved in estimating ecological versatility described in this section show that it is necessary to: (1) enunciate clearly the resource categories that are used, and justify the choice of these categories; (2) interpret the use of resources by organisms in a way that relates to the fitness gained; and (3) provide estimates of resource availabilities at spatial and temporal scales appropriate to the use of the resources. Adherence to these three rules would greatly improve knowledge and understanding of the patterns of ecological versatility in field populations.

2.4 How is versatility measured?

In general, the measurement of resource utilization is hard to relate directly to fitness increments. This is despite the crucial role that fitness plays in being the only common unit among a range of different resource types. For example, food translates into fitness increments primarily by the use of the energy and nutrient content of the food for individual maintenance and the production and nurture of the offspring. Physical shelter affects fitness in various ways, by diminishing the energy requirements posed by the stresses of hostile physical conditions or by reducing exposure to predators.

The analysis of versatility is hindered by the conceptual separation of resources from the underlying basis of fitness. Versatility should be measured relative to fitness increments associated with various categories of resources, but typically this is not done and versatility becomes a measure of the way in which the resources themselves are utilized.

Interpretations of data on resource exploitation depend sensitively on the scales that are used to compute breadth and overlap statistics (Case 1984). Case cited instances in which prey volumes (e.g., Roughgarden 1974a, Pianka 1975, Stamps 1977), numbers of prey items (Stamp and Ohmart 1978) or both (Schoener and Gorman 1968) have been used to analyse versatility. Case (1984) concluded that, at least for food items, mass or some other measure proportional to the nutritional content of the food should be used (see also Simberloff and Dayan 1991: 124). Such measures can plausibly be linked to fitness gains in most circumstances.

Assuming that meaningful fitness increments can be estimated for various resource categories, the next issue is to summarize patterns of exploitation with respect to those resource categories. But how can these patterns be characterized? The short answer is by using indices of niche breadth, which I refer to here as *exploitation indices* for convenience.

Exploitation indices include statistics that summarize (niche) *breadth* and also (niche) *overlap*. Measures of breadth indicate the degree to which resource exploitation is evenly spread among alternative resource categories, and may or may not be related to availability (Petraitis 1981, 1983). Low values of breadth imply that a small number of potential resource categories is utilized, so that the population is 'specialized' in this sense. The argument is made more convincing when based on resource availabilities because a narrow range of resource utilization may merely reflect a biased distribution of availabilities (Petraitis 1979). Overlap has not been referred to in detail in this chapter, but has formed a significant part in ecological thought for many years (e.g., Levins 1968, Pianka 1974, May 1974b, Slobodchickoff and Shulz 1980, Turelli 1981, Rummel and Roughgarden 1985; see Chapter 7).

Following Hill (1973), Petraitis (1981) noted that many existing indices of niche breadth (Levins 1968, Schoener 1974b, Hurlbert 1978, Petraitis 1979) were members of the same family of mathematical functions. These statistics differ in the relative degree to which utilization and availabilities are weighted within the summary index, and the influence that rarer and more common resource categories are assigned (see also Colwell and Futuyma 1971 and Hanski 1978). The general function is:

$$g(b,c) = \left(\sum p_i (p_i/a_i^c)^{b-1}\right)^{1/(1-b)};\ b, c \in 0, 1, \ldots \qquad (2.1)$$

where p_i is the observed proportional utilization of resource state i, a_i is the proportional availability of resource state i, and b and c are the two coefficients that differ between indices.

Expression (2.1) is useful for summarizing the qualities of breadth measures, in particular, for discriminating between two main classes of breadth index. Petraitis (1981) pointed out that the two indices proposed by Levins (1968) do not refer to resource availability; thus, availability is implicitly set to the inverse of the number of resource categories (i.e., $a_i = 1/r$, where there are r resource categories). Any index in which $c = 0$ has this property. An expression of the form $f(g(b, 0))$, where $f()$ is a monotonic function (usually natural logarithm), can be regarded as a *nonconformance* breadth index. Nonconformance breadth measures are not functions of availabilities – there is no test of the degree of matching between utilization and availability. On the other hand, any index in which $c > 0$ is a *conformance* breadth index.

One of the criticisms of the use of exploitation indices over the years has been the difficulty in providing statistical statements for a measured value. If two values differ, is that difference meaningful, or just a chance deviation? With the advent of resampling schemes such as the bootstrap and jackknife (Efron 1982), methods for specifying confidence intervals have been developed (e.g., Mueller and Altenberg 1985) and applied successfully to field data (e.g., Llewellyn and Jenkins 1987). Thus, there seem to be no major obstacles to the detection of changes in versatility (due to season, stage, location, etc.) by using the appropriate conformance statistics.

All indices of the form of (2.1) are limited by being explicitly related to patterns of resource use summed over entire populations. Given the focus on individual exploitation strategies adopted here, it is important to develop more comprehensive statistics that are capable of dealing with differences within populations. Wissinger (1992) recently addressed this issue by adapting the L index of Hurlbert (1978) for use with stage-structured populations. He showed that inferences concerning potentially important interactions within complex faunas can be more profitably explored by using indices that are based simultaneously on utilization and conformance within and between the components of differentiated populations.

It is worth reiterating the desirability of basing both breadth and overlap indices on exploitation rather than utilization, where the former accounts for differential fitness returns of alternative resources. At least if workers think in terms of fitness, they, like Case (1984), most likely will translate raw utilization measures into terms more consistent with fitness returns.

2.5 Local and global pictures of species' versatilities

Recall from Chapter 1 that there is a need to distinguish between levels of versatility shown by populations of species in a local sense, and versatility displayed over all populations of a species (see Fox and Morrow 1981, Nitao *et al.* 1991). There also is a need to differentiate ecological versatility from the degree to which species can occupy different types of habitat. Some of the ideas related to this issue are expanded upon in this subsection.

At scales larger than habitats there are *landscapes*, *regions*, and *biomes*, the first two of which are considered later in this book. Landscapes are regarded as mosaics of different habitat types. Sometimes, they are defined with respect to dispersal distances of particular organisms. Thus, two habitat patches are part of the same landscape for a species if the populations in the patches communicate with one another through migration (Danielson 1992). This means that the populations resident in such habitats can potentially affect each other's dynamics. Regions are larger again, encompassing many landscapes and usually are convenient geographic domains that a researcher wants to use (e.g., Williams 1988).

The concepts of *cosmopolitanism* and *ubiquity* (Hesse *et al.* 1951: 153) were raised in Chapter 1. A ubiquitous species can successfully occupy a relatively large number of distinct habitat types, habitats that differ in *physiognomy* (i.e., physical structure, Whittaker 1975) and/or *floristics* (Rotenberry 1985). Ubiquity can be equated directly with habitat versatility. A formal definition of ubiquity is the degree to which a species maintains the same densities across habitats (McNaughton and Wolf 1970, Rice *et al.* 1980). On the other hand, a cosmopolitan species is one that occurs over a large geographic range (see Rapoport 1982, Brown 1984, Bock 1987), but not necessarily in a large number of structurally or floristically distinct types of habitat. Many species of terrestrial animals are cosmopolitan because they are strongly associated with floristic associations that themselves are cosmopolitan, so that a large geographic range is not necessarily indicative of habitat versatility or ubiquity. Conversely, some species may occupy limited geographic ranges yet display relatively great versatility by occupying many distinct habitat types (May and Southwood 1990). Of course, many widespread species also are versatile with respect to habitat (Pagel *et al.* 1991).

These contrasts are shown in Fig. 2.2, where the positions of four extreme patterns of geographic versatility are shown in relation to the

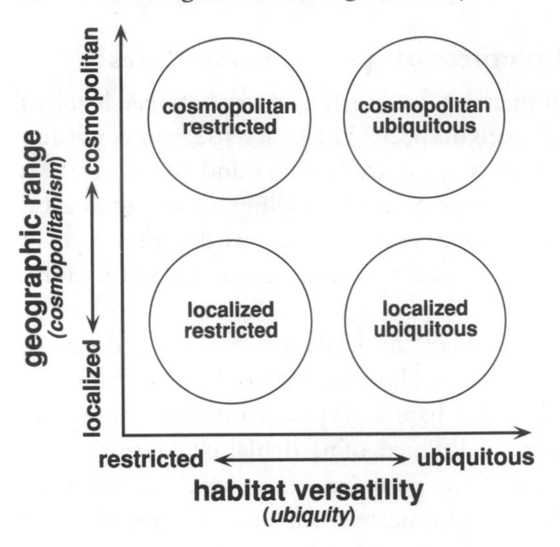

Fig. 2.2 The contrast between cosmopolitanism (the expansiveness of geographic range) and habitat versatility or ubiquity. Four extreme conditions are shown. See also Hesse *et al.* (1951).

cosmopolitan and ubiquity axes. Thus, there may be taxa that can be viewed as cosmopolitan generalists (occupying many distinct habitat types over a wide area), cosmopolitan specialists (few habitats distributed widely), localized generalists (many habitats in a small geographic range), and localized specialists, whose populations, in the extreme, may consist of only a few individuals living in a single valley or shore with very specific habitat requirements. Several notable cosmopolitan generalists were cited by Hesse *et al.* (1951). The cosmopolitanism of a protozoan, *Cyphoderia ampulla*, is striking, ranging from the Arctic Ocean and central Europe to Argentina. Its ubiquity also is pronounced as it occurs in most aquatic habitats.

Although habitats themselves are *not* true resources, patterns of habitat versatility or ubiquity nevertheless are among the most intensively studied features of ecological specialization and generalization. Thousands of research hours have been spent by community ecologists on classifying habitats and recording which species occur where and when (e.g., James 1971, Holmes *et al.* 1979, Wiens *et al.* 1987, Mac Nally 1989). The implicit aim of much of this work has been to infer just which resources and/or constraints actually control population dynamics and distributions, which is most effectively done by performing perturbation experiments (although see Mac Nally 1983a, Roughgarden 1983, Yodzis 1988, Wiens 1989b and Quinn and Keough 1993 for some of the limitations of this approach). Nevertheless, niche theory has depended upon specific responses to habitats as types rather than functionally significant resources within habitats (e.g., MacArthur *et al.* 1966, Rosenzweig 1981). I return to issues related to habitat versatility in Chapter 8.

2.6 A case study

It would seem to be advantageous to draw together some of the main themes of this chapter by using an example drawn from the literature. This work shows how a lack of precision in terminology and methodology can be misleading when one tries to determine from published material a measure of the ecological versatility of a field population. I hasten to add that this particular study is no worse than many others, but it does encapsulate several of the more important points that I have tried to make in this chapter.

The study involved the dietary versatility of a species of colubrid snake of Hispaniola, *Darlingtonia haetiana*. Data were obtained by gut-contents analysis of 131 specimens from five collections. The author contended that 'The results were surprising and indicate that *D. haetiana* has a very narrow trophic niche', and that '*D. haetiana* is stenophagous ... Of 46 prey items found in digestive tracts, 45 (97.8%) were *Eleutherodactylus*' (Henderson 1986: 529). (Note that stenophagy means having a narrowly based diet, *stenos* – narrow and *phago* – to eat.) These rather bald statements would lead one to believe that this snake indeed is an archetypal dietary specialist, in this case on an anuran amphibian, with virtually no dependence on lizards, the staple of most West Indian snakes (at least of the families Boidae, Tropidophiidae, and Colubridae). Yet this picture is misleading for several reasons, especially in relation to the issues of utilization and availability and aliasing that have been discussed in this chapter.

To see this, consider some of the points raised by Henderson (1986) himself in this paper. He noted that *Darlingtonia* is an exceptional snake on Hispaniola because it occupies upland areas in which *Eleutherodactylus* most commonly occurs (other species of snakes do not). Henderson pointed out that lizards also are present in the upland areas (e.g., *Anolis*, *Chamaelinorops* and *Sphaerodactylus*), but contrary to the secretive, retiring *Darlingtonia*, the lizards generally are diurnally active and scansorial. Thus, the lizards are spatially and temporally segregated from *Darlingtonia*, and in this sense, are not 'available' to it. On the other hand, the nocturnally active *Eleutherodactylus* seek cover during the day in much the same microhabitats as *Darlingtonia*, causing Henderson (1986) to conclude that *Eleutherodactylus* is 'in the right place at the right time' to act as fodder for *Darlingtonia*. Thus, by this analysis of availability and utilization, the degree of stenophagy exhibited by *Darlingtonia* would not appear to be very extreme.

Perhaps the stenophagy conclusion could be justified if *Darlingtonia* concentrated disproportionately on only one or a few species of the available *Eleutherodactylus*? But, here too, *Darlingtonia* appeared to be unselective, although availabilities of different species of *Eleutherodactylus* were not known. Henderson (1986) found the remains of (at least) eight species of *Eleutherodactylus* in gut contents, and 13 egg masses as well. These results suggest not only little species-specific specialization, but no stage-specific specialization either. This is an example in which conclusions based on aliased resource categories yield a misleading picture of the true degree of dietary specialization. Perhaps these internal contradictions led to the final sentence, in which, notwithstanding the previous quotations, Henderson (1986: 530) concluded that 'Hispaniolan colubrid snakes are, apparently, *opportunistic* predators, exploiting those prey that are the most often encountered' [italics added].

There can be little question that the vague use of terms like 'stenophagous' and 'frog specialist' can arise if availability (rather than abundance) is not considered, and if resource states are aliased severely. This can be the only explanation for concluding in the same paper that a species is both specialized and opportunistic. Lumping by prey taxon is a particularly insidious form of aliasing (see Chapter 4 and Maiorana 1978b), and should be avoided.

2.7 Summary

Ecological versatility is a measure of the degree of matching between the fitness gained by utilizing resources (≡exploitation) and the relative availabilities of those resources. In simple terms, generalists derive fitness returns from a set of resources in proportion to the relative availability of those resources. More specialized organisms derive larger fitness gains from some resources than would be expected on the basis of the relative availability of those resources, and disproportionately less from others. To make these comparisons requires a spectrum or set of resources against which the exploitation–availability relationship can be gauged.

A perennial problem is how to define and distinguish between resources. There is a welter of existing definitions for the term 'resource', which seems to reflect the underlying impracticality of trying to cater for all organisms and all situations in the one framework. The best that can be done is to make clear statements of what the resource categories are, and to try to justify these choices. This allows other workers to gauge whether the derived versatility estimates are reasonable or not based on the stated set of resources.

Utilization, the 'use' of a resource, is different from exploitation, which is the fitness return corresponding to that utilization. The effective difference between utilization and exploitation often may be slight if the resources are similar in returns (e.g., similar nutritional content and risks associated with gathering the resources, if foods). However, differences sometimes may be so great that completely different interpretations of the versatility displayed within populations may ensue if the distinction between utilization and exploitation is not made. Patterns of versatility need not be governed by the availabilities of resources alone, but also will be affected by environmental components impeding the free use of resources, which here are called constraints. The latter can be either abiotic or biotic in nature, and may force some organisms to be more generalized or more specialized in resource use than they would otherwise be in the absence of the constraints.

Several technical issues are discussed in this chapter. First, ecological scale is important because perceived the versatilities in populations usually will change with different scales of observation (both spatial and temporal) – resources and the capabilities of organisms are scale-dependent. Second, measurements of versatility are influenced by the biases of researchers (conscious or otherwise), particularly in terms of their proclivities to split or lump resource categories in arbitrary ways. The use of a set of resources that is biologically inappropriate is connected with several themes considered here (i.e., definition of resources, ecological scale), and is referred to as *aliasing*. Third, field measurements of resource utilization and availability present their own sets of practical and operational problems, which need to be fully understood if versatility is to be estimated consistently and correctly. Fourth, the degree of versatility expressed in populations at the 'local', or population, scale must be distinguished from between-population versatility (Fox and Morrow 1981, Singer and Parmesan 1993; see Chapter 1). And last, there is a need to distinguish between cosmopolitanism (the geographic range of a species) and ubiquity (the diversity of distinct habitat types occupied by a species).

3 · *Studies of versatility in natural populations*

My prior impression of our knowledge of patterns of ecological versatility in natural communities was that it was limited, incoherent, and inconsistent. This impression I can trace at least in part to several critical papers published in the 1970s and early 1980s, especially those of Schoener (1974b), Hurlbert (1978), Petraitis (1979), and Feinsinger *et al.* (1981). These workers argued strongly that the characterization of niche breadth is rather pointless if resource use is not linked to the relative availabilities of the resources. Perhaps Petraitis (1979: 709) said it most succinctly, by concluding: 'Yet in order to understand the relationship of niche breadth and overlap to community structure, knowledge of available resources is imperative.' Thus, monitoring the availability of resources as well as their utilization (i.e., conformance) is seen as being the *crucial* part of estimating ecological versatility in the field.

However, given the semantics and operational problems outlined in Chapter 2, the issue of availability is just one of the difficulties that has to be contended with when one tries to quantify ecological versatility. We also need to consider, among other things, the difference between utilization and exploitation (i.e., fitness returns from utilizing resources), the proper identification, enumeration and restriction of resource categories (Real 1975), availability as opposed to abundance for each consumer taxon (Wiens 1989a), spatial and temporal scales of monitoring and variation (Chesson 1985, Powell and Richerson 1985), and differentiation within populations and its effect on versatility (Polis 1984, Wissinger 1992; see Chapters 5 and 6 for a detailed treatment). My impression was that, for all of these aspects, knowledge was even poorer than for the conformance issue.

But impressions can be subjective and misleading, as Wiens (1989a: Chapter 14) conceded in his analysis of trends in the nature of studies of bird communities between 1950 and 1984. So, like Wiens, I chose to determine the objectivity of my impressions. To do so, a set of papers describing work on natural populations in which the pattern of resource use was the principal theme was evaluated in terms of how studies of

Table 3.1. *The criteria used in the analysis of studies of ecological versatility in natural communities*

1. Resource specification	Was the set of resources *clearly* defined? Were heterogeneous resource dimensions considered?
2. Fitness	Were differential fitness effects related to utilization of alternative resources considered?
3. Constraints	Were the effects of constraints on patterns of resource use considered?
4. Individual patterns	Were patterns of resource use pooled for all members of populations or were individual differences monitored?
5. Stages/subsets	Were differences between stages (including age-classes) or other population subsets (e.g., by gender) recorded?
6. Conformance	Were availabilities of resources considered as a key feature in patterns of resource use?
7. Temporal variation	Were studies conducted in such a way that temporal variability in patterns (especially by season, month and within-day) could be detected?
8. Spatial variation	Were studies conducted so that spatial variability in patterns could be detected?
9. Versatility indices	Were summary statistics used to characterize the ecological versatility of populations?

ecological versatility have been conducted in recent years. Therefore, Chapter 3 involves an analysis of 145 studies (see §3.2 for selection criteria, and Appendix A.1 for a complete listing) that were mostly published during the 1980s and early 1990s. Each study in the survey was appraised to determine the degree to which it satisfied the criteria outlined in Chapter 2. This review clearly is an indicative rather than a definitive survey of the ways in which studies of ecological versatility have been conducted. Nevertheless, the survey should be a relatively representative sample of recent studies of ecological versatility.

3.1 Survey parameters

In this section, I briefly outline the criteria against which each study in the survey was judged, which are summarized in Table 3.1. The first issue is whether a study provided a clear expression of the resource bases upon which versatility assessments could be made. Specification of resource categories is imperative, particularly if the problem of ecological scale is to be negotiated. A related topic involves the estimation of versatility with respect to heterogeneous resource dimensions, for example, food and microhabitat. Clearly, a comprehensive picture of versatility ultimately will depend on several resource axes, so that definitive estimates should try to marry together versatilities with respect to heterogeneous dimensions.

Fitness plays an important role in assessing versatility. Resource exploitation inherently differs from resource utilization because exploitation encompasses the costs and benefits to fitness in utilizing particular resources. Moreover, if the origins and maintenance of patterns of versatility are to be understood in evolutionary terms, then fitness (or more precisely the variance thereof) *is* the principal criterion (Futuyma and Moreno 1988, Brown and Wilson 1992). A well-focused study therefore should attempt to view patterns of resource use in terms of the fitness returns involved in using alternative resources.

Limitations on the exploitation of resources often will determine the actual patterns of versatility that can be adopted. A certain pattern of exploitation may yield the best possible return with respect to fitness, yet be unattainable because of the influence of external factors, namely, constraints. Studies also were evaluated as to whether any attempt was made to interpret observed patterns of resource use in light of the actual or potential effects of constraints.

The importance of assessing differences in patterns of exploitation between individual organisms cannot be overemphasized. An understanding of the evolution and maintenance of patterns of versatility hinges on knowledge of differences between individuals within populations (Roughgarden 1972, Polis 1984, Wissinger 1992). Therefore, the degree to which interindividual differences were monitored in studies was determined.

In practice, differences between individuals rarely have been considered in studies of resource utilization. More attention seems to have been paid to the properties of different stages or other subsets (e.g., sexes, clones) within populations (see Chapter 5). Therefore, whether studies registered differences due to stage or subset also was noted.

The significance of measuring resource availability and of conformance indices was discussed at some length in Chapter 2. Therefore, each study was assessed for whether resource availability was considered, and also, whether conformance between utilization and availabilities was used to interpret ecological versatility. The survey also is a timely review of the conformance issue, given that Petraitis (1979) believed that up until the late 1970s few field studies considered availabilities when estimating niche breadth. Has this situation changed in the past 15 years or so? This question also is addressed.

Patterns of versatility within populations may vary, particularly in response to changes in the availabilities of resources or under the impact of changing exogenous regimes (especially climate and predators or parasites). Complete descriptions of versatility for a local population require estimates to be made at different times. The two most obvious temporal domains are seasonal and daily, although monthly cycles may also be evident in some organisms. The survey included judgements as to whether studies took into account seasonal and/or daily variations in resource utilization, or at least, recognised the limited nature of estimates made for certain restricted times (e.g., breeding seasons of migrant birds).

Similar comments can be made for the influence of spatial heterogeneity of resource utilization within populations. That is, did studies take into account within-habitat spatial variability, particularly in coarse-grained situations (i.e., where the scale of spatial patchiness is large compared with typical activity ranges of organisms; see Emlen 1973)? The problems associated with estimating versatility with respect to habitat were raised in Chapter 2. There are two issues here: (1) are 'local' patterns of versatility displayed by populations of species *X* repeated in other habitats (see Fox and Morrow 1981)? and (2) what are the tolerances of species with respect to different types of habitat? The final parameter recorded for each study was the type, if any, of measure used for estimating ecological versatility (see §2.4).

The chapter concludes with a section on information on the distribution of generalized, specialized, and intermediate species in the surveyed studies. The perceived distribution of versatilities of collections of populations represents a significant constraint on the development of a theoretical understanding of versatility in general. Therefore, some comments are offered on this issue in §3.4.

3.2 Survey domain

Literature sources and selection criteria

The number of studies that might be included in a survey of ecological versatility is potentially immense. The number of journals publishing results of ecological research is growing rapidly, as are the numbers of pages in each. Therefore, several criteria were used to select a manageable number. The first was recency, specifically, studies published since 1977, so that some of the ideas emerging during the 1970s and early 1980s might have been expected to have been incorporated. Issues such as resource availability and utilization (i.e., conformance), raised in relation to the measurement of versatility by Schoener (1974b), Hurlbert (1978), and Petraitis (1979), and the importance of differentiation with respect to stage (e.g., Polis 1984, Werner and Gilliam 1984), fit this criterion.

A second criterion was the phraseology used in titles of papers. I selected papers in which the titles bore certain combinations of key words or phrases indicating that patterns of resource use by (and sometimes in relation to availability) one or more species were recorded. Typical combinations are listed here, with alternatives in brackets separated by the 'or' symbol '|'.

- (resource | food | dietary | habitat | microhabitat) (use | preferences | choice | availability | specialization | generalization) of . . .
- (resource | food | dietary | trophic) (partitioning | division | selection | ecology | relationships | divergence) among . . .
- niche (breadth | overlap | characterization | separation | partitioning | shifts | relationships | dynamics) (of | among) . . .
- (guild | community) (structure | organization | coexistence) of . . ., and
- (ecological | ecotypic) (segregation | differentiation) among . . .

(Although a digression, I think that it is worth pointing out that it remains relatively easy to select 145 papers with these terms in their titles, many of which are evocative of the impact of interspecific competition on resource use (and of course, many more than this number could have been analysed given greater endurance on my part!). This seems to confirm that interspecific competition still is regarded as a potentially crucial structuring agent in natural communities, as shown by Wiens 1989a: Chapter 14. Indeed, 74 of the studies explicitly mentioned, or framed programs, to study this process.)

The third aspect relates to the source of studies, which I deliberately biased towards mainstream ecological journals. Several more-specialized journals also were examined for pertinent studies, although the intensity of review was less than for the mainstream journals because papers in the latter tend to be several years more advanced in absorbing and applying new approaches and ideas (Wiens 1989a). The most extensive searches were of the main North American journals, particularly *Ecology* (1977–1991), *Ecological Monographs* (1977–1991), *American Naturalist* (1974–1991), *American Zoologist* (1970–1991), and the *Canadian Journal of Zoology* (1978–1991). The more recent volumes (1985–1991) of the principal European journals were scanned (i.e., *Journal of Animal Ecology*, *Journal of Ecology*, *Oikos*, *Oecologia (Berlin)*). Other general ecological journals included were the *Australian Journal of Ecology* and *Evolutionary Ecology* (both up to and including 1991). The marine journals *Marine Biology* (volumes 100–107, 1989–1990), *Journal of Experimental Marine Biology and Ecology* (volumes 26–144, 1977–1990), *Bulletin of Marine Science* (volumes 42–45, 1988–1989), and *Estuarine Coast and Shelf Science* (volumes 22–31, 1986–1990) were examined, as was the limnological journal *Hydrobiologia* (volumes 192–203, 1990). Volumes of avian (*The Condor*, *Auk*, *Ibis*, *Emu*), mammalian (*Journal of Mammalogy*), herpetological/ichthyological (*Copeia*, *Herpetologica*), and insect (*Ecological Entomology*) journals published between 1985 and 1991 (inclusive) also were scanned for relevant work. The volumes of journals listed in this and preceding paragraphs were searched most thoroughly, but this does not mean that appropriate papers that came to my attention from other sources necessarily were excluded from consideration.

Having identified the sources of the surveyed material, I now present a statistical overview of the types of studies, taxa, and habitats covered by the surveyed work. This information should help to place the conclusions of the survey in perspective. Note also that the tallies of all alternatives in some cases may exceed 145 because certain individual studies considered aspects that fell into several categories (e.g., birds regarded as pollinating resources for plants, but the plants also were viewed as nectar-bearing resources for the birds; hence such a study would fall into two taxonomic domains).

Taxonomy

The taxonomic distribution of studies probably reflects the greater attention devoted to ecological versatility by vertebrate ecologists, with 79 of 151 taxa overall based on this group (Table 3.2). Birds (mostly terrestrial) constituted 46% of studies on vertebrates. There were at least eight studies for each of the other vertebrate classes. Another 52 studies were on invertebrates, with the majority on insects (Table 3.2). Many invertebrate phyla were not represented, particularly those of marine environments. The number of studies on plants also was relatively small,

Table 3.2. *Taxonomic distribution of organisms investigated in studies that wee included in the survey (numbers of studies)*

Vertebrate animals	**79**	**Invertebrate animals**	**52**
birds	*36*	*insects*	*28*
forest birds (general)	8	insects (general)	2
frugivores	4	larvae (general)	3
insectivores	5	Hymenoptera	8
granivores	4	Hemiptera	5
nectarivores	7	Odonata	2
raptors	3	Ephemeroptera	3
ducks	2	Diptera	5
seabirds (general)	1	*other arthropods*	*10*
penguins	2	spiders	5
mammals	*13*	mites	2
rodents/small mammals	8	scorpions	2
bats	4	crabs	1
marsupials	1	*molluscs*	*9*
reptiles	*10*	marine gastropods	8
snakes	3	marine pelecypods	1
lizards	7	*annelids*	*2*
amphibians	*8*	polychaetes	1
salamanders	6	leeches	1
anurans	2	*zooplankton*	*2*
fishes	*12*	*bryozoans*	*1*
Plants	**20**		
angiosperms	*17*		
annuals	5		
biennials	3		
perennials	4		
shrubs	5		
mosses	*1*		
algae	*2*		

with just 20 representatives, the majority of which were on angiosperms (Table 3.2).

Habitats

Most studies were of continental systems (Table 3.3). Terrestrial systems on continents dominated the survey, with 97 studies. Fifty-two studies

Table 3.3. *Distribution of major habitat types investigated in studies included in the survey (number of studies)*

Continental terrestrial	*97*	*Marine margins*	*22*	*Continental waters*	*29*
forests and woodlands	52	coasts/nearshore	4	lakes	7
		rocky shores	8	pools/ponds/potholes	7
grasslands and prairies	27	coral reefs	4	streams	6
		kelp forests	2	wetlands/swamps	7
savanna/scrub	5	embayments	2	salt marshes	2
deserts	10	estuaries	2		
caves	1			*Islands*	*11*
rocky outcrops	1				
heathlands	1				

Table 3.4. *Zonal and geographic distributions of studies included in the survey (numbers of studies)*

Zonal	*145*	*Geographic*	*145*
tropics and subtropics	42	Africa	7
warm temperate	22	America, North	81
cool/cold temperate	72	America, Central	19
polar and subpolar	9	America, South	4
		Antarctica	2
		Asia	5
		Australia	10
		Europe	10
		Oceania	7

were set in forests or woodlands, and another 27 in grasslands or prairies. Twenty-nine studies were of taxa occurring in fresh or brackish waters, these studies being evenly spread among lakes, ponds, streams, and wetlands (Table 3.3). Only 22 studies were marine, none of which were of deep-water or abyssal taxa – studies of marine margins were predominantly of rocky shores or coral reefs. Another 11 studies were of insular terrestrial systems (Table 3.3).

Climatic zones and geography

About half of the studies were of cool or cold temperate systems (Table 3.4). Forty-two studies were located in the tropics or subtropics, another 22 in warm temperate zones, but just nine in polar regions (Table 3.4).

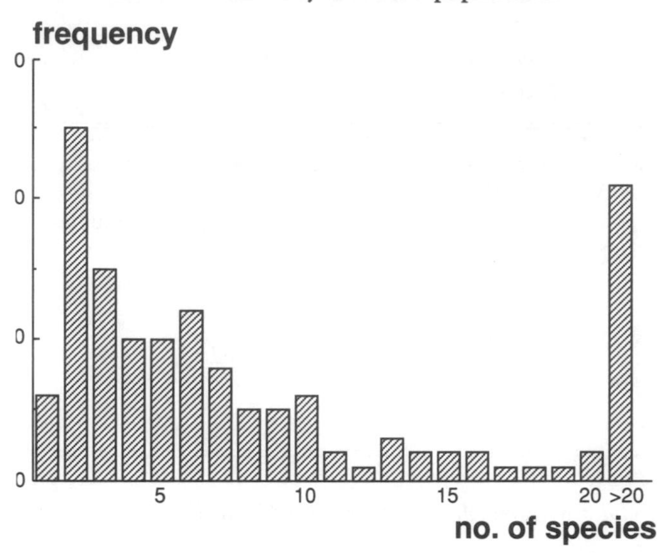

Fig. 3.1 Distribution of numbers of taxa investigated in survey studies. Some studies were based on several locations, each of which is included separately. A few reports involved plant–pollinator systems in which there were sets of both animals and plants. These sets are also counted individually here even though they arose from the one study.

Studies of North and Central (including Caribbean) American systems dominated the survey, constituting just over two-thirds of the total (Table 3.4). There were relatively few studies located in Asia, Africa, Antarctica and South America. Most reports from Oceania concerned systems in the Hawaiian and Galapágos Islands.

Numbers of species per study

The distribution of numbers of species per study is shown in Fig. 3.1. More than 50% of the studies involved fewer than six species, while less than 6% investigated systems of more than 35 species. Many studies involving small numbers of species tended to focus on specific subsets (e.g., nectarivorous birds; annuals) from a much broader taxonomic background (i.e., woodland birds; grasses and forbs). This means that the spectra of resources against which versatility was gauged were restricted relative to the total arrays of resources that may have occurred locally.

The frequent practice of limiting studies to smallish sets of similar species has a major impact on perceived versatility. Thus, scales and circumscriptions of studies may end up having more influence on the evaluation of ecological versatilities than do the actual resource utilization strategies of the species for the reasons outlined in Chapter 2. The versatility of a species will depend greatly on the degree of resource aliasing (i.e., inappropriate splitting and lumping), which is more likely to become serious if small subsets of species are considered in isolation. These problems may be severe in the survey because of the large numbers of limited sets of species considered. Nevertheless, the survey almost certainly reflects general patterns in versatility research, and probably is not biased greatly in this respect.

3.3 The survey

The results of the survey are presented below. For most of the criteria discussed above, statistics are provided for grades of adequacy. Thus, for example, the proportions of studies using conformance statistics are given, as are numbers of studies in which some reference to the importance of availability is made. The results of certain studies of particular interest also are described where appropriate.

Clear specification of resources

Resource categories were either well, or moderately well, defined in most of the surveyed studies, with only 14 reports providing unclear descriptions of the resource bases (e.g., Parsons 1978, Wheelwright 1985). Resources were viewed as belonging to one of four main groups: (1) space; (2) food; (3) time; and (4) biotic resources. 'Spatial' resources fell into two main subgroups, habitat, and microhabitat. Problems with treating habitats as resources have been discussed in Chapter 2, but, nevertheless, habitats frequently were regarded as true resources in surveyed papers. Microhabitats could be further divided into substrates, nutrient, light, or water gradients, and vertical grades such as bathymetry (e.g., distributions along kelp fronds, Lowry *et al.* 1974) or heights within woodlands. Food was treated reasonably consistently. Time consisted of patterns at the order of a year (phenology, seasonality), or at the within-day scale (activities). The biotic group of resources involved two broad divisions, hosts and vectors. Hosts may have been plants or animals, but implied a more crucial dependence than the use of the host for just food, for example. Vectors are organisms that transmit things for other organisms, and are typified by insect or avian pollinators or fruit and seed dispersers (see Howe 1984).

Table 3.5. *Combinations of major resource groups used in surveyed studies*

Group	S	F	T	B	Total
S	32 (5)	36 (8)	10 (3)	4 (2)	90 (22)
F		25 (1)	2 (1)	0	71 (14)
T			4 (2)	5 (3)	29 (13)
B				8 (4)	17 (9)

Notes: S, space; F, food; T, time; and B, 'biotic'. Figures in the body of the table are numbers of studies of the given combination of groups, while associated numbers in parentheses are numbers of 'types' of the given major group of resources (see text). The leading diagonal are numbers of homogeneous resource groups (i.e., just spatial, just food, etc.), while figures in the upper triangle relate to pairs of groups (e.g., food and time). The only set of three groups was space, time, and food, with figures of 8(4). These values are included in totals for the S, F, and T rows.

The distribution of studies with respect to the four main resource groups is presented in Table 3.5. Spatial resources were considered separately in 32 studies, and in 58 other studies with one or two other major resource categories (Table 3.5). Similarly, food formed the single resource group studied in 25 reports, but appeared in conjunction with spatial and/or temporal resources in another 46 studies (Table 3.5). The analogous figures for temporal resources were four and 25, and for biotic resources, eight and nine. There were studies in which all pairwise combinations of major groups were examined, except for the food–biotic resource coupling (Table 3.5). The only combination of three major resource groups was space–food–time, of which there were eight kinds (see below), while no single study encompassed all four major groups of resource (Table 3.5). Overall, spatial and food-based resources were much more intensively investigated than were temporal and biotic resources (Table 3.5).

Figures in parentheses in Table 3.5 provide details on the number of distinct combinations of major resource groups used in surveyed studies. These values reflect the diversity of types of study with respect to each major resource group or combination. For example, there were eight forms of space–food combination: (1) food and microhabitat; (2) food, habitat, and microhabitat; (3) food and habitat; (4) food and feeding substrate; (5) food, habitat, and feeding substrate; (6) food and nesting locations; (7) food and bathymetry; and (8) food, habitat, and bathymetry.

Fitness bases

Fitness is central to understanding patterns of ecological versatility, yet its determination requires rigorous sampling programs, plus substantial interpretation of the probable or actual effects on fitness of different patterns of resource utilization. Few field studies could be expected to meet these requirements. Nevertheless, some of the studies included in the survey did address fitness to some degree. This subsection provides a brief overview of the sorts of fitness variables used by authors who viewed aspects of fitness as being of direct significance to explaining patterns of resource utilization.

The expression of effects on fitness can take many forms, some of which are evident at the level of the population (e.g., growth in numbers of individuals, see Endler 1986), while others are more directly associated with the fitness of individuals. Differences in resource utilization can lead to changes in population characteristics such as density, colonization capacity, and stage-structure, or alter properties of individuals, such as fecundity, germination, growth rates, or other aspects of the nutritional status of individuals (e.g., energy or nutrient stores) (Mac Nally 1983a).

Analyses at the population level obscure many of the subtle although significant components of versatility discussed in Chapter 2, but, nevertheless, are substantially more illuminating than studies in which fitness is not considered at all. Investigations of the effects on individual fitness are most desirable – these fall into two distinct classes. The first involves an assessment of the constraints and putative nutritional or energy returns of utilizing different resources, and often is couched in terms of optimization theories (e.g., Pulliam 1985). That is, no measurements are made of the actual fitness effects *per se*. Such effects are inferred from indirect measurements, such as handling time or nutritional and energy content of foods (see Chapter 4). Studies in the second class were based on individual fitness, thereby addressing certain characteristics of individuals that can be related reasonably directly to individual fitness. Fecundity, individual survivorship, time to maturation, germination, growth, propagule production, and the like fall into this category.

About one in six (24) of the surveyed studies considered some aspect of fitness. Eight of these studies either measured or interpreted their systems in terms of the impact on one or more population variables that could be viewed as aspects of fitness. The most commonly reported effects were changes in demographic patterns, usually stage-structure (e.g., Polis 1984, Lishman 1985, Trivelpiece *et al.* 1987, Riechert 1991). For instance, Spence (1983) and Hairston *et al.* (1987) recorded changes in the densities of populations, while Werner and Platt (1976) and Platt and Weis (1977) monitored effects on propagules in relation to colonization ability.

Six studies analysed resource utilization in terms of the energy and/or nutritional costs and returns to individuals. These reports involved either birds (e.g., Pimm and Pimm 1982, Pulliam 1985) or arthropods (e.g., Harder 1985, Riechert 1991, Tanaka 1991).

More informative analyses from the fitness standpoint have been published for several groups of organisms. Plant ecologists frequently related resource utilization patterns to critical fitness variables, such as germination (Parrish and Bazzaz 1979, 1985, Gerard 1990), flower production (Lechowicz *et al.* 1988), somatic and reproductive growth rates (Parrish and Bazzaz 1979, Parrish and Bazzaz 1985), and seedling survivorship (Kephart 1983). Fecundity and maturation times have been documented by entomologists (Addicott 1978, Yu *et al.* 1990, Brakefield and Reitsma 1991, Via 1991), while general life-history effects have been reported for some groups of birds (e.g., penguins, Trivelpiece *et al.* 1987, and raptors, Steenhof and Kochert 1988).

One example of the use of fitness-related quantities in relation to ecological versatility is a study of use of cactus rots by species of *Drosophila* in north-western México (Ruiz and Heed 1988). In this work, Ruiz and Heed measured three indicators of larval performance: viability, size (indexed by thorax length), and development time (Fig. 3.2). These indicators were combined into a single measure that was deemed to summarize the fitness of larvae under different experimental conditions (viability multiplied by size, all divided by development time). The degree of density-dependent inhibition of larval fitness was gauged by rearing larvae at low and at high densities. The level of inhibition depended on the species of cactus involved, with little effect for cina (*Stenocereus alamosensis*) and prickly pear (*Opuntia willcoxii*), but a substantial reduction of fitness when larvae were raised on agria (*S. gummosus*) and organ pipe (*S. thurberi*) rots.

In general terms, larval performance in each species of *Drosophila* was better on their usual host plant, but a surprising result was that all species, even those that typically use *Stenocereus* (columnar) cacti, performed well on prickly pear. Ruiz and Heed (1988) suggested that *Opuntia* was the original resource used by the ancestral species of this *Drosophila* complex, and that each species remains capable of using it. However, the columnar-cactus breeders now occupy habitats that have become too arid and hot for *Opuntia*, and their 'specialization' is not so much physiological as one of availability (columnar cacti are readily available,

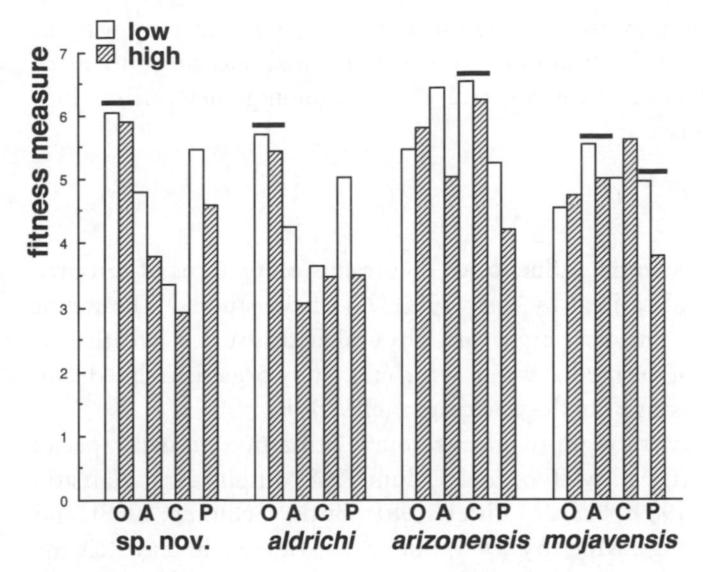

Fig. 3.2 Performance of larvae of four species of *Drosophila* on rots of four species of cactus from north-western México (after Ruiz and Heed 1988). Fitness is a function of larval viability, size, and development time (see text). Bars overlie the main host(s) of the species involved. Experiments were conducted with larvae at low and high densities. Cacti: prickly pear (*Opuntia wilcoxii* – O), agria (*Stenocereus gummosus* – A), cina (*S. alamosensis* – C), and organ pipe (*S. thurberi* – P).

Opuntia is much reduced in availability or absent altogether). On the other hand, the *Opuntia* breeders perform poorly on rots of columnar cacti because of physical and biochemical differences in the rotting processes that make them less amenable substrates than *Opuntia*.

Thus, by looking at the relative fitnesses associated with using different resources, Ruiz and Heed (1988) demonstrated that some species of *Drosophila* (*arizonensis, mojavensis*) show more limited versatility than they are capable of for ecological reasons, that is, due to the unavailability of certain species of cactus at some locations. Other species of *Drosophila* (*aldrichi*, sp. nov.) are physiologically limited in that they suffer reduced fitness on their non-usual hosts. This difference between ecological and physiological 'specialization' surfaces in several other contexts later in this book, especially in relation to the range of host plants used by herbivorous insects and differences in the versatility of host use by some species of animal parasites.

The survey shows that many workers have been aware of the central role that fitness plays in relation to patterns of resource use. However, none of the studies was conducted in a completely satisfactory way. In particular, few of the studies looked at differential effects on fitness of utilizing alternative resources, and then combining these effects into estimates of versatility.

Constraints

The actual or potential influence of constraints on patterns of resource utilization was described by only 19% (27) of the studies. The major classes of constraints were predators, physical factors such as weather or characteristics of the media within which the study organisms lived, and chemicals or toxins of biological origin (*allelochemicals*).

Predators were thought to affect resource utilization in many species of vertebrates (fishes, McIvor and Odum 1988; amphibians, Hairston 1980, Alford 1989; lizards, Lister 1981; birds, Schluter 1988) and invertebrates (bugs, McEvoy 1986; butterflies, Brakefield and Reitsma 1991; limpets, Mercurio *et al.* 1985). Polis (1984) and Polis and McCormick (1987) documented the effects of both cannibalism and intraguild predation on resource use.

Several physical agents were implicated in the control of ecological versatility. These included the effects of weather such as drought (Nudds 1983) and temperature (Schoener 1970), and physical and chemical attributes of both aquatic (salinity and water temperature, Flint and Kalke 1986, Thorman and Wiederholm 1986, Bergman 1988, Sogard *et al.* 1989; wave action, Leviten and Kohn 1980) and terrestrial habitats (Werner and Platt 1976, Mac Nally 1985). Adolph (1990) correctly identified temperature as a constraint rather than a resource and recognized its role in controlling the capacity of individuals to occupy microhabitats and habitats. A major class of constraints was biologically derived chemicals such as phenols, which appear to influence patterns of resource use because of differential palatability or toxicity (Chew 1981, Folt and Goldman 1981, Barker and Chapman 1990).

An important distinction must be drawn between broad effects on the distributions of species and local alterations of resource utilization apparently caused by the impact of constraints. In other words, some workers thought that factors such as salinity or temperature could result in the absence of a species from particular habitats or microhabitats, whereas constraints in the sense used here tend to compress or change patterns of resource usage at the local scale. Occupation of microhabitats frequently was deemed to be controlled at least in part by the effects of predators (Lister 1981, Hallett 1982, Mercurio *et al.* 1985, McIvor and Odum 1988). The clearest cases of moderation of local resource use were instances involving deterrent chemicals in plant–herbivore systems (Coley 1983, Barker and Chapman 1990) (see §4.1 for a detailed look at chemical interactions between plants and herbivores).

Results of the survey showed that the role of constraints in governing ecological versatility is not particularly well documented. Although 20% (29) of studies did view constraints as significant features of systems under investigation, few of these reports quantified the magnitude of the effects. The majority involved anecdotal discussions of the likely influence of constraints on patterns of resource use. There appears to be much scope to increase understanding of this relationship in a quantitative way, either by observational or experimental means (e.g., Denno *et al.* 1990).

Differences between individuals

The classification of versatility within populations depends upon knowledge of the similarity of resource utilization among individuals (or genetically identical clones). Studies were assessed as to whether interindividual differences were monitored, and the degree to which differences were reported or used in the analysis. I divided studies into four main categories, studies in which there were: (*A*) individual-based analysis and interpretation; (*B*) explicit recognition of interindividual differences, usually noted in the methods used, but not specifically reported or analysed in an individual context; (*C*) some level of implicit recognition; and (*D*) no mention or no apparent recognition of interindividual differences. There were six instances of each of the first three categories (e.g., *A*–Ehlinger 1990, Via 1991, and see below; *B*–Harder 1985, Johnson 1986 and Steenhof and Kochert 1988; and *C*–Findley and Black 1983, DuBowy 1988, and Mahdi *et al.* 1989). This indicates that only about one in eight studies (18) regarded interindividual differences as being an important component of the interpretation and understanding of versatility. Four studies of the first class are briefly described below to highlight the gains involved in gathering individual-specific information.

Trowbridge (1991) investigated whether a population of herbivorous sea slugs of the species *Placida dendritica* consisted of polyphagous individuals or collections of differentiated oligophagous individuals. In a series of experiments, she showed that the population consisted of sets of individuals that exhibited specific dietary preferences. Individuals did not appear to change their preferences through time, irrespective of hunger level, algal quality, or different social circumstances. Trowbridge also reported that individuals would perish when maintained on an unfamiliar algal species, even though that food was commonly consumed by other conspecific individuals. Thus, the level of generalization in food use in this population of *P. dendritica* is produced by a differentiation between individuals within the population rather than by each individual being generalized itself.

Fruiting phenologies of 45 species of plants in Florida hammock communities were studied by Skeate (1987), where, from this stance, frugivorous birds (and to a lesser extent, mammals) formed the resource base as dispersal agents for the plants. Skeate (1987) marked individual plants from all species and monitored fruit production of individuals through time. From these data, he computed a variety of indices that can be interpreted as between-individual and overall population variation in fruit ripening periods (see also Thompson and Willson 1979). These data then were interpreted in relation to the 'availability' of fruit-dispersing birds.

Dietary analyses of species of surface-gleaning bats from Panama were reported by Humphrey *et al.* (1983). Faecal pellets of 86 individuals of nine species of bats were used to assess diet (Humphrey *et al.* 1983). These data were used to cluster individual bats to determine their guild allegiances (Fig. 3.3). This study demonstrated that interindividual variability in resource (food) utilization can lead to conspecific individuals being distributed among different guilds, each guild therefore consisting of heterospecific collections of individuals. There was no repeated sampling of individuals so that one cannot say, on the basis of this work, whether results for particular individuals are representative of their diets through time. It is tempting to speculate that these bats may be so opportunistic that individual bats may 'change' guilds if sampled repeatedly through time.

Much the same form of analysis was conducted by Sherry (1984) on species of neotropical flycatchers (Tyrannidae) in Costa Rica, except that stomach contents rather than faecal pellets were used. Diets of some 126 individuals of 16 species were analysed. Prey taxa derived from stomach analyses indicated that well-defined sets of prey occurred, largely discernible in terms of detectability, substrate and location, and escape behaviour. Sherry (1984) computed indices corresponding to between-individual (within-species) heterogeneity of diet. By distinguishing

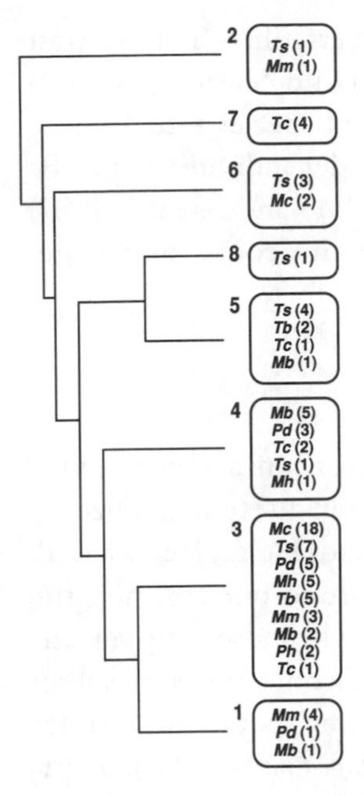

Fig. 3.3 Assortment of 86 individuals of nine species of surface-gleaning Panamanian bats on the basis of faecal pellet contents (after Humphrey *et al.* 1983). Numbers of individuals of each species in each cluster are shown in brackets. The dendrogram structure is a qualitative representation of the order of fusion of clusters. Note the occurrence of individuals of *Tonatia sylvicola* (*Ts*) in six of the eight clusters. Key for the other species: *Mm* – *Micronycteris megalotis*; *Mb* – *Mic. brachyotis*; *Mh* – *Mic. hirsuta*; *Mc* – *Mimon crenulatum*; *Tc* – *Trachops cirrhosus*; *Tb* – *Tonatia bidens*; *Pd* – *Phyllostomus discolor*; and *Ph* – *P. hastatus*.

between individuals, Sherry (1984) concluded that: (1) many species were not very variable in diet; (2) different degrees of variability were displayed by species *within* guilds; (3) overwintering migratory species were more variable that year-round residents; and (4) flycatchers of open habitats and of the canopy tended to show greater dietary variation than species of the forest interior. Thus, by distinguishing the patterns of individuals within species, Sherry (1984) provided a richer understanding of patterns displayed by the flycatchers. Note that some of the patterns concerning distributions of individuals among guilds were the opposite of those found by Humphrey *et al.* (1983). In particular, Sherry (1984) found a much more complete assortment of conspecific individuals into the same guilds, suggesting that the birds were less opportunistic and more consistent feeders than were the bats.

These four studies demonstrate the importance of considering interindividual variation as a component of ecological versatility. The contrast between the results of Humphrey *et al.* (1983) and Sherry (1984) in relation to distribution of individuals of different species into feeding guilds is a striking example of the need to consider individual-specific ecological characteristics. Both Trowbridge (1991) and Skeate (1987) provided an extra dimension by considering the time course of changes in versatility.

Population subsets

The apparent ecological versatility of populations of conspecific individuals may also be influenced greatly by differentiation due to stage, size, or sex (see Chapter 5), and in some cases clonal structure (e.g., Jaenike *et al.* 1980). The more dissimilar population subsets are to one another, the greater the need to distinguish between subsets when assessing versatility. Studies of populations showing the most distinct types of subset versatility will be those in which stages in the population differ substantially in somatic organization (*distinctly staged populations*). Typical examples are systems of seeds–seedlings–mature plants (Parrish and Bazzaz 1985) and eggs–larvae–adults (Chew 1981). These somatic differences translate into potentially huge dissimilarities in ecological requirements, and, therefore, inflated versatilities if no distinction is made between subsets. Reduced amounts of subset differentiation are expected in populations in which stages mainly differ in size alone (*size-stepped populations*; e.g., leeches, Davies *et al.* 1981) or populations in which there are gender-specific differences (e.g., anoline lizards, Lister 1981; hawks, Preston 1990). Some workers have attempted to distinguish between phenotypic races within populations (e.g., biochemically, Jaenike and Selander 1979, or by coloration, Brakefield and Reitsma 1991).

Twenty-seven studies (19%) explicitly addressed subset differentiation in their analyses of versatility. Another three recognized the biases involved in lumping subsets but did not distinguish between them (Crowley and Johnson 1982, Bennett and Branch 1990, Gascon 1991). There was some overlap between studies involving size-stepped and sexually differentiated populations, particularly those of *Anolis* lizards (Schoener 1970, Talbot 1979), salamanders (Maiorana 1978a), and scorpions (Polis and McCormick 1987). Counting these studies twice, there were seven studies of distinct-staged populations (including plants, Gerard 1990; invertebrates, Stiling 1980; vertebrates, Diaz and Valencia

1985), 12 of size-stepped populations (four on invertebrates, the remainder on vertebrates), ten investigations of sexually differentiated populations (mainly vertebrates), and the two studies based on phenotypic differences mentioned above.

The proportion of studies in which subsets were considered as separate entities was small. Most workers either did not choose to consider the implications of failing to distinguish between major subsets within populations, or else, for one reason or another, narrowly focused on a restricted set of subsets within populations (e.g., only males, Mac Nally 1983b). The former tack obscures important sources of variation while the latter does not provide a full picture of versatility for the population. Either way, greater attention needs to be devoted to measuring the differences in versatility between population subsets.

One notable study looked at clonal structure in populations of an earthworm, *Octolasion tyrtaeum*, in Tennessee and North Carolina, USA (Jaenike *et al.* 1980). Genetically distinct clones of this parthenogenetic species displayed substantial differentiation in microhabitat use. Some clonal strains were generalized, while others were much more restricted. The generalized clones (*A* and *B* in the authors' nomenclature) showed as much ecological variation as the entire species (in the area studied), and, indeed, as much as many sexually reproducing species of earthworms. These results indicate extreme complexity of resource use within populations of species to the extent that more generalized components of populations may use supersets of the resources used by other, more restricted components. The possible effect of such complex patterns of resource use on population dynamics and evolution is intriguing, but no general framework currently exists to handle it.

Availabilities and conformances

Each study in the survey was allocated to one of the following classes in relation to the degree of attention paid to the resource availability and conformance issues (in order of decreasing desirability): (*A*) full conformance analysis, including availabilities, utilization, and a conformance measure of some kind; (*B*) availabilities and utilization qualitatively compared (i.e., no statistic used); (*C*) resource availabilities monitored; (*D*) the importance of resource availabilities mentioned, although not measured; or (*E*) resource availabilities not considered, either directly or indirectly. A study was classified insofar as it satisfied the criteria appropriate to this classification.

The results of this survey indicated that workers frequently measured resource availabilities, and in some cases calculated conformance statistics. Thirty-four studies (23%) involved measurements of resource availability. Of these, only seven did not explicitly compare availability with utilization. Seven of the remaining 27 studies computed conformance statistics. Another 16 studies in addition to the 34 mentioned above recognized the significance of resource availability without actually measuring it.

One class *A* study was that of Feinsinger *et al.* (1985), who explored the dependence of nectarivorous birds on their floral resources on the Caribbean islands of Trinidad and Tobago. The study involved the variation of resource use with respect to season and location and provided information on the intensity of utilization of nectar. Feinsinger *et al.* (1985) showed marked differences in dietary versatility between conspecific populations on the two islands. The black-throated mango, *Anthrocothorax nigricollis*, seemed to have a much more restricted diet on Tobago than on Trinidad (median diet breadths: 0.13 versus 0.49), while the reverse appeared to be true for the ruby-topaz hummingbird, *Chrysolampis mosquitus* (0.26 versus 0.03). The bananaquit (*Coereba flaveola*) population on Tobago also seemed to be more specialized in diet than its Trinidadian counterpart (0.16 versus 0.50), but the versatility of the copper-rumped hummingbird (*Amazilia tobaci*) did not differ significantly between islands (0.52 versus 0.35). These results indicate that even closely related species may show idiosyncratic changes in the versatility of resource use in different places, even at sites that are not widely separated geographically.

The other class *A* reports were: (1) Adolph (1990), who used the Ivlev index of electivity (Ivlev 1961) to quantify the preferences of two species of lizards for various microhabitats; (2) a rank comparison of the gut contents and the potential invertebrate prey of two species of warblers in an Illinois forest by Raley and Anderson (1990); (3) differences in raptor diets in an Idaho canyon (Steenhof and Kochert 1988); (4) diets of sticklebacks in brackish waters (Delbeek and Williams 1987); (5) an investigation of mycophagous species of *Drosophila* in forests of the eastern USA (Lacy 1984); and (6) Llewellyn and Jenkins (1987), who looked at seasonal variation in microhabitat versatility in two species of *Peromyscus*.

Another facet of the conformance and availability issue merits attention. Versatility with respect to consumable or renewable resources, such as food, might be better understood by comparing

Table 3.6. *Dietary versatilities of six species of South African estuarine fish based on standing crops (≡ availability) and on production of prey species. The conformance statistic is the proportional similarity of measure ($B_{PS} = 1 - \frac{1}{2}\sum_i |U_i - A_i|$, where U_i is proportion of category i utilized, and A_i is the availability or production of class i)*

Species of fish	Standing crop	Production
Atherina breviceps	0.458	0.639
Gilchristella aestuarina	0.235	0.459
Psammogobius krysnaensis	0.264	0.461
Clinus spatulatus	0.704	0.625
Syngnathus acus	0.227	0.380
Caffrogobius multifasciatus	0.492	0.452

Source: Data of Bennett & Branch (1990).

utilization with *production* rather than with availability (i.e., standing crop) (Lawlor 1980). Bennett and Branch (1990) provided data on the diets of several species of estuarine fishes in South Africa. They also published details on the standing crops (availability) and production of the main prey species. These data can be used to interpret the versatility patterns of the fishes (see Table 3.6).

Note that the spread of versatilities based on standing crop (0.227–0.704) is substantially greater than that based on production (0.380–0.639). Also, three of the species (*Atherina breviceps, Gilchristella aestuarina*, and *Psammogobius krysnaensis*) appeared to be much less specialized when their diets were compared with production than with availability. Why is this so? Part of the reason lies with the differential impact of predators on prey. Bennett and Branch (1990) reported that the consumption-to-production ratio for aquatic insect larvae was 88%, but only 41% for amphipods. This means that for these prey taxa, the apparent standing crop will be lessened proportionately more than for prey taxa that are not subject to such severe predation. *P. krysnaensis* was heavily dependent on aquatic insect larvae, *A. breviceps* on amphipods, while a sizeable fraction of the diet of *G. aestuarina* consisted of amphipods and aquatic insect larvae. Therefore, there can be pitfalls in estimating versatility with respect to consumable resources. Conformance measures based on availability can be misleading if prey classes suffer differential amounts of predation, as is likely. It is generally desirable to base versatility measures of consumable resources on conformance between utilization and production (Lawlor 1980), or better still, between exploitation and production.

A related point is the difference between the 'abundance' and the 'availability' of resources, especially of food. This distinction is related to the relative availabilities of alternative resources as *perceived by the organisms* under study (see Chapter 2 and Wiens 1989a). None of the surveyed studies provided sufficient details to derive conformance measurements with respect to perceived availability, although some, such as Sherry (1984) and Holmes and Schultz (1988), were well aware of this problem. Thus, all mention of availability in this chapter refers strictly to relative abundances (as they were sampled) of alternative resources.

In summary, 52 of 145 papers (36%) involving measurements of ecological versatility directly or indirectly addressed the issue of conformance. This suggests that the state of awareness concerning the importance of resource availability to the interpretation of ecological versatility is better now than in the late 1970s (cf., Petraitis 1979), although it still is far from ideal.

Spatial variability

Spatial variation in resource utilization is a complicated issue (see Fig. 3.4). One aspect is whether an investigator distinguishes microhabitats as 'actual' resources, or as 'receptacles' holding true resources. In the former case, which appears to be the prevalent view (e.g., Schoener 1970, Lister 1981, Adolph 1990), microhabitat versatility becomes a measure of the range of versatilities with respect to microhabitats displayed by different populations of a species. That is, species X displays microhabitat versatility V_A in habitat A, and versatility V_B in habitat B. This resembles the local/general versatility picture developed by Fox and Morrow (1981).

Resources also can be discerned within microhabitats within habitat types. For example, food resources for species of *Conus* might be identified for various microhabitats (e.g., limestone pavements, coralline rock, sand-filled depressions) on rocky intertidal benches (see Kohn and Nybakken 1975, Leviten and Kohn 1980). Food-based versatilities then could be evaluated for each of these microhabitats, and a within-habitat statement of microhabitat versatilities made. Clearly, the most complete

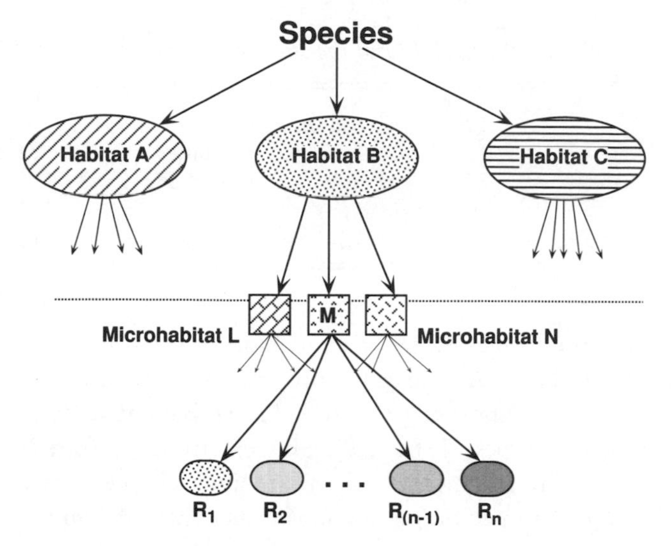

Fig. 3.4 Distinction between within-microhabitat and within-habitat versatility. One species occurs in three types of habitat. In each habitat, there are several microhabitats, not necessarily of the same types or the same number. Similarly, there are suites of true resources occurring in each microhabitat, again not necessarily the same numbers nor types. Usually, microhabitats *per se* are viewed as resources (viz., monitoring ceases at the dotted line), which may be appropriate for microhabitats that provide shelter, for example. Microhabitat versatility (below dotted line) involves another level of nesting, pertinent to different patterns of true resource exploitation within microhabitats within habitats.

picture would involve nesting of resources within microhabitats within habitats (see Fig. 3.4).

Studies were assigned to one of eight categories based on: (1) whether a single set or several sets of species were monitored; and (2) the nature of the study design with respect to differentiation of microhabitats and habitats (see Table 3.7). The first criterion allowed an assessment of the degree to which utilization patterns within specified sets of species have been investigated with respect to spatial variation. The second criterion was divided into four classes, namely: (*a*) microhabitats within two or more habitats were recognized; (*b*) microhabitats within one habitat type were used; (*c*) several habitats studied, with no microhabitat distinction; and (*d*) neither habitats nor microhabitats formed a component of the study.

Eighty-four reports (57%) involved single sets of species, with 12 of

Table 3.7. *Treatment of spatial variability in the study of versatility*

Single set of species	*84*	*Several sets of species*	*64*
Microhabitat in habitats	19	Microhabitat in habitats	22
Microhabitat in one habitat	37	Microhabitat in one habitat	2
Habitats	16	Habitats	27
No distinction/not relevant	12	No distinction/not relevant	13

these studies drawing no distinction between habitats or microhabitats. Just under half of the studies addressed different patterns of microhabitat use at single locations, while another 19 extended microhabitat analyses to several sites or habitat types (Table 3.7). Sixteen studies looked at habitat versatility, which, as emphasized in previous chapters, is a characteristic of a set of conspecific populations rather than of populations *per se*.

Another 64 studies (43%) involved different sets of taxa, although there may have been considerable taxonomic overlap in combinations in some cases (e.g., M'Closkey 1978). Just under half of these studies were of patterns of habitat versatility, another 22 of microhabitats within habitats, and just two of microhabitats within a single habitat. Thirteen studies did not address habitats or microhabitats at all (Table 3.7).

Therefore, most workers collected information with respect to spatial variation, whether at the level of microhabitat (54% of all studies) or habitat (29%). Relatively few studies involved the desired nesting of microhabitat within habitat (28%), and only about half of these were based on the same set of species in each habitat. Therefore, habitat-specific microhabitat versatility has attracted only a moderate amount of attention in the past 15 years or so.

Variation in resource utilization within microhabitats and between habitats has not been widely studied. Only a single example in the survey involved a design to look at this doubly-nested sampling. Delbeek and Williams (1987) collected information on food availability and diet for four species of sticklebacks in two habitats, lake and marshland, in New Brunswick, Canada. Several sampling sites in each habitat were established, differing to some extent in either inundation frequency, bottom substrate, etc. These sample sites effectively were microhabitats within either a lake or marshland setting. Adults and juvenile sticklebacks of all species were treated separately as well. (Data for food and diets were collected seasonally, but were pooled.) Thus, this investigation provided information for a design similar to that depicted in Fig. 3.4, but with the added 'bonus' of stage dependency.

Temporal variability – seasonal trends

Many, perhaps most, resources change in availability or suitability throughout the annual cycle, so that patterns of versatility might be expected to follow suit. Therefore, it is important to monitor fluctuations in versatility with respect to meaningful 'seasons' within the year. Seasonality may involve the familiar temperate sequence, summer, autumn, winter and spring, or the tropical and subtropical cycle of 'wet' and 'dry' (e.g., Smith *et al.* 1978, Carroll 1979, Findley and Black 1983). The term season is used here in a general way embracing both meanings.

There are several issues related to the way in which seasonal variability in versatility is treated. The first is the need to delimit the main relevant seasons for the component of the population being studied. For example, if one were considering the patterns of flowering in adult plants (e.g., Kochmer and Handel 1986) or activity in adults of many insects (e.g., Addicott 1978, Joern and Lawlor 1980), then some parts of the year (particularly winter in temperate zones) would not be pertinent because there would be no representatives of that component during those parts of the year. Therefore, it would be unnecessary to record patterns of versatility over the entire year – a clearly circumscribed subset would do. A second consideration is whether information, although collected over an annual cycle, is not analysed with respect to season (e.g., Snow and Snow 1972, Vitt *et al.* 1981, Hallett 1982). This pooling approach provides a maximum estimate of versatility, but probably occludes important information on seasonal changes in patterns of resource use. Third, one can ask – did a study register information for each season over at least several years to determine the stability of versatilities within seasons with respect to different years (e.g., Skeate 1987, Bell and Ford 1990)? In terms of the reliability and completeness of a versatility estimate for a population, a study involving repeated seasonal estimates over several years represents the best approach (e.g., Llewellyn and Jenkins 1987). Limited studies involving short-term collection of data (possibly from whirlwind visits, e.g., Schoener 1970, Cox 1981, Mercurio *et al.* 1985) do not engender much confidence in the published versatility assessments. In contrast, detailed month-by-month measurements of ecological versatility clearly would yield a good picture of within-year changes for most organisms (e.g., Herrera 1978a, Davies *et al.* 1981).

Some 108 studies in the survey (74%) described the sampling regime within years in sufficient detail to assess the degree to which within-year variability had been tackled. Sixty-three of the studies did not cover the whole annual cycle, ranging from around three weeks in total at the most brief (e.g., Schoener 1970, Lowry *et al.* 1974, Cox 1981, Ford and Paton 1982, Mercurio *et al.* 1985), up to nine months or so (e.g., Gascon 1991). Some studies were constrained to parts of the year by deliberate design (e.g., Ebersole 1985, DuBowy 1988), while many mainly focused on the spring and summer months (e.g., Lechowicz *et al.* 1988, Moyle and Vondracek 1985, Trowbridge 1991) and/or during breeding (e.g., Whittam and Siegel-Causey 1981, Lishman 1985, Mountainspring and Scott 1985).

In forty-five studies, information was collected either throughout the year or for all pertinent seasons. A third of these reported information at monthly or bimonthly intervals (e.g., Herrera 1978a, Feinsinger *et al.* 1985, Levey 1988), while almost half did so for seasonal composites (e.g., Meserve 1976, Llewellyn and Jenkins 1987). The remaining studies either did not explicitly segregate information within the year, or chose to select certain months as being 'characteristic' of the annual cycle.

What is to be gained by using season-specific information? Several authors have shown a contraction of versatility (i.e., increased specialization) with respect to season as resource availabilities change. For example, Wagner (1981) investigated versatility with respect to foraging substrates of gleaning insectivorous birds. She reported that four of the six year-round resident species had narrower versatilities in autumn and winter than in spring. Moreover, within-year variation in versatility was substantially less than between-year variation. Smith *et al.* (1978) showed a similar contraction in the versatility in diets of Galapágos ground finches in the dry season, as the availability of smaller soft seeds and fruits declined. Llewellyn and Jenkins (1987) described between-season variation in microhabitat utilization in two species of *Peromyscus*. There was a distinct decrease in versatility during winter in one species (pinyon mouse, *P. truei*), but not so for the other (deer mouse, *P. maniculatus*) (Fig. 3.5). However, the seasonally collected information showed that the absence of a large change in microhabitat versatility in deer mice was not a function of constant relative preferences for microhabitats, but rather a

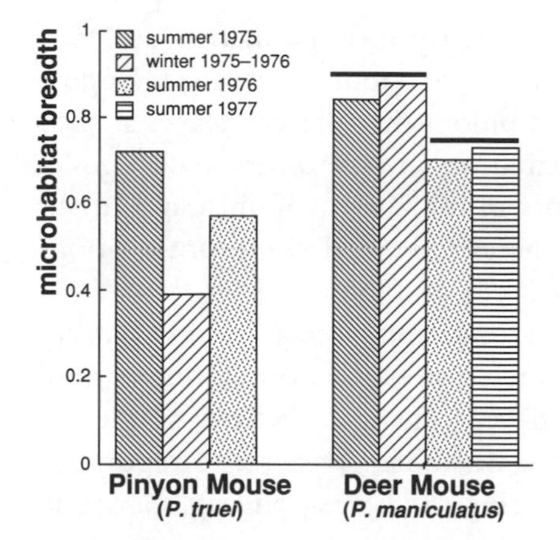

Fig. 3.5 Microhabitat niche breadths calculated for four seasons for two species of *Peromyscus*, the pinyon mouse (*P. truei*) and the deer mouse (*P. maniculatus*) (after Llewellyn and Jenkins 1987). All between-species comparisons within seasons were significantly different; all between-season comparisons for *P. truei* were significantly different; seasons with bars overlain by horizontal lines in *P. maniculatus* were *not* significantly different (statistical significance computed by using the bootstrapping method of Mueller and Altenberg 1985).

realignment of electivities from season to season (Llewellyn and Jenkins 1987).

Feinsinger *et al.* (1985) explored the versatility of nectarivorous birds on Trinidad and Tobago with respect to plant use. There appeared to be diet convergence on profusely flowering species of plants, and divergence when resource availability declined. Thus, the versatility of nectarivores was highly dynamic and linked to phenological changes in the production of different species of plants through the year. Versatility in this system was further complicated by different availabilities of nectar for long-billed and short-billed species of birds. Depending upon which flowers were blooming, the availability of nectar for long-billed species might be relatively high, but for short-billed species, which could not utilize the flowers with long corollas, availability could be quite low. The mix of flowers with short and long corollas depended upon the time of year.

Herrera (1978a) intensively studied the community dynamics of birds occupying an evergreen oak woodland at a site in south-western Spain. He computed versatilities with respect to foraging substrate throughout the year. From these data, Herrera deduced that year-round resident species were efficient specialists at utilizing the most permanent and least seasonal resources, while migrant species were more generalized opportunists that followed emergent food blooms in different places as they occurred. He related monthly information on versatilities to changes in the densities of putative competitors in the current month and in the previous one or two months. This analysis revealed significant relationships for resident species for lags of up to two months.

These examples show that the interpretation of ecological versatility must take into account the dynamic nature of resource fluctuations associated with seasonal or within-year changes. Nested studies of seasons within years are even more desirable given some reports of significant changes in versatilities during the corresponding seasons in different years (e.g., Wagner 1981).

Temporal variability – daily trends

Whereas most species with lifespans of the order of a year or more will be influenced by (seasonal) variation in resources, this will not necessarily be so at the daily timescale. For example, many predators capture prey relatively infrequently, perhaps only once in several days or months (e.g., Polis 1984). The scale of daily variation in food availability will not be pertinent to versatility for these organisms. Other species, particularly homeothermic herbivores and nectarivores, need to feed almost continuously to maintain themselves (e.g., Feinsinger *et al.* 1985). Daily variation in resource utilization is potentially significant for them. There are two issues involved here. First, do species differ in their daily versatilities? And second, do versatilities change throughout the day?

Relatively few studies monitored versatility components associated with changes throughout the daily cycle. Only 12 studies built daily variation into their study programs; another two mentioned possible contingencies related to daily variability (Parsons 1978, Hines 1982). All surveyed studies addressing daily versatility did so in relation to the first issue. Vitt *et al.* (1981), in a study of three species of lizards in Arizona, pooled information seasonally to develop foraging activity patterns with respect to the rise of the sun. From these data, daily activity breadths were computed. Anderson *et al.* (1979) viewed 'zonation' in terms of substrate and time of day as a component of the specialization of breeding birds in deciduous forests of eastern USA. Brown and Kodric-Brown (1979) showed that two species of hummingbird-pollinated plants, *Penstemon barbatus*, and *Ipomopsis aggregata*, produced nectar throughout the day, although the former did appear to secrete nectar more evenly throughout the day and at night. Parrish and Bazzaz (1979) provided a detailed analysis of daily flowering breadths of early and late successional plants at several sites in Illinois, USA. They observed that mean daily breadths did not differ much between communities of winter and summer annuals, perennials, and prairie species. Feinsinger *et al.* (1985) showed that the availabilities of nectar for hummingbirds (and the bananaquit) contracted greatly at some times during the year. At these times, nectar sources were seriously depleted by dawn and remained so throughout the day. That is, demand rivalled production. During more profuse phases, nectar remained available at dusk. Clearly, temporal variability in dietary versatility in these birds involves an intricate relationship between seasonal and daily effects.

There appears to be little information concerning daily variation in resource utilization *per se*. Most workers used daily patterns of activity as a niche dimension in its own right. More studies on comparisons of daily variability in patterns of resource utilization are needed to assess this component of versatility.

Measures of breadth and overlap

Breadth indices were used in 39 studies and overlap statistics in 53. Thirty-five of the studies using a breadth index were published after 1980, while the corresponding figure was 40 for overlap measures. Both breadth and overlap were calculated in 26 studies, although only twice for plants (Kephart 1983, Parrish and Bazzaz 1985). This indicates that there continues to be some credence attached to these sorts of measures to summarize and interpret the results of studies of resource utilization.

An almost bewildering array of measures turned up in the course of the survey. Thirteen breadth indices (counting standardized versions of other statistics separately) and nine overlap measures were used. Formulae, numbers of studies, and a brief comment on the usefulness of each measure are provided in Appendix A.2.

The most common breadth measures were the standardized or unstandardized inverse Simpson diversity index (Levins 1968), used in 16 studies, and the standardized or unstandardized versions of the Shannon–Wiener information statistic (Levins 1968), which appeared in another 12 reports. Some of the more unusual breadth measures were based on the standard deviations of resource use in canonical spaces (Werner and Platt 1976, Hoffmaster 1985), none of which is a conformance-based measure.

However, some workers acknowledged the need to compare utilization with availability to calculate versatility. Both Feinsinger *et al.* (1985) and Llewellyn and Jenkins (1987) used the adapted proportional similarity measure (Feinsinger *et al.* 1981) for comparing utilization and availability. Delbeek and Williams (1987) was the only surveyed work in which the Petraitis (1979) *W*-statistic was used. Adolph (1990) employed the Ivlev (1961) selectivity index for each resource state, which can be translated easily to yield the Canberra dissimilarity metric (Lance and Williams 1967), which can be viewed as being a conformance-based versatility statistic. Several papers followed Colwell and Futuyma (1971) and Clarke (1977) when estimating breadth (e.g., Landres and MacMahon 1983, Lacy 1984).

Overall, however, the use of conformance breadth statistics has not followed the increased measurement of resource availability, and nonconformance measures of breadth still dominate. Neither Schoener (1974b), Hurlbert (1978), nor Petraitis (1979) seem to have been heeded much. This is shown by noting that all but four of the papers using nonconformance indices were published since 1980, and 18 since 1985.

The distribution of studies using overlap measures was more skewed than for breadth measures. Thirty studies used one version or another of the proportional similarity measure (Schoener 1970), and another 12 applied symmetric or asymmetric forms of the Pianka (1974) overlap measure. Another nonconformance statistic used to quantify overlap was the Morisita (1959) and Horn (1966) measure (Kohn and Nybakken 1975, Vitt *et al.* 1981). Landres and MacMahon (1983) again followed Colwell–Futuyma tenets, while just two studies (Crowley and Johnson 1982, Delbeek and Williams 1987) employed the Hurlbert (1978) *L*-statistic. The Pianka and proportional similarity measures clearly have maintained their preeminent position as overlap indices, except that the latter seems to have surpassed Pianka's index (see Polis 1984: 543), perhaps due to the commendation of Linton *et al.* (1981).

Overview

Although not an exhaustive review of studies of ecological versatility, the survey does provide a good picture of the approaches and limitations of studies. The lack of operational consistency between studies was striking, although some of this can be attributed to different purposes for which studies were conducted. No single study combined all aspects that might lead to a definitive statement on the ecological versatility of a species – the work of Delbeek and Williams (1987) perhaps came closest to the mark. More consistency between studies is called for, as is greater recognition of fitness-related effects, individual ecological behaviour, and conformance-based estimates gathered throughout the annual cycle (where applicable) and over different years.

3.4 The distribution of versatility in nature

The distribution of observed versatilities represents a major constraint on the relative capacities of rival ideas to account for origins and maintenance of versatility. Here I briefly review some of the patterns of versatility reported in papers included in the survey. Again, these results are hardly exhaustive, but they do provide a basis for scrutinizing theoretical predictions. None of the work was based on fitness values, so one has to rely on indirect measures of resource exploitation, namely, utilization.

Studies considered here must have satisfied the conformance criterion. Comparisons of conformance and nonconformance statistics for natural systems in which details of availabilities were provided can be used to demonstrate the biases in interpretation that can arise by using nonconformance measures. Unless otherwise stated, the conformance measure was the proportional similarity adapted for measures of breadth (e.g., Feinsinger *et al.* 1981), while the nonconformance statistic was the standardized Levins (1968) breadth ($\beta_{1/\lambda} = 1/Q\sum_{i=1}^{Q} \pi_i^2$, where there are Q resources or states, and π_i is the proportional utilization of the ith state). Different interpretations can be derived by using conformance and nonconformance indices – some differences in interpretation are illustrated with examples of variation in versatility due to stage and space (Delbeek and Williams 1987), to stage alone (Talbot 1979, Preston 1990), and to season (Lister 1981).

A study of brackish-water sticklebacks by Delbeek and Williams (1987) was referred to earlier in relation to spatial variation in resource utilization. Patterns of versatility (based on the Petraitis 1979 W statistic; see Table 8 of Delbeek and Williams 1987) for two of the species, which occurred in all microhabitats, are reproduced in Fig. 3.6. Several points are worth mentioning. First, dietary versatility in juveniles exceeded that of their conspecific adults in almost all cases, usually by substantial

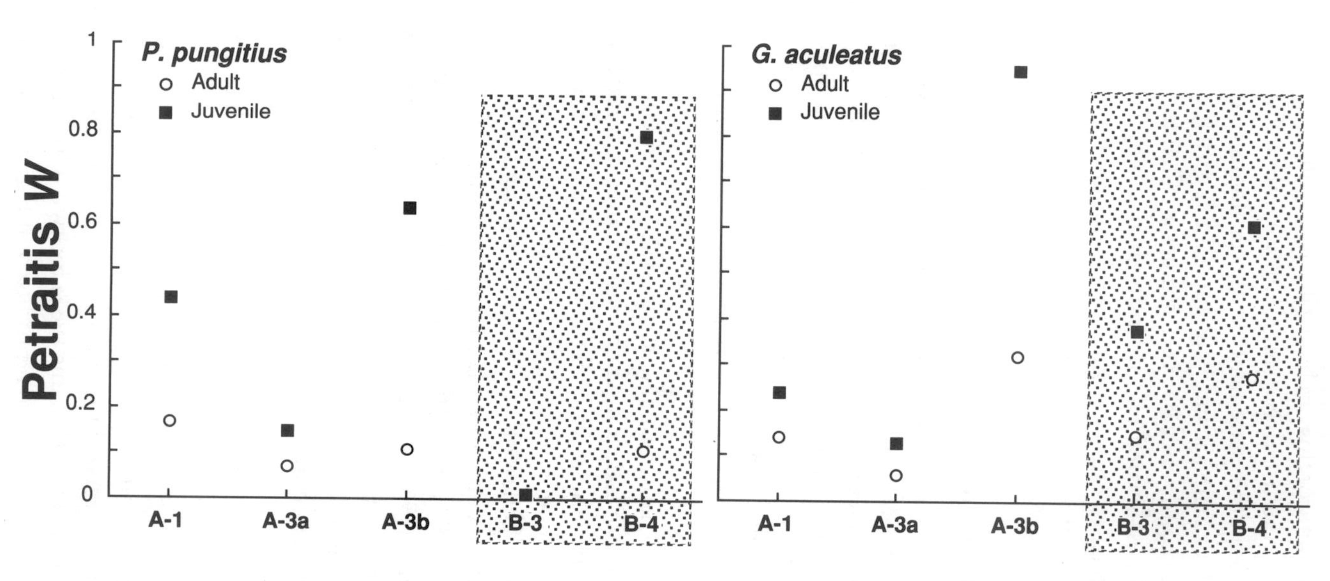

Fig. 3.6 Versatility statistics (based on Petratis' (1979) *W* measure) for two species of sticklebacks from New Brunswick, Canada (data of Delbeek and Williams 1987). Several microhabitats were sampled in a lake (A sites) and in marshland (B sites, stippled). Both adults and juveniles of *Pungitius pungitius* displayed no versatility at the B-3 study site.

amounts. Adults of *Pungitius pungitius* consistently were trophically limited, as were the adults of *Gasterosteus aculeatus* (Fig. 3.6). An assessment of dietary versatility for juveniles of both species depended upon where one looked. Variation in versatility in marshland populations of juveniles of *P. pungitius* was very marked, and similarly for juveniles of *G. aculeatus* in the lake habitat (Fig. 3.6). These observations suggest that although versatility in microhabitat use may be a meaningful measurement in its own right, within-microhabitat resource utilization often involves a significant share of the overall versatility displayed by a population.

Abbott *et al.* (1977) reported variation in dietary versatility of six species of Darwin's finches in the Galapágos Islands, in which there were distinct island-specific patterns. The small ground-finch, *Geospiza fuliginosa*, occupied six of the islands, usually displaying generalized diets with respect to seeds and fruits (conformance versatilities of 0.62 to 0.93). On Rabída, however, the dietary breadth of this species was just 0.17. Similar variation was evident in other finches. One or another of the island populations of four other species of *Geospiza* were moderately or extremely generalized (versatility ≥ 0.60), yet conspecific populations on other islands were quite restricted in diet (versatility ≤ 0.30). This amount of inter-island variation presumably reflects real differences given that the same methods of data collection by the same investigators were used throughout. This is a stark indication of the degree of spatial variation that can occur between conspecific populations, although these levels may be exaggerated by isolation and varying amounts of subspeciation.

Several other workers considered stage-based and sexual differentiation in resource use. Preston (1990) showed that males, females, and juveniles of the northern harrier, *Circus cyaneus*, displayed similar breadths with respect to habitat use despite different preferences between males on the one hand, and females and juveniles on the other. In this case, the use of conformance and nonconformance statistics did not greatly affect the conclusion, although nonconformance values were a little inflated. An investigation of food use by the sexes of two species of *Anolis* (Talbot 1979) yielded much the same sorts of results – males and females of *A. humilis* and *A. limifrons* did not differ greatly in dietary versatility. However, conformance values showed that the versatilities of males of the two species were similar, but nonconformance statistics implied that the versatility of males of *A. limifrons* was 34% greater than that of males of *A. humilis*.

The potential interpretational problems by using nonconformance indices can be highlighted by plotting corresponding nonconformance and conformance values for studies in which both sets can be calculated. If patterns of versatility are to be interpreted by using a statistic such as the standardized Levins index, then the corresponding conformance value must be consistently estimated by the nonconformance value. As an example, consider the work of Lister (1981), who looked at diets and food availability in three species of anoles in neotropical rainforests in Puerto Rico. Nonconformance measurements indicated that two of the species, *Anolis evermanni* and *A. stratulus*, expanded their dietary versatilities moderately in the dry season compared with the wet season (Fig. 3.7). Both species could be viewed as displaying intermediate dietary versatility. There appeared to be relatively little seasonal difference in the versatility of *A. gundlachi*, which seemed to be a dietary specialist (Fig. 3.7). Conformance statistics yielded a rather different picture. Although versatilities during the dry season were increased in *A. evermanni* and *A. stratulus*, the magnitude of the expansion was lessened. Rather than being a dietary specialist displaying no seasonal change in versatility, *A. gundlachi* actually was the most generalized of the three species when food availability was included in the equation, and showed an *increase* in versatility from the dry to the wet season of about 30% (Fig. 3.7). In this case, the use of nonconformance statistics yielded not just an imprecise view of the system, but rather a picture that was diametrically opposite to that based on conformance measures.

To show further the problems associated with use of nonconformance measures, I combined information from several studies irrespective of population subsets, season, and site (Fig. 3.8). These data show that the Levins measure generally is a poor estimator of conformance statistics. Nonconformance estimates of dietary versatility of *Anolis evermanni* in the dry season (Lister 1981), and of an estuarine fish, *Gilchristella aestuarina* (Bennett and Branch 1990 – availability based on standing crop for consistent comparison with other studies), were substantially greater than the corresponding conformance values (Fig. 3.8). As reported above, both dry- and wet-season nonconformance measurements of diets of *A. gundlachi* (Lister 1981) were much too small, as were the estimates of the versatility of food utilization in another species of estuarine fish, *Clinus spatulatus* (Bennett and Branch 1990), microhabitat occupancy in the prosobranch gastropod *Conus chaldeus* (Kohn and Nybakken 1975), and microhabitat distributions within *Solidago* stems of the spittlebug *Philaenus spumarius* (McEvoy 1986). Note that these latter populations would have been viewed as specialized based on nonconformance statistics, but turn out to be intermediate, or, in some

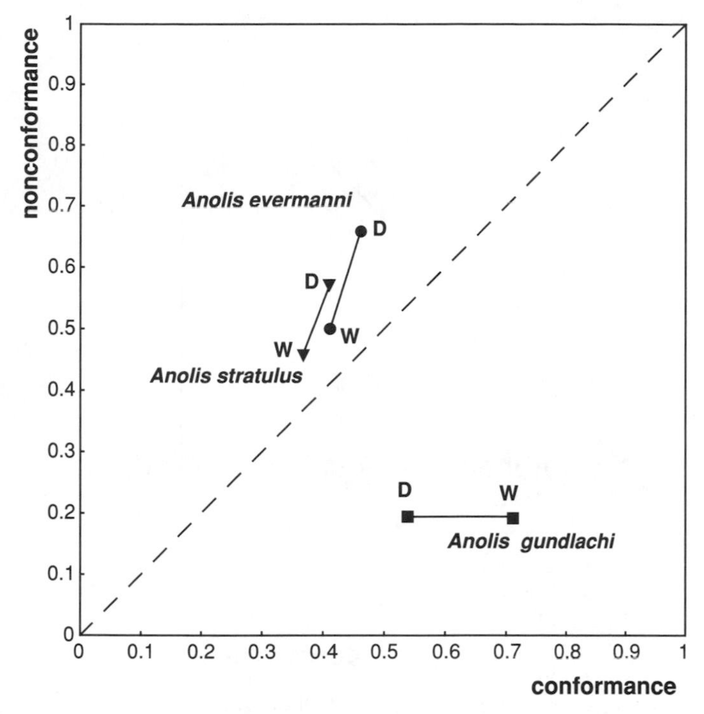

Fig. 3.7 Estimated dietary versatilities of three species of anoles in the dry (D) and wet (W) seasons in some rainforests of Puerto Rico. The ordinate is the nonconformance estimate (standardized Levins 1968 statistic), while the abscissa is the proportional similarity conformance statistic (Feinsinger *et al.* 1981). Data of Lister (1981).

cases, moderately generalized (Fig. 3.8). These results reinforce the view that testing theoretical predictions of patterns of resource use in community ecology cannot be done reliably by using nonconformance statistics (Hurlbert 1978, Petraitis 1979).

The distribution of all conformance measurements in the sample of studies is presented in Fig. 3.9. There was a reasonably even spread of versatilities, which suggests that neither generalized nor specialized taxa are particularly prevalent in natural communities. Of course, there are several major conditions on these conclusions. First, the data are based only on studies from a relatively limited survey. Second, resource aliasing almost certainly influences results, and presumably there will be different degrees of aliasing in the studies. Third, several major groups of resources were mixed to generate this overall result, including food and microhabitat. Fourth, much of the information comes from just a few studies, particularly Abbott *et al.* (1977), Feinsinger *et al.* (1985), and

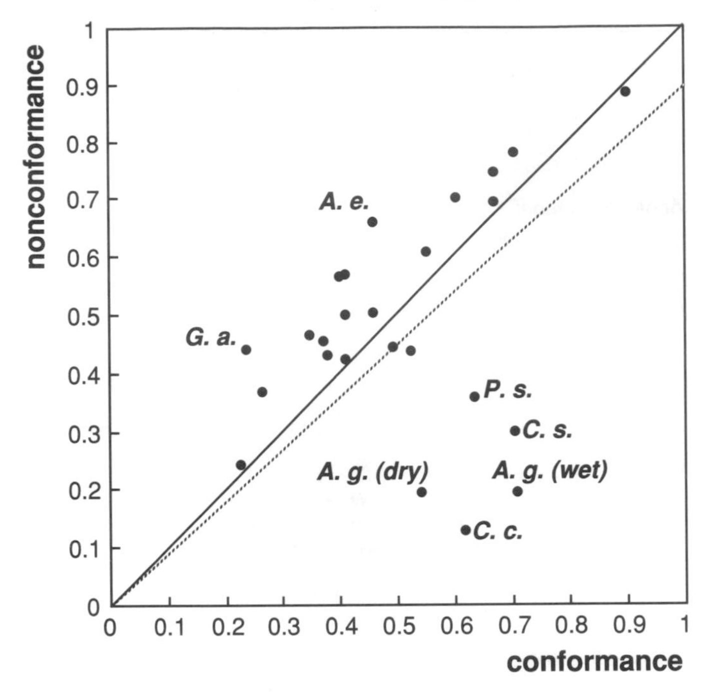

Fig. 3.8 Nonconformance (standardized Levins 1968 statistic) versus conformance (proportional similarity, Feinsinger *et al.* 1981) estimates for several species. Solid line is the line of equality, while the broken line is the regression slope (0.89). Acronyms: *Anolis evermanni* (*A. e.*) and *A. gundlachi* (*A. g.*) (Lister 1981), *Clinus spatulatus* (*C. s.*) and *Gilchristella aesturania* (*G. a.*) (Bennett and Branch 1990 – availability assessed as standing crop), *Philaenus spumarius* (*P. s.*, McEvoy 1986, on *Solidago* stems), and *Conus chaldeus* (*C. c.*, Kohn and Nybakken 1975).

Bennett and Branch (1990). And last, spatially discrete, seasonal and stage-dependent estimates of single populations also are merged together in the data set. These limitations make the conclusions rather weaker than one would like, but in any event there seems to be little evidence for numerical dominance of generalized or specialized strategies.

3.5 Summary

A detailed analysis of a set of 145 studies on ecological versatility in natural populations is presented. The criteria considered here address many issues raised in the previous chapter, including the clarity of the definition of resource states, the fitness basis of versatility, constraints,

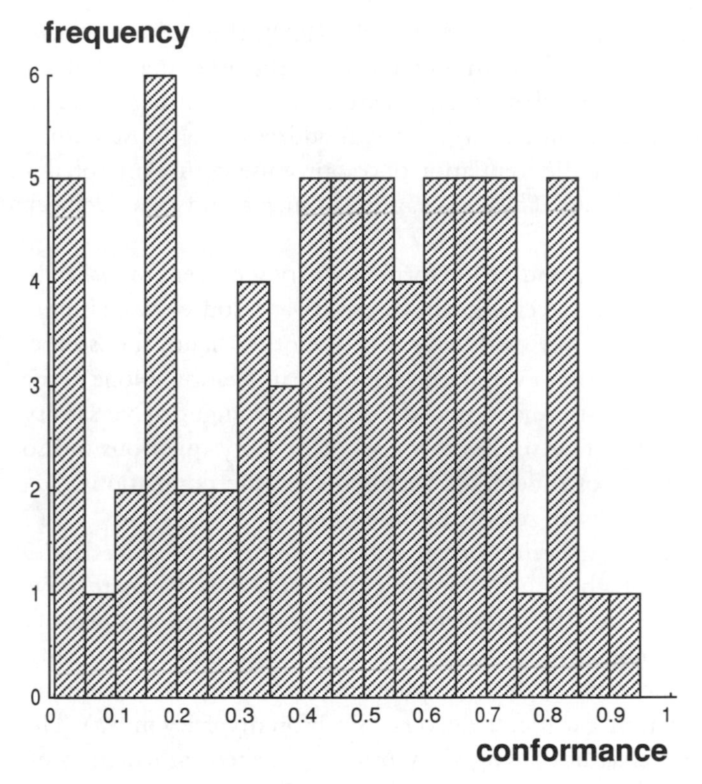

Fig. 3.9 The distribution of conformance values of versatility for surveyed studies.

differentiation of resource use in populations, spatial and temporal patterns, and conformance between utilization and availability.

Most studies provided reasonably clear pictures of resource categories. Comparatively few gauged relative fitness gains associated with the use of alternative resources. Botanists (and some entomologists) generally used fitness-related quantities to a greater extent than zoologists. The influence of constraints on resource use generally was not well documented. Few workers looked closely at differences in resource utilization between individuals, but those who did derived important insights into reasons for similarity or differences within populations and possible effects on population dynamics. Another facet of differentiation within populations, subsets such as the sexes, or different stages, also was poorly studied.

Interpretation of versatility by using conformance between utilization and availability seems to have improved in the period since Petraitis (1979) correctly lamented the lack of attention to this crucial aspect.

Results of one study, Bennett and Branch (1990), showed the dependence of interpretations of versatility on production rather than availability (i.e., standing crop) when resources are consumed during use (e.g., food). Spatial variation in resource use was addressed by most studies, but few have looked at differentiation of resource use within microhabitats and between habitats. The best example of such a study was Delbeek and Williams (1987).

Seasonal changes in resource use formed a strong current of research. Many studies documented changes in versatility as resource availabilities varied throughout the seasonal cycle. On the other hand, few studies considered daily patterns of variation in resource utilization. None of the investigations in the survey provided details of changes of versatility during the day. The few studies that looked at daily questions did so from the viewpoint of differences between syntopic populations (i.e., resource segregation with respect to time of day).

There were many measures of versatility – 13 breadth indices and 9 overlap statistics. Although monitoring of availability seems to have increased in the past 15 years, the use of availabilities to quantify versatilities by using an appropriate statistic has not followed suit. Most workers continue to use nonconformance measures (e.g., the Shannon–Wiener information statistic, the inverse Simpson diversity index). The distribution of versatilities gauged by using published information on utilization (note, not exploitation) and availabilities showed that there seems to be no tendency for specialization or generalization to dominate. In fact, the distribution was comparatively 'flat'. This result is indicative rather than definitive because few studies provided sufficient information to compute conformance statistics. The result nevertheless shows that communities may typically consist of mixtures of populations displaying specialized, intermediate, and generalized patterns of resource use in roughly equal numbers.

One concludes that there has been no consistent protocol for measuring ecological versatility. Ideally, studies should account for all of the components discussed here, particularly spatial and temporal patterns of exploitation and resource availability, or productivity where appropriate. The influence of constraints on resource use also is significant, as is differentiation between individuals. Only when all of these facets are combined together routinely will a more complete picture of patterns of resource in natural populations emerge. A few studies have combined several of these aspects together, and the richness of interpretations associated with these relatively complex analyses reflects the importance of more complete sampling designs.

4 · *The influence of interspecific interactions on versatility*

Of course it is one thing to develop a semantic framework and judge the state of knowledge of ecological versatility in these terms (Chapters 2 and 3), but quite another to begin to understand or explain the evident variation in versatility that we see in nature. I think that the differences in the quality of studies revealed by the survey are related rather directly to the variety of ecological mechanisms involved and also to taxonomic characteristics. Thus, the importance of constraints (e.g., allelochemicals) on resource use emerged as a key issue in studies involving plants and their herbivores, as did aspects of fitness, which were seen as fundamental measures of plant responses to resource availability. Neither aspect was as well studied by ornithologists or mammalogists. On the other hand, in taxa in which populations are differentiated in an obvious way, such as many species of *Anolis* and the sticklebacks of Delbeek and Williams (1987), data often were collected in such a way as to allow the intrapopulational differences in versatility to be gauged. In species in which differentiation is less obvious, data for individuals often were pooled together with little regard. Thus, differences in the way in which studies of resource use are done seem to depend upon: (1) the apparency or 'obviousness' of possible factors, which depends upon processes and taxonomy; (2) the difficulty of measurement; and (3) (almost certainly) the cultural norms and channelized thinking of each ecological subdiscipline (e.g., herbivory, predation, polymorphism).

So when one tries to account for or explain the reasons for different degrees of ecological versatility, one must contend with a rich and diverse array of ideas from each of the subdisciplines. This is the purpose of this chapter and the next. In Chapter 4 I review many of the ideas on how versatility has been thought to be determined in interactions between populations. The review is designed to be comparatively brief because of the immense existing literature and variety of ideas on many of the topics – for example, there are well over a dozen explanations for the prevalence of restricted use of plant species by herbivorous insects. Nevertheless, the general flavour of most of the main arguments will hopefully be conveyed. Similar sentiments hold for Chapter 5 in which I consider some of the theories involved in the effects of differentiation within populations on ecological versatility, which, again, has an extensive literature with connexions to population genetics and developmental biology.

As just mentioned, the topics covered in this chapter involve various forms of interspecific interactions, especially trophic interactions. These include: herbivory, predation, parasitism (and hyperparasitism), omnivory and symbioses (including mutualisms). Much of the theory on ecological versatility has been derived from ideas on the potential influence of interspecific competition on resource use in syntopic populations. Although this topic clearly could be covered in the current chapter, I prefer to treat it in a much more detailed way in Chapter 7 because the discussion there depends upon the modelling methods of Chapter 6. I hope that the reader will bear with me in this apparent dislocation but I am confident that the structure works best the way I have chosen to present it.

The versatility of resource use by detritivores and decomposers cannot be considered in the context of interspecific interactions even though decomposition is a crucial part of food web function (Pahl-Wostl 1993) – the 'interaction' involves a nonliving component. The situation for the diversity of inorganic and organic nutrients used by plant species is similar. As I noted in Chapter 1, neither of these issues is covered in this book.

The interactions are considered in a systematic but arbitrary order, beginning with interactions that are +/− in terms of the effects of the interaction on the participants. That is, the interaction benefits one member to the cost of the other. These include herbivory (herbivore +, plant −), parasitism (parasite +, host −), and predation (predator +, prey −). Then follows omnivory, which also is a +/− interaction, but one spanning several trophic levels. The last of the interactions is symbiosis, which comprises a diverse range of relationships, many of which are +/+ interactions (e.g., mutualisms, mutualistic symbioses), while others (including parasitism) are +/−.

Some caveats

The framework used in this chapter is designed to allow a systematic examination of the effects of interactions on the versatility of resource use. However, there are several limitations to using this traditional classification of interspecific interactions. First, the outcomes of an interaction between populations of two species need not invariably be the same – interactions may vary between mutualism and antagonism, say, for a variety of reasons (Thompson 1988b, Bronstein 1994). In addition, interacting populations may be involved simultaneously in several forms of interaction with one another and an interpretation of the factors governing the evolution of the relationship between the populations cannot be undertaken without a balanced appraisal of the effects of each form of interaction. Thus, although I recognize the possibility of spatial or temporal variation in the interactions between the participant populations, or the concurrent operation of several types of interaction, the discussion in this chapter is restricted to the traditional view of the interactions mainly for simplicity.

Second, and not unrelated to the first point, is that some workers now view a strict delineation between trophic levels as being an abstraction, or phenomenological description, rather than a functional description of food webs (Cousins 1987). They recognize a 'trophic continuum' rather than the well-defined trophic levels with which most ecologists are familiar. In trophic continua, taxa operate in varied roles possibly spanning several of the traditional trophic levels and presumably in similar ways to omnivores (see §4.4). This view of food webs certainly is not universal (e.g., Oksanen 1991b), but it is worth keeping in mind when reading this chapter.

And last, it will become evident that 'ecological versatility' in this chapter (and indeed the next) does not have the same semantic precision as in the previous chapters. This is because much of the published work does not consider the relative availability of alternative resources, the difference between utilization and exploitation, or temporal and spatial variation in the patterns of resource use. Thus, versatility in Chapters 4 and 5 usually refers to crude measures of the variety of resources used rather than a precise estimate of the actual versatility of resource use.

4.1 Herbivory

Herbivory is the complete or partial consumption of plants by animals. The process has been studied intensively for many years for several reasons: (1) insect herbivores wreak heavy damage on agricultural and forest industries; (2) patterns of forage use in vertebrate herbivores (mainly ungulates and rodents) is a critical factor in the management of rangelands and some woodlands (Owen-Smith and Novellie 1982) and in animal husbandry (Provenza and Balph 1990); and (3) herbivores probably control rates of nutrient cycling and resources available to detritivores and decomposers, especially in aquatic ecosystems (Cyr and Pace 1993). Thus, apart from their inherent ecological interest, food use, oviposition behaviour, and population dynamics of herbivores have been investigated in detail for their impact in agriculture, forestry, and environmental management. The current section cannot fully cover the literature in this vast field but some aspects relevant to ecological versatility are considered (for detailed reviews see Rosenthal and Janzen 1979, Crawley 1983, Futuyma 1983, Futuyma and Peterson 1985, Hay and Fenical 1988, Berenbaum 1990, Jaenike 1990, Andow 1991, Rosenthal and Berenbaum 1991).

Some workers have regarded herbivory (i.e., animal–plant predation) as being not fundamentally different from animal–animal predation, and optimal diet selection arguments have been applied freely to herbivores (e.g., O'Dowd and Williamson 1979). In predators, diet selection has been regarded as an optimization problem in which the rate of energy gained is maximized (Verlinden and Wiley 1989). The body of theory relating to this has been called the *contingency model* (Belovsky 1984; see §4.3). However, there are significant differences between herbivory and predation (Westoby 1974, Owen-Smith and Novellie 1982, but see Dade *et al.* 1990). For example, some common issues arising in models of optimal foraging in predators are of minor or no importance in herbivory, particularly of generalized mammalian herbivores. In addition, the equivalent of predatory pursuit time is negligible for most herbivores, especially 'large' ones (Owen-Smith and Novellie 1982), the probability of 'capture' is one (Westoby 1974), processing time and gut volume are crucial (Belovsky 1986, Verlinden and Wiley 1989), the victim (the plant) often is not consumed fully (Westoby 1978) and, related to the last point, deterrent allelochemicals or defences must be evident prior to the consumption of animal prey to be effective (e.g., Ball 1990, Vanderploeg *et al.* 1990) but often are dealt with following consumption by herbivores (Westoby 1974: 293). Some workers recognize that the optimization approach is just one of many mechanisms for explaining diet selection in herbivores (e.g., Provenza and Balph 1990). These points more than justify a separate section on herbivory, quite apart from the inherent ecological interest of this field.

The herbivory section is divided into two parts because of a dichotomy recognized by many students of herbivores, namely between the so-called *generalized* and *specialized* herbivores (Caughley and Lawton 1981, Belovsky 1986). Therefore, the first part of this section provides a brief overview of some of the ideas involved in dietary versatility in generalized herbivores, which are animals capable of eating many species and/or parts of plants. The determination of diets in generalized herbivores appears to be related to criteria akin to optimal foraging in predators, such as feeding time minimization or energy–nutrient maximization (Belovsky 1986). The second part explores differences in host range in small herbivores with a particular emphasis on phytophagous insects. The reasons for the prevalence of limited numbers of hosts in herbivorous insects is a topic that has seen fierce debate over the past 30 years since the pioneering ideas of Ehrlich and Raven (1964). There appears to be a complex interaction of many effects, including sequestering of secondary compounds by insects from plants for defence, enemy-free space (Gilbert and Singer 1975), mutualism, selectivity, food finding, insect foraging mode, and digestive and detoxification capabilities (Thompson 1988a, Jaenike 1990). Some reference is also made to diet in other small herbivores including marine invertebrates.

Dietary versatility in 'generalized' herbivores

Westoby (1974) noted that the optimal foraging models in vogue in the late 1960s and early 1970s were not suitable for diet selection in large, generalized herbivores. Westoby believed that these animals optimize diet according to nutrient mix but he conceded that optimization may be imperfect for various reasons, such as deficiency-induced thresholds, over-consumption of nutrient-enriched items, and (undetectable) lethal toxins.

One of Westoby's main predictions was that foods will be consumed for nutritional optimization rather than for reasons related to how difficult they might be to obtain. This led him to conclude that the proportion of a specific plant in the diet of a given herbivore ought not change much over the range of availabilities of that plant. That is, the nutritional quality of a plant becomes more important than its availability if digestive capacity rather than search/pursuit time is limiting. He postulated threshold availabilities below which plants gradually would be lost from the diet. Above the threshold, the amount of a plant in the diet would change little (Fig. 4.1). A corollary is that rarity attracts an increased pressure of herbivory, which might destabilize plant communities (Westoby 1974). These predictions are relevant only to potentially generalized herbivores and when the herbivore's ability to gauge the nutrient content of a range of foods by sensory means is limited.

How does this picture relate to versatility? Recall that my operational

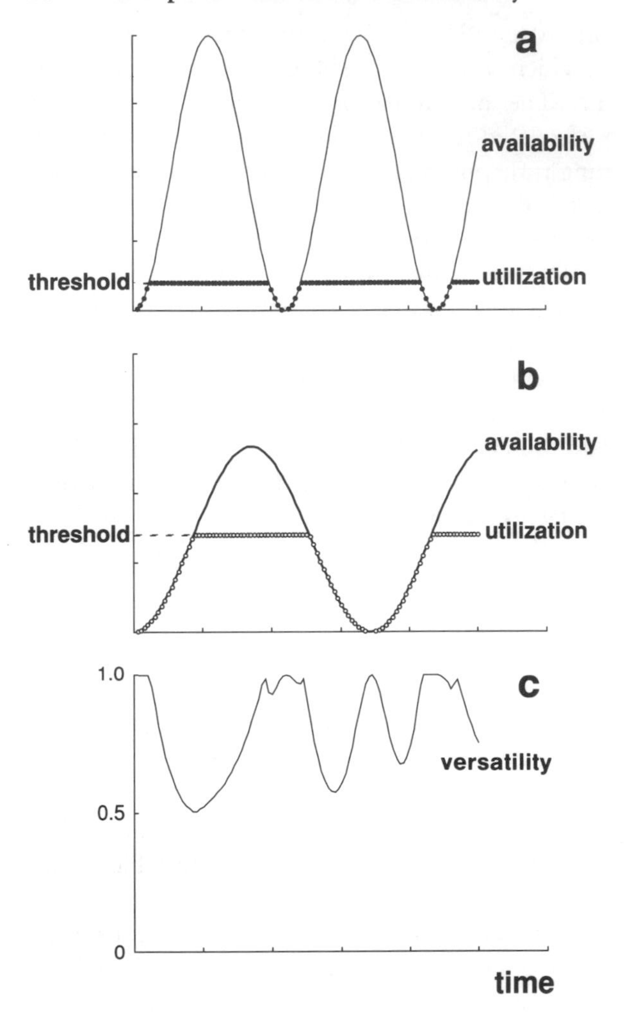

Fig. 4.1 'Passive' variation in versatility in an organism that uses resources in proportion to their availability up to a certain threshold value. The threshold for resource *a* is comparatively low compared with that of resource *b*, but the maximum availability of resource *a* is twice that of resource *b*. Versatility (*c* – proportional similarity measure) is a complicated function of the respective availabilities of the resources and the thresholds.

definition of a generalized organism is one that dynamically matches utilization (actually exploitation) with availability (Chapter 2). According to Westoby, this close correspondence would not be evident if the optimization hypothesis holds for large herbivores because of the insensitivity of consumption to availability above nutritional thresholds. As relative availabilities change, the diet may stay more or less the same except when few plant species are available. Is there a solution to this apparent paradox, namely, that there may be no consistent relationship between consumption and availability in the diets of 'generalized' herbivores?

If a particular diet does show so little sensitivity to the changes in relative availability then that diet can be regarded as a form of specialization, albeit one extended to some or most plants encountered by the animal. The distinction may be a semantic one based on how versatility is gauged – by the number of different resources used (hence variable conformance) or by the dynamic mixture of resources utilized by organisms as functions of changing availabilities. The latter seems to be more consistent with prevailing views on niche breadth (e.g., Hurlbert 1978, Petraitis 1979, Wissinger 1992).

In another approach to the modelling of diet selection in large herbivores by using optimality criteria, Owen-Smith and Novellie (1982) attempted to simulate a 'clever' ungulate in a model of diet in a species of browsing antelope in southern Africa, the kudu (*Tragelaphus strepsiceros*). Cleverness was couched in terms of the short-term optimization of foraging performance. Owen-Smith and Novellie noted that the protein content of forage is normally regarded as the main limiting nutrient for ungulates. However, they showed that gross protein intake greatly exceeded the maintenance requirements so that energy consumption appeared more likely to be the limiting factor. They attributed a marked lack of correspondence between relative acceptability and protein content of alternative plant taxa to this effect. The degree to which this lack of conformance was expressed was more pronounced during the dry season than at other times of the year. However, Verlinden and Wiley (1989) offered an alternative explanation for the poor performance of the clever ungulate model by noting that this model was a variant of the contingency model of diet selection (see §4.3), and that this model may not be adequate for herbivores (see below).

Like Westoby (1978), Owen-Smith and Novellie (1982) noted that diet of kudu was substantially more varied than their optimal model suggested it should be. They cited possible causes such as phenological change in plant quality (hence animals need to sample), allelochemicals and other feeding deterrents, and a potentially crucial feature (especially for ruminants) – the gut microbiota (see also Barbosa *et al.* 1991). Owen-Smith and Novellie suggested that the maintenance of a diverse microbiota by having a varied diet may be adaptive but not optimal *per se*. They also noted that many microbes are substrate-specific, and that this specificity may act to curb sudden changes in diet, which might prove detrimental or lethal to the herbivore. If these ideas are correct then the state of the microbiota may act inertially to maintain a more diverse diet than might be anticipated by using optimality arguments. The inertia also may lead to effects like those described by Westoby (1974) in terms of the relation between availability and consumption (i.e., an availability threshold and more or less constant consumption above that threshold).

Belovsky (1986) eschewed optimal modelling, preferring to view diets of generalized herbivores in terms of minimal acceptabilities. In one aspect of his study, Belovsky constructed models of diet selection for moose (*Alces alces*) and snowshoe hare (*Lepus americanus*) in Michigan and for two species of grasshoppers in Montana. These models were based on plant qualities and herbivore digestive characteristics. For example, minimum food quality was deemed to be a function of the following quantities:

$$\frac{\textit{requirement} \times \text{food bulk}}{\textit{digestive throughput} \times \textit{capacity} \times \text{nutrient content} \times \textit{digestibility}}$$

where italicised quantities are characteristics of herbivores and roman quantities those of plants.

Belovsky's models provided a good agreement between the predicted and observed diet selection in terms of both minimum plant quality and minimum plant size. Belovsky conceded that his models represented suboptimal decision rules in dietary selection, but that they may represent a strategy by which selection errors are minimized. From this and other analyses, Belovsky (1986) concluded that generalized herbivores employ a selection method maximizing the rate of energy and nutrient consumption rather than minimizing feeding time.

Also, if availabilities were rescaled according to the models of plant quality and herbivore characteristics, then a good match between consumption and 'useable' availability existed (Belovsky 1986: 59). Intake appeared to be unrelated to raw or unscaled availabilities except for the elk, *Cervus elaphus*. Therefore, differences between consumption based on raw availabilities and those derived from rescaled availabilities (which are specific to each herbivore) yield important mechanistic information on why different species of herbivores display different levels of dietary versatility. There appears to be scope to extend these results of detailed studies on particular herbivores to other generalized herbivores, on which such intense effort has not been expended, by appealing to allometric relationships with body size as the key link (see also Peters 1983). Belovsky (1986) suggested that this relatively crude approach can account for much of the variation in diet in a broad array of generalized herbivores.

A more recent model for diet selection is the digestive rate model (DRM) proposed by Verlinden and Wiley (1989). They noted that feeding consists of two phases, ingestion and digestion, and that diet selection should be constrained by whichever of these phases is slowest. That is, energy or nutrient assimilation will be rate-limited by either ingestion or digestion. The DRM is based on the principle of the maximization of the *rate of absorption* of energy by the digestive system rather than the maximization of the *rate of ingestion* of energy, which characterizes the contingency model. If food items differ little in digestibility (e.g., nectar/pollen, flesh, grain), then the ingestion rate becomes the key factor and the contingency model is expected to explain better patterns of diet selection. If food items differ substantially in digestibility (e.g., plants, chitinous and nonchitinous insects) then Verlinden and Wiley (1989) believed that digestion would control diet selection and that the DRM should apply. Others have arrived at similar conclusions (e.g., Taghon *et al.* 1990).

The DRM has several characteristics of direct relevance to versatility. First, Verlinden and Wiley (1989) noted that the expected appearance of items in diets generally is binary (used or not used) in the contingency model but is continuous in the DRM. Unpreferred foods will be consumed to the extent that preferred foods are available under the DRM. Thus, the DRM needs no special conditions to explain the common observation of nonpreferred food types in diets, whereas the contingency model does (Verlinden and Wiley 1989). On this basis, the DRM generally predicts greater dietary versatility that the contingency model, apparently in accordance with observation. And second, Verlinden and Wiley argued that size-related selectivity evident in many mammalian herbivores is explained well by the DRM. They suggested that large herbivores such as the moose (*Alces alces*) require a substantial amount of time just to fill their digestive systems, allowing little flexibility in search time. Smaller mammalian herbivores, such as many rodents, require little time to fill their guts, which may permit much greater selectivity. Verlinden and Wiley believed that this greater selectivity allows smaller herbivores to have less efficient digestion than larger herbivores rather than (relatively) inefficient digestion forcing dietary specificity (cf. Demment and Van Soest 1985). The DRM clearly makes the prediction that dietary versatility should correspond to the ratio of time required to handle foods compared with search times (the T_h/T_s ratio). Generalization should be associated with high T_h/T_s ratios, while specialization should typify low T_h/T_s ratios.

Although diet selection (and hence breadth) in generalized herbivores seems to be reasonably well described by arguments related to processing time, other factors also may be important. Several attempts have been made to interpret diet selection subject to constraints, such as the need for large herbivores to be vigilant of predators (e.g., Stephens and Krebs 1986: Chapter 5, Lima 1987, Godin 1990).

In summary, current ideas on diet in large herbivores involve the effects of both the acquisition and processing of food. There are many factors to be considered, some of which are gut volume, rates of throughput, fodder digestibility, nutrient content, chemical and physical deterrents and food availability and distribution. A succession of models from the early 1970s through to the present time has included more of these quantities leading to an increased ability to account for the observed patterns. It seems that large to very large herbivores (e.g., ruminants) must spend a significant fraction of their foraging time just filling their guts. In the main, this persistent need for ingesting bulk limits the time available for selective foraging leading to greater digestive efficiency and dietary generalization. On the other hand, smaller herbivores require less time to fill their guts, which may afford them time for dietary selectivity and a reduced imperative for digestive efficiency *per se*.

Host-plants of small herbivores, especially phytophagous insects

Given that this is intended to be a relatively brief review only, I have chosen to consider limited host-plant range in small herbivores in relation to the following questions: (1) how prevalent is narrow range (*stenophagy*)? (2) what are some of the main ecological factors contributing to stenophagy? (3) which mechanisms are involved in restricting the range of hosts used? (4) how are evolutionary relationships (cladistics) used to assess the problem of stenophagy? and (5) are phytophagous insects are 'good' model for other small herbivores, including marine invertebrates?

(1) How prevalent is narrow range?

Monophagous or oligophagous specialization is common in phytophagous insects, particularly in those that pass their larval lifetimes within the confines of an individual host plant (Lawton and Pimm 1978, Price 1980, Thompson 1982). Perhaps fewer than 10% of phytophagous insects in intensively studied systems use plants from more than a couple of plant families (Price 1983). Eighty per cent of North American butterflies are restricted to a single plant family while fewer than 1% use five or more families (Ehrlich and Murphy 1988). Some researchers believe that the degree of monophagy may be exaggerated to some extent by insects being unable to locate hosts or the unavailability of alternative hosts at some times or in some places (Fox and Morrow 1981, Michaud 1992). Thus, potentially polyphagous (*euryphagous*) herbivorous insects may appear to be relatively specialized due to proximate ecological effects. This has been referred to as *ecological monophagy* (Bernays and Graham 1988).

However, not all insect herbivores show extremely limited host-range. Certain adaptations help some herbivorous insects to exploit a wide range of plants despite sophisticated host defences. For example, many species of plants have evolved canalized defences such that certain defensive chemicals (e.g., latex, resins) are restricted to channels within leaves under hydraulic pressure. When a herbivore severs a canal, exudates are released, often gumming up the mouthparts and leading to the starvation of the herbivore. Species of four families of insects (noctuiid and sphingid moths, nymphalid butterflies and species of Danainae, Ithomiinae, and Heliconiinae, Doussard 1993) are able to circumvent this plant defence mechanism by cutting a trench across the leaf. This action eliminates the hydraulic pressure emanating from the rest of plant and allows the insect to feed unimpaired on the distal end of the leaf. Doussard and Denno (1994) have shown that nontrenching species are capable of growing as well on the distal ends of leaves modified by trenching species (and also on detached leaves, which lack hydraulic pressure) as on leaves lacking canals, which indicates that their host range is limited by an inability to trench. Thus, the absence of a behavioural trait leads to limited versatility in nontrenching species and greater versatility is possible from the capacity to trench. Ehrlich and Murphy (1988) also observed that some insects display remarkable generalization in host range (jacks *and* masters of many trades, in their terms), particularly *Heliothis* and the bug *Leptoglossus phyllopus*. The latter feeds on xylem and young plant embryos, a strategy so successful that it can exploit species of almost 30 families of plants. Some recent work suggests that monophagy or oligophagy may have been overstated in relation to insect herbivores of tropical rainforests (see Gaston 1993).

(As a peripheral comment, it is important not to be misled by the taxonomic breadth of host range *per se* (Barbosa 1988, Futuyma and Moreno 1988). Some phytophagous insects select hosts of quite distinct taxonomic origins yet those hosts share common chemical (Barbosa and Krischik 1987) or anatomical (Powell 1980) features. Maiorana (1978b) was particularly critical of the use of taxonomic breadth as a direct indicator of niche breadth and attributed a lack of correspondence between these quantities as the source of problems in the conclusions drawn by Futuyma (1976). The ability to 'order' resources in terms of their respective similarities helps to characterize versatility better but the taxonomic similarity of hosts may not be a useful dimension to use.)

Despite these counter-examples, it does seem that host-plant specialization, or *stenophagy*, is characteristic of herbivorous insects, so much so that most species tend to extreme monophagy and have trouble effectively utilizing even close relatives of their principal hosts (Ehrlich and Murphy 1988).

(2) Which ecological factors contribute to stenophagy?

Jaenike (1990) reviewed the factors that influence levels of host specificity in herbivorous insects, with an emphasis on general characteristics and ecological constraints. He concluded that learning and aspects affecting thresholds for oviposition behaviour may be the most likely proximate determinants of host selection.

Jaenike (1990) treated general ecological factors in terms of density-dependent and density-independent mechanisms that might produce narrow or broad host ranges (Table 4.1). The ideas presented in Table 4.1 are largely self-explanatory, so that only a brief overview of Jaenike's conclusions seems necessary.

Of density-independent hypotheses, one appealing idea is that selection leading to adaptation for one host species should lead to reduced performance on other potential hosts – the performance compromise

Table 4.1. *Some of the proposed ecological factors that might affect host range in phytophagous insects*

	Restricted host range[1]	Broad host range[1]
Density-independent	Host-specific adaptations (hence performance compromise) Hosts abundant, consistent and easily found by adults Interspecific competition (food or enemy-free space) Resistance to generalized predators Similarity of hosts to unsuitable plants (see Fox and Lalonde 1993)	Hosts rare, unpredictable, or hard to locate by adults Physically small hosts hence larval grazing favoured Risk spreading Genetic correlation – selectivity (adults) versus performance (larvae) Ant mutualists
Density-dependent	Mate location Overwhelming plant defences	Intraspecific competition Predator functional response Pathogen numerical response

Note: [1]The entries in corresponding rows of the tables are not contrasts with each other as such – the columns are independent lists.
Source: After Jaenike (1990).

hypothesis. Few experimental studies appear to support this idea, and sometimes produce opposing conclusions (Futuyma and Philippi 1987; see also Abrams 1990, Thompson and Burdon 1992). However, Jaenike (1990) concurred with Rausher (1988) by concluding that there are several weaknesses in experimental programs designed to test the idea, which may mean that the influence of performance compromises should not be eliminated as a potentially important factor just yet.

Bernays and Graham (1988) suggested that the breadth of host range is controlled by natural enemies such as generalized predators, parasites or parasitoids. Although there seem to be some examples in which this is supported (e.g., Bernays 1989), Jaenike (1990) concluded that this is unlikely to be a general phenomenon.

There appears to be some support for the idea that the abundance and predictability of preferred hosts are correlated with monophagy, as is the ease with which preferred hosts can be found. This has been invoked as a contributory factor in the occurrence of latitudinal gradients in the frequency of monophagy within higher taxonomic units. The greater diversity of plant species in the tropics and a consequent reduction in the average density of any one plant species locally mean that finding a particular host is more difficult in the tropics than in temperate regions. Thus, less specificity is to be expected (and is found) in some groups of insects. However, other factors such as the degree of adult mobility and longevity may obliterate this trend because long-lived, mobile adults have a greater capacity to find rare hosts.

Mutualisms with ants may reduce host specificity in some insects, with the occurrence of ants overriding host quality as a signal to oviposit, although this effect may be relatively uncommon (Jaenike 1990). Risk-spreading and the correspondence between adult host preference and larval performance also seem to be of limited importance.

Density-dependent factors are conjectured to work in a variety of ways. A general class of mechanisms involves increasing the likelihood of encountering conspecific individuals, especially mates. Jaenike (1990) believed that critical tests of this idea have not been conducted and that circumstantial evidence indicates that mate-finding is unlikely to be a major factor leading to narrow host range. However, it is possible that the concentration of larvae on particular hosts allows them to overwhelm host defences but whether this itself accounts for stenophagy *per se* is unclear. Alternatively, by spreading out over a diversity of hosts, polyphagous species might avoid the negative effects of predators preferentially attacking aggregations or the contagion of pathogens. Another obvious density-dependent mechanism is intraspecific competition, the action of which would be expected to lessen host specificity (see Chapter 5). Ovipositing females might avoid preferred hosts that already have been attacked by conspecific individuals, possibly being forced to use less suitable ones. Jaenike (1990) believed that intraspecific competition should not be discounted yet as a potentially important agent in controlling the host specificity of herbivorous insects.

(3) Which mechanisms restrict the range of hosts used?

Jaenike's (1990) review provided a general framework for considering how host range may be controlled by ecological factors. However, it is worth while reviewing in greater detail some ideas either not considered by Jaenike or that have been the centre of recent active debate. Perhaps the most widely used explanation is chemical coevolution or 'arms races' between herbivores and plants (see Rosenthal and Berenbaum 1991). Bernays and Graham (1988) argued that the concentration on coevolutionary issues between insects and their hosts (e.g., plant chemistry and detoxifying capabilities) may have overshadowed the possible impact of other factors in controlling host range. They cited evidence suggesting that adaptation to new hosts can be accomplished relatively quickly so that restricted host use can occur during intervals far shorter than the time thought to be needed for coevolution. Apparent aversion in some cases may be overcome by habituation without apparent effects on the fitness of the insects. Bernays and Graham (1988) also noted that specialized morphologies for clinging to hosts against abiotic threats (e.g., wind, storms) might contribute to restricted host ranges (see Denno *et al.* 1990).

On the basis of these observations Bernays and Graham (1988) argued that host range selection is too ecologically dynamic to support the widespread coevolution of specific pairs of plants and insect herbivores (see Strong *et al.* 1984, Bowers *et al.* 1992). They believed that the lability of factors affecting host range should inhibit coevolution through mutual selection on host and herbivore and that predation and parasitism of herbivorous insects may be the principal reasons for restricted plant host range. They suggested that the normally low densities of herbivores compared with their hosts (Strong *et al.* 1984) would impart little selective force on plants to develop specific weapons to deal with individual species of insects. Bernays and Graham (1988) proposed that generalized predators and parasites have driven the process of host specialization by forcing phytophagous insects to use particular hosts, which may confer an advantage on the herbivores. Laboratory studies conducted by Bernays (1989) appeared to support the idea that generalized or polyphagous insect herbivores suffer significantly heavier mortality from several insect predators (a vespid wasp, *Mischocyttarus flavitarsus*, the Argentine ant, *Iridomyrmex humilis*, and a coccinellid beetle, *Hippodamia convergens*) than monophagous herbivores.

Jermy (1988) countered the claims of Bernays and Graham (1988) by suggesting that stable host ranges may be at least as common as labile ones, and he provided numerous examples in support of his statement. Thompson (1988a) and Courtney (1988) each noted that phytophagous insects must contend with plant chemistry irrespective of a strict coevolutionary coupling. The origins of secondary compounds and other features in plants may not even be due to herbivory as such (at least by insects) but may incidentally have a deterrent effect causing strong selective pressure on insects (Rausher 1988). From the plant perspective, allelochemicals may represent a general defence rather than a specific one against particular species of herbivore – some workers have contended that allelochemicals are primarily directed against microbial infection anyway (Courtney 1988). Thus, plant chemistry may influence host range in phytophagous insects with or without bilateral or specific coevolution between a species of plant and a particular insect herbivore. Host-plant specialization does not *require* coevolution, only that there are differential rewards or costs associated with alternative hosts (Courtney 1988). Apart from these points, the Bernays and Graham position drew criticism from many workers who, like Jaenike (1990), noted that a variety of other mechanisms would be expected to influence host range in insects (Thompson 1988a), some of which are reproduced in Table 4.2.

Specific experiments have been conducted to explore the interaction between various factors in determining host range of particular species of insects. One example is the work of Denno *et al.* (1990) who considered two species of leaf beetles (*Phratora vitellinae* and *Galerucella lineola*) utilizing three species of willow in Sweden (*Salix fragilis*, *S. dasyclados*, and *S. viminalis*). The comparison included contrasts between (more) specialized (*P. vitellinae*) and (more) generalized (*G. lineola*) herbivores, and between host plants with distinct differences in allelochemistry, nitrogen content, pubescence and leaf toughness. Denno and his cow-orkers found that the larvae of the specialist (*P. vitellinae*) sequester salicylate compounds to produce a secretion for defence against predators. Although both *S. fragilis* and *S. dasyclados* contain such salicylates, the latter also contains high concentrations of simple phenolics that prove to impair larval performance. Thus, oviposition preference in *P. vitellinae* is consistent with the absence of salicylate compounds in *S. viminalis* (hence no defensive capabilities) and high levels of phenolics in *S. dasyclados* (poorer larval performance) leading to a disproportionate use of *S. fragilis*. On the other hand, the larvae of the generalist (*G. lineola*) do not employ allelochemicals for defensive purposes, and oviposition preference was inversely proportional to the concentration of salicylate (Fig. 4.2). In neither species of beetle did oviposition preference conform to patterns of nitrogen content, pubescence or leaf toughness.

In summary, there are many mechanisms or factors that can influence host range in herbivorous insects. The intensity of feeling on the importance of particular explanations for limited host range probably reflects the importance of a given factor (e.g., natural enemies, chemical coevolution) in systems with which a proponent is most familiar (e.g.,

Table 4.2. *A selection of mechanisms or factors that have been proposed to produce restricted host range in herbivorous insects*

Mechanism	Explanation
1. Specific insect–plant (chemical) coevolution	Escalated 'arms races' between particular species pairs (insect herbivores and plant host), with the coevolutionary process excluding other (nonspecific) herbivores and making the insect herbivore less able to attack other plant species (the compromise hypothesis)
2. Ecological monophagy	Potential alternative host plants unavailable locally leading to apparent restriction (Bernays and Graham 1988)
3. Microclimate	Gross- and micro-structural differences between hosts may produce a range of suitabilities to which insects respond (e.g., Dobkin 1985)
4. Host plant ecology (including resource concentration and associational resistance, see O'Dowd and Williamson 1979, Andow 1991)	Abundance and spatial and temporal distribution of plants, including phenology (Barbosa 1988, Thompson 1988a)
5. Insect feeding mode	Grazing versus parasites that complete development within a host (Thompson 1988a)
6. Genetic contention	Adults and larvae may perform significantly better on alternative hosts leading to contention (Futuyma and Peterson 1985; see also Chapter 5)
7. Herbivore variability	Differences due to stage, sex or physiological responsiveness (Jaenike 1990)
8. Host variability	Individuals or parts of individual host plants differ substantially in suitability, both in terms of palatability (Damman 1987), camouflage or refugia (Walter and O'Dowd 1992a, 1992b). NB, polyphagy or plasticity of host use may be favoured if there is spatial (between habitats) or temporal (between generations) switches in relative suitability of hosts (Michaud 1992)

Table 4.2. (*cont.*)

Mechanism	Explanation
9. Fauna associated with plants	Protection afforded to herbivores by fauna associated with particular plants (e.g., myrmecophilous lycaenid butterflies, Jordano and Thomas 1992, Jordano *et al.* 1992)
10. Generalized predators and parasites	Predation and/or parasite pressure forces insect herbivores to use particular hosts (Bernays and Graham 1988)
11. Behavioural limitations	Limited recognition or neural capabilities of most insect groups (Jermy 1988, but see Fox and Lalonde 1993)
12. Radiation of stenophagous sibling species	Formation of sibling species from existing ones (Jermy 1988) via host race stages (Bush 1975)

Bernays 1989). It is more likely that a few or several factors interact in any one system and that no mechanism is a general or most influential one in herbivorous insects as a group. For example, Jaenike (1990) analysed frequency distributions of host range in four families of insects from Canada, Great Britain, and Costa Rica. He found that the results of this analysis were consistent with the independent or weakly interdependent, simultaneous impact of several factors. Thus, limited host range may be an epiphenomenon arising from the concurrent influence of a set of mostly independent forces. The main issue then, is the degree to which each factor determines host range in specific instances. By recognizing that host range in herbivorous insects probably is affected to a lesser or greater extent in any one species by several of the factors listed in Tables 4.1 and 4.2 (Courtney 1988, Thompson and Pellmyr 1991), research clearly needs to be framed in terms of how factors interact with one other to determine the diversity of hosts used within a population.

(4) How are evolutionary relationships used to assess reasons for stenophagy? The predominance of stenophagy may arise from the formation of novel stenophagous genotypes from existing ones (Jermy 1988), which then become sibling species through an intervening host race stage (Bush 1975). An increasingly important tool for gauging the importance of

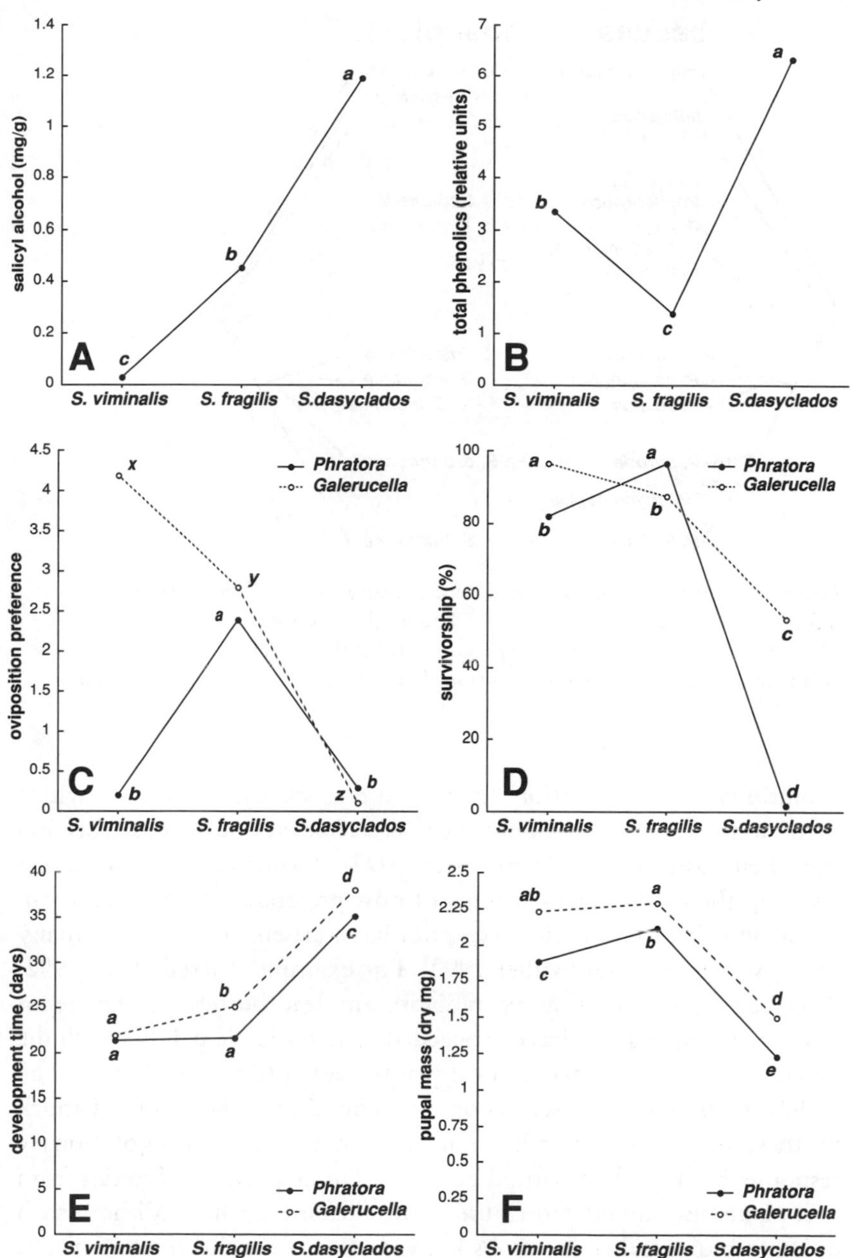

Fig. 4.2 Chemical characteristics (A – salicyl alcohol concentration, B – total phenolics), oviposition preferences (C – egg masses per shoot), and measures of larval fitness (D – survivorship, E – development times, and F – pupal dry mass) for two species of beetles (*Phratora vitellinae* and *Galerucella lineola*) using three species of willow (*Salix viminalis*, *S. fragilis*, and *S. dasyclados*) (after Denno *et al.* 1990). In all plots except C, means with different small letters were significantly different. In C, the same holds, except that comparisons were within-species (hence the a–b–c and x–y–z sequences).

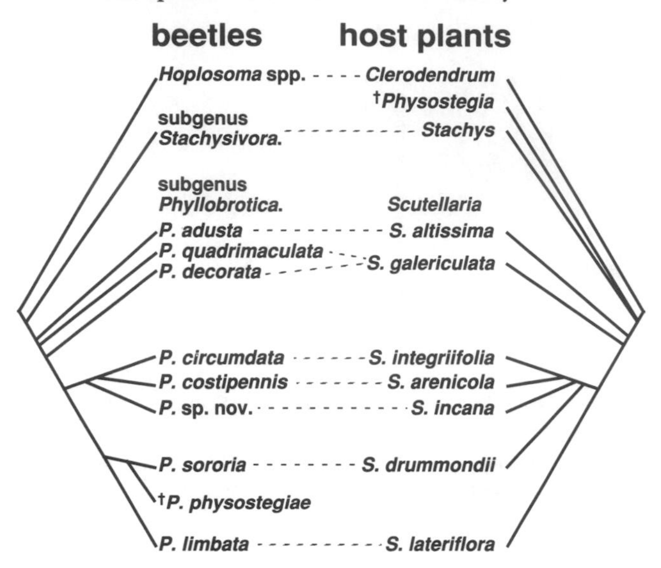

Fig. 4.3 Cladograms of leaf beetles and their host plants. Beetle species are linked to their hosts by a dashed line. The single exceptional mismatch, *Phyllobrotica phrysostegiae* and the perennial mint *Physostegia* (marked by †), may be a recent colonization from an ancestral annual species in the same habitat. After Farrell *et al.* (1992).

evolutionary diversification to host use is cladistic analysis, which involves looking at the coupling between lineages of insect herbivores and their host plants (Armbruster 1992). Cladistics are critical for assessing the evolutionary origins of host preference because they can reveal whether certain characteristics have arisen just once, or many times, within lineages (Miller 1992). For example, Farrell *et al.* (1992) described how evolutionary radiation in leaf beetles of the genus *Phyllobrotica* appears to have proceeded in remarkable parallel with the radiation of their herbaceous host plants, Lamia (Fig. 4.3). Farrell *et al.* (1992) attributed this close, stenophagic coupling to the nature of attack by these beetles, which is liable to be so damaging that evolutionary responses by host plants would be likely. The acquisition of toxins from host plants also should contribute to the extreme fidelity. Without such intimate biological connections between the insect and the host plant, one need not expect such extreme stenophagy (Farrell *et al.* 1992). On the other hand, Miller (1992) discussed host plant use in notodontid moths in which different clades appear to have taken quite different evolutionary paths *vis-à-vis* host use. Some lineages are restricted to the same host (taxon) while other clades attack a broad range of plants that are neither taxonomically nor chemically similar.

Of course, cladistic analysis is concerned with the breadth of host use in insect lineages rather than with the versatility of individual species. Nevertheless, these forms of analysis seem to show two important things. First, host use appears to be governed mainly by sequential evolution rather than by coevolution (Miller 1992). That is, evolution of insect lineages has occurred against an existing background of plant taxa (Jermy 1984). And second, plant chemistry seems to be an important factor in some insect clades but, generally speaking, chemical constraints are not the primary determinants of host use (Miller 1992). Host characteristics may be partly responsible for this effect. As Miller (1992) pointed out, herbivores that are reliant on trees as hosts face a quantitative rather than specific, or qualitative, chemical defence, which is much less likely to be associated with extreme specificity (see Feeny 1991). Thus, cladistic analysis can provide a general context within which differences in host use between closely related herbivores can be understood.

(5) Are phytophagous insects a good model for small herbivores?

There are both interesting parallels and differences with respect to host range in small herbivores other than insects. In recent times, much work has been done on marine herbivores in which varying degrees of specificity for algae and seagrasses have been found (e.g., Paul and Hay 1986, Hay and Fenical 1988). Some species show relatively great versatility by being able to sequester and store allelochemicals but others may be limited by plant toughness and calcification (e.g., the sea hare, *Dolabella auricularia*, Pennings and Paul 1992) or algal polyphenols (Tugwell and Branch 1992). Some species face selection for specificity for certain algae, which have their particular retinues of secondary chemicals, to avoid deliberate or incidental predation (Hay *et al.* 1989, Hay *et al.* 1990). Yet other marine herbivores show facultative diets depending upon the availability of preferred plants (Carefoot 1987). Recent work has shown that small mammalian herbivores also respond to deterrents produced by plants. Gali-Muhtasib *et al.* (1992) showed that the concentration of silica in grasses affects plant selection in the prairie vole (*Microtus ochrogaster*). Individuals of Abert's squirrel (*Sciurus aberti*) appear to choose particular individual host plants of ponderosa pine

(*Pinus ponderosa*) on the basis of plant chemistry, notably specific monoterpenes, nonstructural carbohydrates, and heavy metal ions (Snyder 1992). Do these observations mean that some of the arguments put forward to explain plant use by insects also apply in other small herbivores?

Hay's (1991) recent comparison of the trophic ecology of small-sized herbivores of marine and terrestrial origins suggests that there may be significant differences between the two groups in relation to trophic specificity. Hay noted that the oviposition behaviour of adult females of several major groups of insects appears to be at least partly responsible for the specificity shown by the larvae of those insects. Marine herbivores generally have planktonic larvae and the opportunity for them to 'select' or 'sample' particular algae before settling may be rather limited. Nevertheless, the adults of many marine herbivores do appear to prefer chemically-rich algae that are distasteful to predators. This use of certain host plants may reduce their susceptibility to predation (or parasitism) either by helping to avoid incidental consumption by large herbivores (via the distastefulness of the plant itself) or by sequestering plant metabolites to increase the distastefulness of the small herbivore itself. Although this mechanism may have driven host-plant specialization in many insects, Hay (1991) argued that generalist marine herbivores also use sequestering for defence. Thus, the sequestering of plant chemicals that some workers think has had a major effect on insect host-plant specificity cannot be readily extended to marine situations and does not seem to explain differences in the dietary versatility of small marine herbivores.

Herbivory – an overview

The length of the current section reflects the interest evoked by plant–herbivore systems. The division of the section into two parts, one on large, essentially 'generalized' herbivores and the second on host range in small herbivores, particularly insects, may be artificial to some extent, although the mechanisms that are considered to be most important in each case seem to be quite different. Whether this reflects a fundamental biological difference attributable mainly to the size of the herbivore is unclear at present. Ideas on diet selection in generalized herbivores focus on optimality issues, including compromises between the processing and acquisition of food, while dozens of factors have been raised to explain various degrees of host use in smaller herbivores and especially insects.

4.2 Parasitism

Price (1980) has written that, gauging by the number of species, parasitism is more common than any other feeding strategy making it the most successful way of life (see Kennedy 1984). There are various definitions for what constitutes a parasite. For example, Price (1980: 4) defined parasitism to be 'living in or on another living organism, obtaining from it part or all . . . organic nutriment, commonly exhibiting some degree of adaptive structural modification, and causing some degree of real damage'. Phytophagous insects often have been thought of in terms of being plant parasites. Given the attention already paid to them, the current section will focus mainly on parasitism in animals with just a few comments on plant parasites such as mistletoes.

Parasitoids are a subset of parasites. They kill just one host individual during a given life-history stage (Kuris 1974). Kuris and Norton (1985) suggested that parasitoids are physiologically similar to parasites but have ecological effects more like those of predators. Species of both forms of parasite have been regarded as being extraordinarily specialized, particularly in relation to the hosts used and/or the locations of attack on hosts. In the following sections, I describe some of the ideas that have emerged to account for patterns of versatility of host use, first in parasitoids, and then in parasites.

Parasitoids

There are several key factors that appear to affect the versatility of parasitoids, some of which are related to characteristics of the host organisms and others to the biologies of the parasitoids themselves. The main characteristic of hosts entering into the question is that of feeding biology: are hosts exophytic feeders, leaf miners, gallers, borers, root feeders or mixed endophytic–exophytic foragers (Mills 1992, 1993)? The degree of exposure to attack, ease of location of hosts by parasitoids and probabilities of hyperparasitism of the parasitoids themselves all depend upon host trophic ecology (Hawkins *et al.* 1992, Hawkins and Gross 1992).

Some parasitoids arrest the development of the host individual upon attack. This is termed *idiobiosis* and parasitoids operating in this fashion are called idiobionts. The alternative strategy is to allow host development to continue following infection (*koinobiosis*, hence koinobionts), but this strategy requires physiological adaptations on the part of the parasitoid to contend with the counter-responses of hosts. Idiobionts appear to be more likely to attack endophytic hosts because of the incapacity of immobilized hosts to avoid predation and inclement conditions (and thus 'protect' the parasitoid). Idiobionts generally are ectoparasites, another feature favouring the exploitation of endophytic hosts. Thus, the habits of the host often afford a level of protection for the parasitoid. On the other hand, koinobionts seem to be more common on exophytic hosts because of an apparent competitive inferiority to idiobionts. The crucial issue relating these alternative strategies to ecological versatility is that the continuing interaction following attack between a koinobiont and its host should make the relationship more specific to a given pair of species than in the idiobiont case. The koinobiont parasitoid must also have sophisticated capabilities to locate specific hosts (Kuris and Norton 1985). Therefore, idiobionts are expected to be more generalized in host exploitation than koinobionts.

Another feature of herbivorous insect–parasitoid systems is the phylogenetic differentiation of physiological adaptations to counter host defences in koinobionts. Hawkins *et al.* (1992) reported that dipteran koinobionts may be relatively more generalized in host use than might otherwise be expected. However, they noted that many species of the family Tachinidae, the dominant parasitoid dipteran family in North America, have physiological and morphological adaptations that can counter the immune responses of developing hosts (Hawkins *et al.* 1992: 65). Some species of tachinids employ the koinobiont strategy but circumvent its restrictions thereby ameliorating the need for extreme host specificity. Hence, idiobiosis and koinobiosis do not necessarily correspond consistently to a generalized–specialized dichotomy.

This brief review shows the complexity of interactions between factors controlling versatility. Host trophic biology and stage, relative predation risk, and the densities and distributions of host populations together affect the appropriateness of rival parasitoid strategies. Although koinobionts usually can be regarded as being relatively specialized and idiobionts more generalized in host use, characteristics of different parasitoid groups can blur the general validity of this distinction (e.g., some koinobionts can generalize use of hosts by virtue of special physiological adaptations; specificity to particular host stages, Mills 1993).

Animal parasites

Many parasitologists have listed what they believe to be the advantages to parasites of using many host species (i.e., 'generalized' parasites,

Garnick 1992). Some of these benefits are: (1) reduced susceptibilities to fluctuations in host densities and resistances; (2) higher genetic differentiation; and (3) more extensive geographic ranges. Despite these apparent advantages, a high fraction of parasites appears to be obligatorily dependent on single host species ('specialized' parasites). One solution to this paradox, as mentioned above in relation to parasitoids, is that the specialization of a parasite on a particular host engenders an intense selective pressure on the parasite to specialize on that host. Pressure for close adaptation leads to certain hosts acting as primary reservoirs despite the use of other hosts (Garnick 1992). The exploitation of other hosts probably reduces the potential for finely tuned adaptation to the reservoir host and if hosts are too different from one another, generalization will not be favoured. This result may be consistent with the observation that evolutionarily specialized hosts support specialized parasites (Cameron 1964, Jones 1967).

It seems that host specificity can be pronounced in certain groups of parasitic organisms. For example, Jones (1967) noted that the cestode genera *Moniezia*, *Cittotaenia*, and *Anoplocephala* are specific to the ruminants, lagomorphs, and equines respectively, and that the cestode family Hymenolepididae consists of about 800 species, most of which parasitize a single species of bird or mammal. Rohde (1979) documented impressive host specificity in some groups of marine parasites, particularly in the platyhelminth order Monogenea. In these trematodes, almost 80% of species parasitize single species of fish (Fig. 4.4 a). Only 2% of species attack hosts from more than one order. Sometimes, this phylogenetic specialization is replaced by an ecological specificity in which certain species of trematode afflict hosts having similar ecologies rather than phylogenies. Extreme specificity may also be expressed in relation to 'microhabitat', which corresponds to anatomical locations on host fishes such as gill opercula, filaments and arches, internal organs, and skin. It is even possible that chiral specificity on gills occurs (e.g., only left *or* right gill filaments in a particular species of fish, Rohde 1979). At the other extreme, the parasitic worm *Polymorphus minutus* has been recorded in more than 80 species of birds (Crompton and Joyner 1980).

As with insect herbivores and their plant species (see Fig. 4.3), parasitologists have developed cladistic representations that have been taken to indicate concurrent or parallel coevolutionary speciation between certain groups of parasites and their hosts ('association by descent,' Brooks 1988). Such systems imply extremely high host specificity. However, Downes (1990) has shown that some parasitic mites of freshwater molluscs in Florida develop host-specific morpholo-

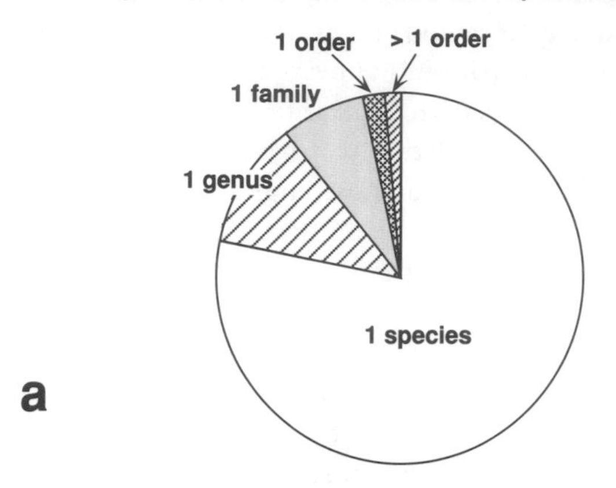

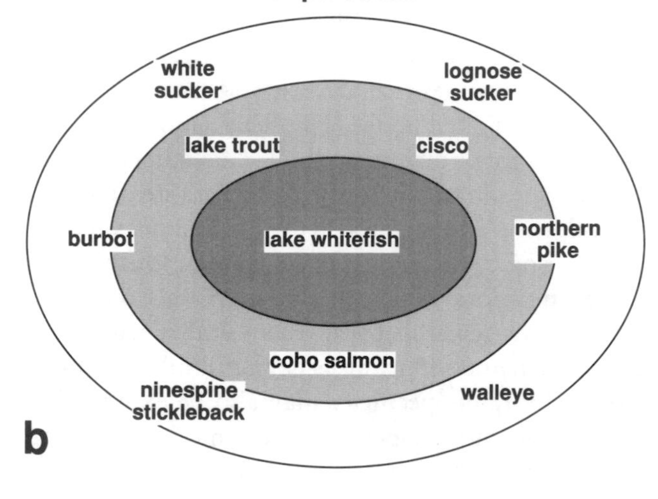

Fig. 4.4 (a) distribution of host specificities of helminth parasites on marine fishes (order Monogenea). Data from Rohde (1979). (b) distribution of fish species among categories recognized by Holmes (1979) for the acanthocephalan parasite *Metechinorynchus salmonis* in Cold Lake, Alberta, Canada: inner oval (heavy stipple) – required host(s); middle oval (light stipple) – suitable hosts; and outer oval (clear) – unsuitable hosts. After Holmes (1979).

gies when colonizing different hosts. In other words, known nymphs of the one parasite species develop different morphologies in different host species and in the process may mislead parasitologists into believing that different species of parasites are involved. This has led, in Downes' view, to a situation in which the high cladistic correspondence between 'species' of parasitic mites and molluscs may be mostly an artefact of host–specific morphological development. If the latter were a widespread phenomenon (Downes cites several groups in which it is known to occur, such as acarines, trematodes, cestodes, and protozoans), then at least some of the perceived host specificity in parasites may be spurious. The solution to this dilemma is the development of independent measures of specific differentiation other than morphology (Downes 1990), possibly by using genetic markers or DNA sequencing.

Some parasitologists adopt fine-grained distinctions in relation to host specificity. For example, Holmes (1976, 1979) recognized these gradations: (1) *required* host – host(s) to which a parasite has the highest degree of adaptation and without which it cannot survive (there may be a further division between *definitive* and *intermediate* required hosts); (2) *suitable* hosts – hosts in which the parasite can mature and reproduce, but reproduction is at rates that are too low to sustain the parasite population; and (3) *unsuitable* hosts – hosts in which the parasite may become established but within which it cannot mature or reproduce. Holmes (1979) illustrated these distinctions by using the data collected by Leong (1975) on the hosts of an acanthocephalan parasite, *Metechinorynchus salmonis*, in Cold Lake, Alberta, Canada (Fig. 4.4 b). The required definitive host was the salmonid lake whitefish (*Coregonus clupeaformis*) of which 99% of 863 individuals were found to be infected (Holmes 1976: Table 1). A required intermediate host was the amphipod *Pontoporeia affinis*. Several other salmonid fish were suitable hosts, namely, lake trout (*Salvelinus namaycush*, 100% infection), cisco (*Coregonus artedii*, 32%), and coho salmon (*Oncorhynchus kisutch*, 100%). Non-salmonid fishes generally were unsuitable hosts (e.g., white and longnose suckers, *Catostomus commersoni* and *Ca. catostomus* respectively; ninespine stickleback, *Pungitius pungitius*; walleye, *Stizostedion vitreum*) apart from two relatively uncommon predatory species (burbot, *Lota lota*; northern pike, *Esox lucius*), which Holmes (1979) regarded as bordering on the verge of 'suitability'. Thus, host specificity in many parasites may be significantly more complicated than simple measures of host use might suggest.

Hearkening back to Chapters 1 and 2, it is evident that if we were to interpret host use by parasites in relation to exploitation (fitness returns) rather than utilization (raw usage patterns) then the differential suitability of hosts in terms of the gains in parasite fitness would reflect Holmes' (1979) scheme well. In particular, the versatility of host use in the Cold Lake example would be rather less than the bald number of host species in which the parasite was found (eleven species).

There are two aspects to versatility that need to be emphasized in relation to parasites, namely, (1) physiological versus ecological specificity (Baer 1951: 136–155), and (2) stage-specificity (Jones 1967). Physiological specificity arises when a parasite is incapable of manufacturing nutritional components, or lacks certain enzyme systems, and so uses the host to supply them. Immunological factors also contribute to physiological specificity. Ecological specificity is a manifestation of the availability of hosts. The parasite may be capable of exploiting a wider range of hosts than it seems to do but those hosts are generally not exposed to the parasite.

The majority of parasites have complex life cycles, typically dispersal, growth, and sexual phases. This means that they may have multiple hosts, depending upon stage, and stage-specificity varies markedly. In some parasites, one or more phases show little host specificity, opportunistically infecting hosts. Other phases have a limited set of hosts. Together, these observations suggest that niche breadth in many parasites may vary substantially between stages, and in some cases, there may be gender-specific differences in host specificity as well (e.g., *Unionicola* mites, Downes 1989)

The physiology/ecology and stage-specific ideas generally are not independent of one another. For example, Jones (1967: 362–363) noted that the miracidial stage of the widespread Asian blood fluke *Schistosoma japonicum* is specific to single species (sometimes races) of aquatic snails and, when exposed to other snails, is incapable of successfully infecting them. This specificity is due to immunological, nutritional and other characteristics of the snails and parasite. On the other hand, the cercarial stage can infect almost any vertebrate with which it comes into contact, so that its specificity in the field is a function of opportunity and so is ecologically controlled. Another example is the cestode *Schistocephalus solidus* of North America. This cestode has a free-swimming coracidium, a procercoid phase that utilizes copepods, a cercoid stage that uses the three-spined stickleback (*Gasterosteus aculeatus*) and the adult, which is capable of afflicting more than 40 species of fish-eating birds (LoBue and Bell 1993).

The specificity in many parasites therefore has to be couched in terms of the versatility of host use of different stages (see Chapter 5) and this is intimately connected with physiology and ecology. Some stages are physiologically specific, and their range of host use would remain the same independent of the respective availabilities of potential alternative hosts. These stages are ecological specialists. Other stages that show much more opportunistic infection would be expected to use alternative hosts in relation to their availability, thus showing ecological generalization.

Much of the preceding discussion concerns generalizations based on extensive observations of patterns in groups of parasites, especially in relation to the epidemiology of parasitic infections in humans. There also is an extensive theoretical literature on the population dynamics of hosts and parasites that attempts to identify the reasons for differential host specificity in parasites (e.g., Esch 1977, Kennedy 1984, Esch and Fernandez 1993).

For example, Garnick (1992) recently looked at the conditions that might determine the evolutionary stability of a parasite population consisting of specialized and generalized phenotypes attacking a reservoir host and an 'incidental' host. The specialized morph was regarded as being obligatorily dependent on the reservoir host, while the generalized morph could facultatively infect both hosts, although less efficiently than the specialized morph on its host. Given these assumptions, Garnick (1992: 71) showed that there are conditions under which either the specialist or the generalist phenotype prevails, or coexistence occurs. The critical variables in which conditions for these different outcomes can be expressed are the basic reproductive rates. These are defined as average numbers of secondary cases that an infected host generates during its infectious lifespan if set among a large population of susceptible hosts. The specialist persists alone if its reproductive rate exceeds that of the generalist, while coexistence occurs if the total reproductive rate (over both hosts) of the generalist exceeds that of the specialist (on the reservoir host). There are conditions in which the generalized strategy persists alone despite being incapable of persisting on either host species separately. High rates of cross-host transmission are needed in this case. Thus, the degree of host specificity may be dependent on the rates of transmission in all pathways through host species utilized by a population of parasites. Garnick (1992) concluded that host generalization in parasites is likely to be more common in indirect, free-living, and vector modes of transmission. Specialization seems more likely in direct transmission unless there is a prolonged, close association between the host species. Unfortunately, there appears to be a paucity of data with which to test these deductions because transmission rates have rarely been measured directly in the field (Downes 1995).

Animal parasitism – an overview

These reviews show that the determinants of versatility in parasitic lifestyles encompass many processes. Rarity and dispersion (Rohde 1979) and physiological similarity and cross-infectiousness between hosts may be important effects. Sparse distributions of hosts, especially in the tropics, may prevent the expression of extreme specialization in the parasitoids of herbivorous insects, winnowing out specialists and leaving mainly generalized taxa capable of exploiting at least several species of host. The dispersion of parasite populations in the seas and oceans may favour extreme specialization to assist individuals in contacting one another for reproduction although, given that many parasites are hermaphroditic, mate-finding is at best a partial explanation.

The different nature of host exploitation in parasitoids and parasites might be expected to produce different effects. Coevolution appears to play an important role in specialization in koinobiotic parasitoids and their hosts, a race in which some dipteran parasitoids seem to have gained an advantage by the development of characteristics to combat effectively the counter responses of developing hosts. Host specificity in animal parasites involves both ecological and physiological limitations. Some parasites are capable of infecting many host species but do not come into contact with them (ecological specialization) while others cannot infect more than one host irrespective of exposure (physiological specialization). Perhaps the most important observation in relation to host specificity in parasites is that the stages of the complex life cycles exhibited by most parasites show remarkably different levels of specialization. Certain stages in most parasites show little specificity (e.g., any vertebrate will suffice) whereas others display extreme specificity (e.g., a particular species or race of aquatic snail). In other words, host specificity in animal parasites generally must be treated as a problem in which the various life history stages (and in some cases the sexes) are considered separately.

Plant parasites

Although the primary focus of the present section is on parasites of animals, there also is a well-developed body of thought on the specificity of plants, fungi and bacteria parasitizing plants. Callow (1977) reviewed the biochemical basis for host specificity and noted a distinction between parasites that kill host cells, and those requiring living host cells. In a way reminiscent of host specificity in insect parasitoids (the idiobiont–koinobiont dichotomy), the former group generally is much less dependent on specific hosts than the latter group (Day 1974).

A widespread group of macroscopic plant parasites is the mistletoes. Their host specificity has been linked to many factors. Some important constraints on host specificity in this group have been thought to be: (1) vector specificity (i.e., behavioural patterns of birds); (2) the physiological compatibility of a potential host and a mistletoe; (3) host size; (4) environment, such as water availability; and (5) the avoidance of herbivores (Yan 1989). There appears to be a wide range of specificities for different hosts in species of mistletoe with some species exploiting many or most available hosts while others seem to be restricted to single host species. Specificity also appears to vary geographically, perhaps in relation to climate and/or genetic variation in host populations (Yan 1989). Experimental evidence involving inoculations of mistletoes onto a variety of hosts suggests that germination efficiency rather than differential exposure is the main reason for differences in versatility displayed by species of mistletoe (i.e., physiological rather than ecological specialization, Lamont 1985, Hoffmann *et al.* 1986, and references cited by Yan 1989). Host resistance is a function of the ability of the host to resist penetration by the mistletoe haustorium (Kuijt 1969).

4.3 Predation

Predation is used here in the sense of carnivory – the consumption of an animal by another organism (possibly a plant; e.g., Givnish 1989). Consumption often will be complete but not necessarily (see Giller 1980). Predation can be regarded as a relatively short-term interaction leading to the demise of the prey. Strictly speaking, predators are constraints for prey but prey are true resources for predators. Predators have been implicated in controlling the ecological versatility of prey as in, for example, the enemy-free space hypothesis to account for the range of host plants used by phytophagous insects (e.g., Gilbert and Singer 1975, Smiley 1978, Atsatt 1981, Lawton and Strong 1981).

In this section, the relationship between predation and ecological versatility is treated as three topics. First, some general ideas on how the two are linked are discussed briefly. Second, one of the most influential theories in predator–prey relationships is that of optimal foraging by predacious animals and it seems appropriate that a consideration of this idea and how it relates to ecological versatility should appear in this section. And third, the act of predation perhaps is one in which the biophysical capabilities of organisms have the most influence in terms of resource use. Therefore, several examples of the importance of the perceptual and hunting abilities of predators are discussed, and how these affect the realized dietary versatility.

Predation and versatility – general ideas

Emlen (1973: 176–178) suggested that there is a constant pressure for dietary specialization by predators and that the occurrence of more generalized feeding is due to the effects of several factors. He cited food abundance, spatial and temporal variability in the environment and the need for balanced nutrition as reasons why specialization is not universal. If preferred food types become or are scarce, then extreme specialization becomes uneconomic and diets necessarily must be expanded. Spatial variability probably affects small animals more than large ones because the environmental graininess perceived by them is at a finer scale (May 1978). That is, small animals may spend all their time in a single 'patch' while larger animals move across patches, perhaps being largely oblivious to the patchiness to which the smaller animals respond. Thus, the perceived environmental grain may dictate dietary versatility to some extent. For these reasons, dietary specialization may be expected to be more common in small animals than in large, although in at least some lineages of small predators (e.g., insects) more generalized feeding appears to be the primitive and dominant condition (e.g., North American arboreal lacewings, Milbrath *et al.* 1993).

Variation in food availability through time might be expected to hinder dietary specialization. Emlen (1973) argued that one of the most significant factors preventing specialization is that of nutrient balancing. A single food type is unlikely to satisfy all of the nutritional requirements of an organism so that systematic or even occasional dietary expansion will be necessary.

Close links between niche breadth and the effects of predators on prey have been proposed for many years. Glasser (1983) provided a simple theory relating the impact of predators on abundances of prey populations to try to account for the perceived differences in versatility between top-level consumers (i.e., primary carnivores) and species lower in food webs (Fig. 4.5).

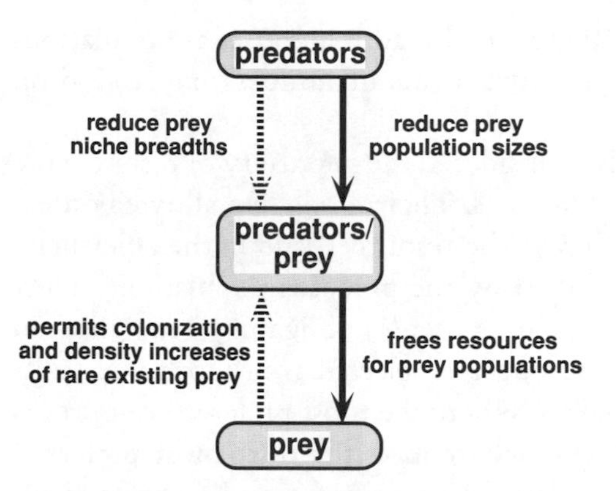

Fig. 4.5 Glasser's (1983) scheme for describing the impact of predators on the densities and niche breadths of their prey species and the implications for niche packing that stem from these effects. Direct effects are indicated by solid arrows while indirect or consequential ones are shown by hatched arrows.

The gist of this argument is that the higher a population is in a food web the less constrained it is by predation. Predators differentially impinge on prey populations, such that the most abundant prey populations are consumed disproportionately more. This has the effect of freeing a greater diversity of resources at that trophic level because abundant populations are regarded as being more generalized (Glasser and Price 1982). A relaxation of exploitation pressure in lower trophic levels then contributes to a contraction of prey niche breadths because overabundance is thought to lead to allow greater concentration on preferred food resources, and hence, narrower breadths (MacArthur 1972). If there is chronic overabundance of resources relative to exploitation pressure, then relatively extreme specialization on small subsets of available resources may occur (Glasser and Price 1982).

Glasser (1983), by using arguments that are discussed more fully in Chapter 5, proposed a species packing mechanism in which obligatory specialization or generalization would be expected under conditions of constant resource superabundance relative to pressure. Facultative exploitation strategies would be more effective if that relationship fluctuates (i.e., niche contraction when resources are overly abundant, and greater generalization under stringency). Thus, this scheme strongly couples the abundance of populations to that of their resources and to position in food webs, providing an expectation of increased generalization at higher trophic levels compared with lower levels. The effect is mediated by the assumption that niche breadth increases as populations grow and contracts when populations shrink in numbers (e.g., Svärdson 1949).

Glasser and Price (1982) also considered the versatility of resource use in relation to predator–prey dynamics. Their modelling allowed differences in both the densities of prey (the resources) and in the efficiencies with which the prey are exploited by the predator populations. They considered three strategies of resource use: (1) obligatory generalists, in which the probability that a resource is selected is proportional to its relative availability; (2) specialists, where the most preferred resource is selected until it is fully utilized, whereupon the next most preferred resource is selected until it too is used completely, and so on; and (3) facultative strategists, which behave like specialists when resources are abundant (hence narrow breadth) but becoming more generalized as resources become rare (hence wider breadth). Trajectories of densities and breadths of the predator populations were shown to depend critically on how similar efficiencies of utilization were on different resource populations. Glasser and Price (1982) found that when efficiencies were equal, facultative strategists reached their highest population densities but generalists exhibited the greatest breadth of prey use (Fig. 4.6). When efficiencies differed markedly, the population density of the obligatory generalist was depressed and both facultative and specialized strategists displayed similar densities and narrow niche breadths.

Optimal foraging – the contingency model

Optimal foraging is one of the prime (and most vibrant) areas of research in behavioural and evolutionary ecology (see, for example, Schoener 1971, Werner and Hall 1974, Charnov 1976, Pyke *et al.* 1977, Caraco 1980, Pyke 1984, Stephens and Krebs 1986, Hughes 1990). It is used to predict the diversity of resources used by heterotrophic organisms, and focuses on fitness-related quantities of individuals such as rates of energy gain, starvation risk and vulnerability to predation (Jaeger and Lucas 1990, Taghon *et al.* 1990). Thus, optimal foraging theory shares much in common with the ideas on ecological versatility considered here. In its original formulation, optimal foraging referred to the factors involved in the three phases of foraging: *search*, *encounter* and *decision* (Emlen 1966, MacArthur and Pianka 1966). Searching means looking for potential food items until one is found. Encounter means the predator's detection of a potential food item and the associated cessation of searching. The

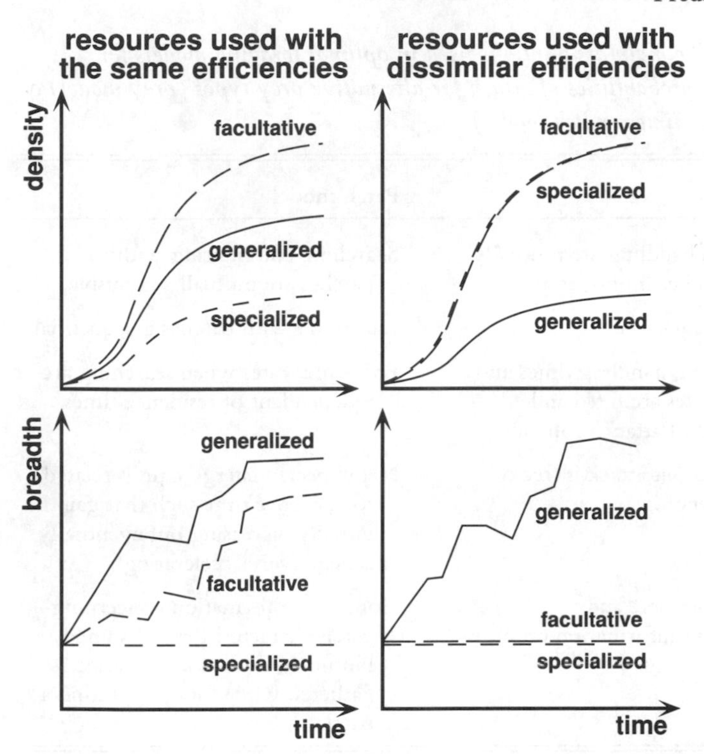

Fig. 4.6 Trajectories of changes in population density (top row) and breadth of prey resource use (lower row) for three predator strategies: obligatorily generalized, facultative and obligatorily specialized. The effects of different degrees of evenness of efficiency of use of resource classes are shown: left – all resources can be used with the same efficiencies; right – resources used with different efficiencies, one particularly efficiently and the rest very inefficiently. These plots were adapted from the modelling results of Glasser and Price (1982).

decision is whether to attempt to attack the encountered food item or not, or possibly whether to move elsewhere to continue searching. Therefore, foraging is seen as a repetitive cycle of 'search–encounter–decision–search . . .' and foraging is treated as an optimization problem in which the average ratio of energy gained to time spent in acquiring that energy is maximized. The time scale over which the average rates are maximized has been controversial and, apparently, has not been fully resolved as yet (see Fairweather and Underwood 1983, Lucas 1990).

There are two basic models, one related to decisions as to which prey to include in diets (the *prey* model) and the other to modelling how long foragers should remain in patches containing food items (the *patch*

Table 4.3. *The main assumptions used in optimal foraging models for predicting the probabilities of attack for alternative prey types (prey model) or patch residence times (patch model)*

Prey model	Patch model
Searching and handling are mutually exclusive and cannot overlap	Searching and hunting within patches are mutually exclusive
Encounter is sequential	Encounter with patches is sequential
Net energy gain, handling times and encounter rates are fixed and independent of attack probability	Encounter rates when searching are independent of residence times
Encounter without attack is free of costs and benefits	Net expected energy gain is related to residence time such that gain is initially increasing but eventually is negatively accelerating
Forager has complete and unambiguous information on prey types	Complete information concerning patch characteristics is assumed but no additional information is gathered while foraging within a patch

Source: After Stephens and Krebs (1986).

model). In the terminology of Chapter 2, patches in this sense are microhabitats that house food resources rather than being resources in their own right. Analytical models are generally used to make predictions about the way in which predators should forage and, because the approach is an optimization one, solutions are found by imposing 'constraints'. Constraints here may or may not be constraints in the sense used in Chapter 2 (true constraints) but 'constraint' is the term that is used in linear programming to solve optimization problems so that I will use it here in this restricted sense and contrast it where necessary with the term 'true constraint'. The basic models are static ones providing no dynamic solutions (see Mangel and Clark 1988 for an overview of dynamic models).

The main assumptions of the basic optimal foraging models are (Table 4.3): (1) encounters with new prey or patches can only occur once handling the current prey, or hunting within the current patch, is completed; (2) encounters with prey or patches are sequential (not concurrent); (3) model parameters are fixed independently of the choice to attack a prey item or stay within or leave a patch; and (4) predators have complete information on alternatives but cannot derive additional information during foraging.

For the prey model, there are three main predictions. First, particular prey types either are always attacked or always rejected depending upon their relative profitability. Second, types are added or deleted from diets according to their relative profitability, with the most profitable included first. And third, whether or not a prey type is included depends upon only its relative profitability and not on the encounter rate of that type (Stephens and Krebs 1986: 23).

Given that encounter rates are identical to availabilities (not abundances), the third result from the basic optimal foraging theory is a disappointing one. It means that dietary breadths are predicted to be independent of relative availabilities so that versatilities (i.e., the respective fitness gains from using a set of resources compared with their relative availabilities) may vary passively (and meaninglessly) as availabilities change. Certain alterations in some of the basic model assumptions are needed for versatilities to vary in a meaningful way.

Of course the assumptions listed in Table 4.3 are often too restrictive and biologically simplistic, which, among other things, has led to criticism of the whole optimality approach (e.g., Stephens and Krebs 1986: Chapter 10). For example, assumption (3) above is violated in situations in which the catchability and handling of prey improve with experience (e.g., Hockey and Steele 1990) and/or when physiological status is important (Croy and Hughes 1990). Much of the theoretical development since the early models has involved the relaxation of one or more of these assumptions, leading to more realistic predictions and conformance with experiments. Some of this work is described by Stephens and Krebs (1986: Chapter 3).

Having considered the basis of optimal foraging theory, it is now time to relate it to dietary versatility. Westoby (1974, 1978) noted that the basic theory generally predicts narrower dietary range than that seen in nature, and that the first prediction listed above (always attack or always reject) rarely is borne out. In other words, diets generally are broader than the basic models predict with partial preferences for types being observed rather more often than absolute attack/reject preferences. Westoby himself suggested several reasons for this but Godin (1990) has compiled a more complete list, which is summarized in Table 4.4.

Most of the factors in Table 4.4 lead to violations of one or more of the

Table 4.4. *Factors leading foraging organisms to display partial rather than absolute preferences for alternative prey types*

Foraging time limitations	Forager hunger status
Nutrient limitations	Learning–dependence (e.g., handling)
Sampling of available prey types	Lags in prey recognition time
Discrimination errors	Simultaneous prey encounters
High value prey rarely encountered	Stochastic inclusion thresholds
Stochastic returns of prey	Discounting of future prey rewards
Energy not the currency of importance	Nonrandom encounters with prey types
Diet selection resulting from averaging over different conditions	Competition
Predation risk	

Source: From Godin (1990).

assumptions of the basic models. Some, like those listed by Westoby (1974, 1978), suggest that the optimization of the rate of energy acquisition is not the appropriate currency of optimization (see McNamara and Houston 1986), while others are based on the effects of true constraints on dietary selection.

Predation risk and intraspecific competition are likely to be two important modifiers of foraging behaviour. Kleptoparasitism is an example of how intraspecific competition can influence dietary selection (e.g., Hockey and Steele 1990). Predation risk for the predator itself has been considered by many workers in both theoretical and experimental terms. For example, Gilliam (1990) generalized the optimal model to include predation risk, and the optimization problem was framed as a maximization of energy gain given the predation risks. In this model, the predator was provided with a safe refuge so that dietary versatility could be modelled in relation to the hazard involved in searching for and handling prey (Fig. 4.7). If searching and handling entailed equal risk, then Gilliam showed that mortality rates were independent of diet selection and the optimal diet solution simplified to a maximization of energy gain (i.e., the basic model, Fig. 4.7 a). If handling were more

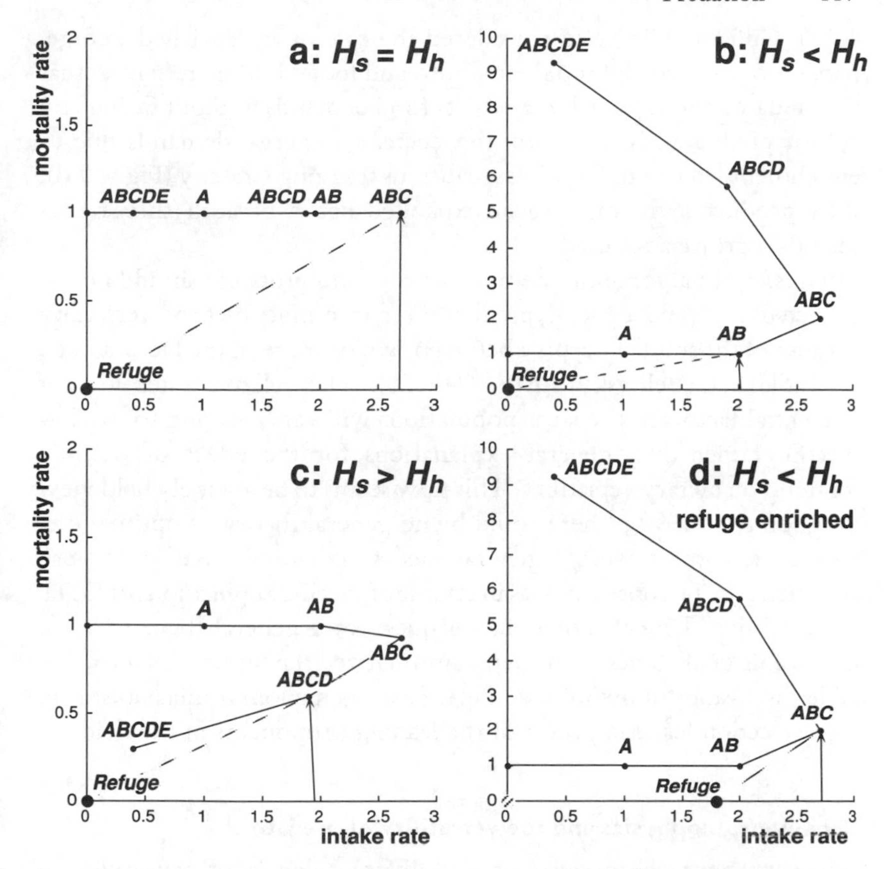

Fig. 4.7 Mortality rates plotted against energy intake rates for four situations involving predation risk: (a) searching and handling equally hazardous (i.e., $H_s = H_h$); (b) handling more hazardous than searching; (c) handling less hazardous than searching; and (d) handling more hazardous than searching but with an 'enriched' refuge (after Gilliam 1990). Five diets were considered (*A*, *AB*, *ABC*, *ABCD*, and *ABCDE*, where *A–E* are dietary classes). The optimal diet, which maximizes energy gain to mortality risk, is one at the point at which the dotted line first touches the solid curve (i.e., the line pivots about the refuge point from the *X*-axis in an anticlockwise direction). In (a), (b) and (c) the refuge yields zero risk *and* zero gain while in (d), food accumulates in the refuge so providing some risk-free gain. Note the different ordinate scales between (a)/(c) and (b)/(d).

hazardous than searching, then the dietary breadth that maximizes energy gain relative to predation risk was narrower than (or possibly equal to) the optimal dietary breadth in the absence of predation risk (Fig. 4.7 b). However, if handling were less hazardous than searching, predation risk increased dietary breadth relative to the basic model (Fig.

4.7 c). Gilliam (1990) also considered the case of an 'enriched' refuge, perhaps one in which food drifted into and lodged in the refuge so that the predator could partially satisfy its food demands without facing any risk of predation. In this case, the decreased energy demands due to enrichment led to a different, less cautious foraging strategy (Fig. 4.7 d). A by-product, in this case, was an expanded dietary breadth (three rather than two prey classes used).

Thus, optimal foraging theory does not (and probably should not be expected to) yield general predictions concerning dietary versatility because of the subtle interplay between two or more of the factors listed in Table 4.4 (Belovsky 1990: 274–276). The relative intensities of influential factors in predator populations will vary, leading to species-specific rather than general explanations for the effect of optimal foraging on dietary versatility. This now seems to be a widely held view (Taghon *et al.* 1990): 'there could be no general theory of optimal diet choice or, if there were a general theory, it would have to be very complex . . . The consensus was in favour of specific applications of ODT [optimal diet theory] rather than a quest for a general theory, . . . an impossible goal'. These comments summarized the findings of a round-table discussion following a conference on behavioural mechanisms in food selection featuring most of the leading proponents in the field.

Behaviour, biophysics and the versatility of predators

Another approach to account for different levels of versatility in predators involves the erection of purely mechanistic explanations. These mechanisms may involve behavioural and/or biophysical constraints that operate to restrict the range of resources used by predatory animals.

One example is that of foraging restriction in migrant wood warblers of North America (family Parulidae). Greenberg (1983) conducted a series of experiments on several species but concentrated on the chestnut-sided (*Dendroica pensylvanica*) and bay-breasted warblers (*D. castanea*). Greenberg's results seemed consistent with the idea that by breeding in coniferous forests, the bay-breasted warbler would be less well adapted for foraging in broad-leaf woodlands during winter than the chestnut-sided warbler, which breeds in broad-leaf woods. This poorer adaptation was deemed to be the source of greater opportunism or generalization of food use by bay-breasted warblers. The proximate mechanism for this effect appears to be less inhibition in approaching novel sources of food and/or microhabitats by the bay-breasted than the chestnut-sided warbler. Thus, coniferous forest specialists such as the bay-breasted warbler appear to be more generalized foragers and more likely to exploit changes in availabilities of food resources when occupying other types of habitats.

The versatility of resource use may also be constrained or modified by morphological characteristics. For example, Freeman (1979) noted that insectivorous bats have been regarded as relatively generalized omnivorous mammals, as opposed to the more obviously specialized carnivores and herbivores. She was able to show that some insectivorous bats may be as specialized in jaw structure and dentition as any mammalian carnivore or herbivore. Freeman assessed jaw characteristics in the light of biophysical performance; her results were consistent with the interpretation that some bats of the family Mollosidae (e.g., *Nyctinomops*) are specifically adapted to seize and process large, soft-bodied insects like moths while others have morphologies suited to crush hard-shelled insects such as coleopterans (e.g., *Molossus*). These results suggest that restricted diets may arise within general categories (e.g., insectivores) through morphological specializations related to feeding.

Carnivorous plants are some of the most fascinating predators. The gains in fitness associated with predation seem to be very great for them (Chandler and Henderson 1976). Givnish (1989: 260–262) discussed how the characteristics of carnivorous plants appear to be associated with 'specialization' on certain insect groups. The entrapment apparatus of butterworts (*Pinguicula* spp.), for instance, seems to be effective in capturing only small gnats and springtails. The coloration, nectaries, trigger hairs and widely spaced cilia of the Venus flytrap (*Dionaea muscipula*) appear more likely to ensnare larger insects. Many pitcher plants (*Sarracenia*) are adapted for preying upon ants, with some exquisite morphological adaptations for this purpose. However, some species of *Sarracenia* have characteristics more suited for attracting and capturing flying insects such as wasps and bees. Givnish (1989) also noted that resource partitioning or character displacement may possibly occur in communities of carnivorous plants in the south-eastern United States, where species are distributed along a height gradient in relation to the ground. However, he noted that in many circumstances, the degree of 'specialization' shown by carnivorous plants cannot be assessed as yet because the availabilities of prey types in the plants' environments generally have not been monitored.

Dietary versatility frequently appears to be greater in larger predators than in small ones, which may reflect the superior capabilities of larger individuals in detecting and capturing prey (e.g., Selander 1966, Herrera 1978b; although see Brönmark and Miner 1992). This need not be a solid rule, however, because biophysical effects might interfere under some circumstances. For example, Barclay and Brigham (1991) explored why most aerial insectivorous bats are so small. Some of the reasons they considered included aerodynamics and manoeuvrability, phylogeny and the sensitivity of echolocation. Barclay and Brigham (1991) ruled out all but the last explanation. They showed that biophysical effects involved in generating and receiving high-frequency sounds prevent large size in aerial insectivorous bats, except where relatively large insects (e.g., moths) are available in moderate to high densities. Large bats either are unable to detect or detect too slowly the much more numerous smaller insects.

These ideas led Barclay and Brigham (1991) to contest the hypothesis that large aerial insectivorous bats actively select large insects and that smaller bats opportunistically use small insects (Swift *et al.* 1985). Barclay and Brigham argued that the large bats exploit those insects that are effectively available to them, while smaller bats have a greater range of options of airborne prey. What does this mean for ecological versatility? If biophysical constraints are not considered then the larger bats might appear to be more selective or specialized than the smaller bats. However, it seems probable that large bats are more or less equally opportunistic as small ones but the range of resources effectively available to them due to biophysical constraints is reduced, leading, falsely, to a conclusion that the larger bats are more 'specialized'. Thus, the biophysical capabilities of species can alter the effective availabilities of alternative food types, which reflects the important distinction made in Chapter 2 between abundance (what is actually there) and predator-specific availability (what the predator perceives).

I have described several cases in which morphological and related biophysical characteristics of predators appear to constrain the breadth of resource use. However, it would be remiss to overlook documented cases in which specific tests for morphological associations with degrees of versatility have been made and in which no relationship has been found. One case is the study by Werner and Sherry (1987) on specialized foraging behaviour in individuals of the Cocos finch, *Pinaroloxias inornata*. They found no relationship between foraging specialization and morphology. Instead, Werner and Sherry (1987) attributed the origin of individual foraging specialization to a cultural transmission (learning by watching experienced individuals), which would allow individuals to reach higher levels of foraging efficiency in less time.

Another instance is a test of the morphological similarities between populations of the yellow-billed cacique, *Amblycercus holosericeus*, from Central and South America (Kratter 1993). Populations of this bird in the highlands of Costa Rica and in the Andes occupy only bamboo forests (predominantly *Chusquea*), and specialize on bamboo as a foraging substrate. However, populations of lowland forests often occur in habitats lacking bamboo and are more generalized in substrate use too. The species appears to be absent from forests at intermediate elevations so that the bamboo specialists are largely isolated from lowland populations. Under these conditions, Kratter (1993) contended that morphological adaptations (hence convergence) might be expected for utilizing bamboo. Morphological analyses did reveal a dichotomy – but the distinction was between populations to the north of Panama and those of South America rather than a generalist-bamboo specialist difference. Kratter (1993) concluded: 'bamboo specialization arose independently in the two populations . . . [suggesting that] bamboo specialization does not require a strict morphology'.

These two examples – and there are many others – show that morphological characteristics do not necessarily relate in a direct way to ecological versatility, although there are often situations in which morphology and biophysical capabilities do interact to limit the spectrum of resources that are effectively available to a predator.

Predation – an overview

Dietary versatility in predators clearly is affected by many factors. Although the contingency model is useful for deriving expectations of diet selection for particular predators, a suite of physical and biological effects appears to invalidate some or all of the basic assumptions of the model. This often leads to observations that are inconsistent with model expectations. Gauging relative prey suitability often is difficult for predators because of the absence of reliable cues or because the sensory capabilities of individual predators may have moderate to severe biophysical limitations (i.e., the detectability of particular prey is predator-specific). Morphology can constrain versatility although not invariably (e.g., Kratter 1993). Learning (e.g., Werner and Sherry 1987), inherent behavioural characteristics (e.g., inhibition) and social interactions (e.g., Morse 1974) may produce apparently anomalous obser-

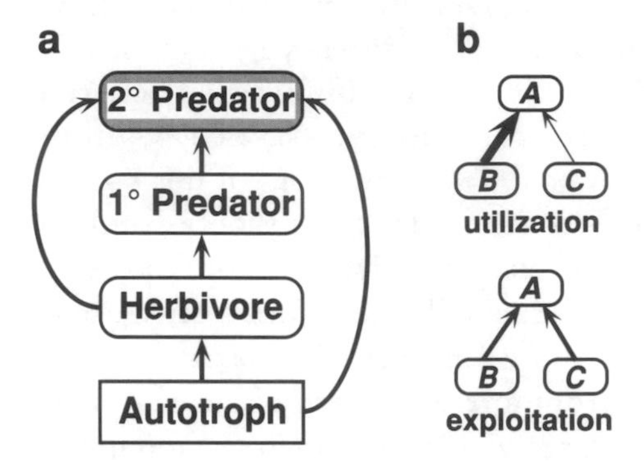

Fig 4.8 (a) schematic representation of a simple food web, where arrows indicate consumption of the organism at the tail by the organism at the head (after Pimm and Lawton 1978). Shaded boxes (in this case, 2° predator) indicate omnivores. (b) the one simple web consisting of three populations, represented in terms of the relative utilization by population *A* of the two prey taxa *B* and *C* (top) and relative fitness gains (bottom).

vations. Different intensities of predation pressure at various trophic levels also have been invoked to explain apparently greater generalization at higher trophic levels than in the lower strata of food webs.

4.4 Omnivory

Omnivory is feeding at several trophic levels (Briand and Cohen 1987, Diehl 1992). Omnivores may utilize prey from one, two or more trophic levels 'down' from where they themselves appear in the food web (Fig. 4.8 a; see also Pimm and Lawton 1978). Some high-order predators practise cannibalism (e.g., Polis 1984, Van Buskirk 1992) and intra-guild predation (e.g., Polis and McCormick 1987, Polis *et al.* 1989, Johanson 1993) or reciprocal predation (Sprules and Bowerman 1988), which also affords them omnivore status because such species gather food from their own trophic level as well as from lower ones. Omnivory and carnivory may exist as alternative ecological phenotypes depending upon prevailing conditions, such as food availability and/or the ephemerality of habitats (Pfennig 1992). Optimal diet in omnivores using both plants and animal prey is not expected to be a simple intermediate strategy between carnivores (maximization of rate of acquisition) and herbivores (throughput maximization), although some experiments suggest that herbivory constraints may predominate (Ball 1990).

Almost by definition omnivores must be regarded as relatively more generalized than organisms depending upon a single, lower-order trophic level (e.g., the omnivore *Mysis relicta*, see Sprules and Bowerman 1988). Omnivores feeding on both plants and animals may face especially difficult problems in contending with the demands of exploiting diverse foods (Yodzis 1984). Presumably, this argument refers to phytophagy coupled with carnivory rather than frugivory, granivory or nectarivory with carnivory, for many birds and mammals seem at ease with omnivorous diets consisting of fruit, seeds, nectar and invertebrates (Lack 1954, Bourliere 1956, Poulin *et al.* 1992; see also Bazely 1989).

Omnivory is a crucial component of food web theory. From an early model, Pimm and Lawton (1978) made two predictions: (1) omnivory should be relatively rare in most natural communities; and (2) omnivory based on nonadjacent trophic levels also should be rare. Pimm and Lawton arrived at prediction (1) from a consideration of the stability of mathematical models of communities in which omnivory is represented. In ecological terms, an omnivore both competes with and preys upon an intermediate species, which presumably makes persistence difficult for the intermediate taxa (Sprules and Bowerman 1988). However, the persistence of populations at intermediate trophic levels might be enhanced by the availability of refugia (Diehl 1992).

Prediction (1) actually is more specific than stated, namely, that omnivory is more rare than would be expected on the basis of what Yodzis (1984) called the 'common-sense [ecological] viewpoint'. Physiological, anatomical, and behavioural requirements that are needed to exploit effectively prey from different trophic levels act to make omnivory comparatively difficult, and hence rare (Yodzis 1984). Yodzis (1984) re-analysed data from the 40 natural food webs considered by Briand (1983). Yodzis' analyses suggested that omnivory indeed is more rare than expected but much of this effect can be attributed directly to the particular difficulties of feeding on both plants and animals rather than on just animals from several trophic levels. Yodzis also showed that although prediction (2) of Pimm and Lawton (1978) is supported by information on natural communities, it might be adequately explained on the basis of the common-sense viewpoint.

Unfortunately, the rarity or commonness of omnivory and its impact in community dynamics is more unresolved than any of these arguments suggest (Kerfoot 1987, Schoener 1989b, Cohen *et al.* 1993, Hall and

Raffaelli 1993). Paine (1988) doubted the conclusions of many analyses inferring ecological pattern from disparate data sets on food webs (e.g., Cohen 1978, Pimm 1982, Briand 1983), emphasizing the sensitive dependence of perceived web structure on the observer, location and time. Strengths of interactions, which are relatively difficult to represent in complex webs (see Figs. 6, 7, and 9 of Winemiller 1990 to see why) and still harder to measure, are critical components of community dynamics, yet they have not been or cannot be incorporated into compendia on food webs (although see Wootton 1994). Another problem is that many field workers and theorists alias lower trophic levels much more than upper strata (i.e., clumping autotrophs and primary consumers to the level of family or order but treating high-order predators at the specific level), which prejudices many conclusions (e.g., Briand and Cohen 1984, 1987, Cohen *et al.* 1990; see Greene and Jaksić 1983, Hall and Raffaelli 1991, Martinez 1991 and Pahl-Wostl 1993 for criticisms).

Omnivory may be more prevalent than has been portrayed in the older archives of published data sets (e.g., Briand 1983) with relatively high levels reported in recent studies in both aquatic and terrestrial systems (Sprules and Bowerman 1988, Vadas 1990, Hall and Raffaelli 1991, Polis 1991; note also that recent theoretical treatments indicate that omnivory need not be as rare as Pimm and Lawton 1978 suggested – see Law and Blackford 1992: 575). For example, analyses of glacial lakes of the Canadian Shield suggest that omnivory is common in pelagic crustacean zooplankton communities (see Table 2 of Sprules and Bowerman 1988). Top-order carnivorous zooplankters in these systems such as the opossum shrimp, *Mysis relicta*, and the copepod *Senecella calanoides*, prey upon phytoplankton and primary and secondary predatory zooplankton (Fig. 4.9). *Mysis*, for example, consumes algae in its early juvenile stages and as adults eat herbivorous cladocerans and rotifers, and secondary (e.g., cyclopoid copepods) and tertiary consumers (e.g., the copepod *Epischura*). Omnivory in the glacial lakes is almost four times as common as in the Briand data set although some of the discrepancy might be due to aliasing in the latter (Sprules and Bowerman 1988).

Omnivory is especially high in lakes inhabited by *Mysis*. The opossum shrimp seems to derive a disproportionate amount of energy directly from herbivorous zooplankton (i.e., a strong interaction from the perspective of *Mysis*), which may explain why *Mysis* can 'afford' to be so omnivorous. Sprules and Bowerman (1988) also hypothesized that there may be a coupling between omnivory and high trophic position. They also commented that the common occurrence of omnivory in pelagic

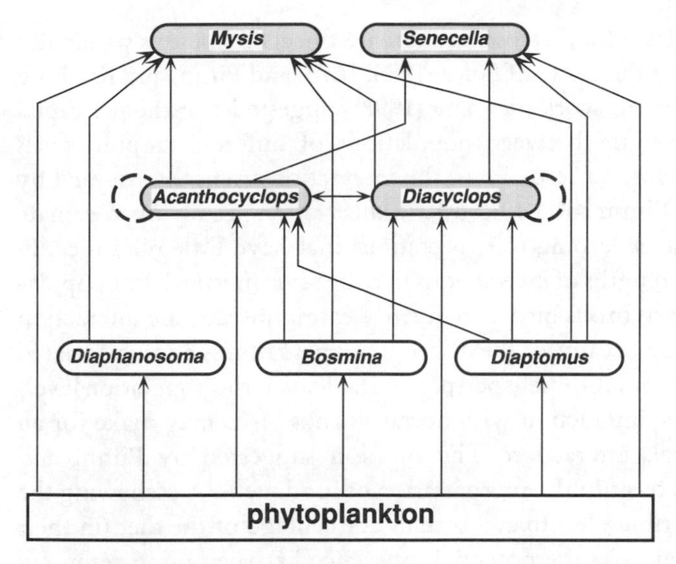

Fig. 4.9 Example food web for pelagic plankton from a Canadian glacial lake (after Sprules and Bowerman 1988). Other conventions as for Fig. 4.8. Dashed, recurved loops indicate cannibalism.

lake systems might be due to the simple mode of feeding, namely, by engulfment. Thus, body size largely determines trophic position and omnivory does not necessarily need to overcome very difficult problems. This may account for the development of the trophic continuum concept by aquatic ecologists (e.g., Cousins 1985). Feeding in lakes or seas often involves rather simple methods of prey capture and processing so that the distinction between trophic levels can become blurred.

The issue then may not be so much whether omnivory is rare or common as such but its importance in community dynamics and the dietary versatility of organisms. Some workers have attributed much importance to omnivory (e.g., Polis 1984, Polis and McCormick 1987, Sprules and Bowerman 1988, Wissinger 1992). Many essentially granivorous or nectarivorous birds (and carnivorous plants for that matter, e.g., Givnish 1989, Karlsson *et al.* 1991) take quantities of insects for essential compounds that they themselves cannot or do not synthesize, or that are in short supply (Remsen *et al.* 1986, Poulin *et al.* 1992). Relative utilization of resources (i.e., seeds, nectar, and light versus insects) does not reflect the impact of omnivory in this case but relative exploitation would (Fig. 4.8 b; see also Polis 1991).

Paine (1988) advocated paying more attention to the strengths of interactions within food webs although he conceded that this would not be easy (contrast the views of Peters 1988: 1675, and Pimm and Kitching 1988: 1671). Recent work by Paine (1992) suggested that the per capita interaction strengths between populations of different trophic levels often may be close to zero. Thus, the interaction strengths assumed by theorists (e.g., Pimm and Lawton 1978) may often grossly overestimate natural intensities, leading to expectations that have little relevance. In addition, the strengths of interactions may be asymmetrical. For populations of a nectarivorous bird infrequently eating insects, the interaction may seem strong because of the disproportionate fitness boost associated with the link. From the insect perspective (at least at the population level) the rarity of the interaction with nectarivorous birds may make for an effectively weak interaction. The solution suggested by Pimm and Lawton (1978) to simplify interpretation of food webs by eschewing the links that contribute less than a certain per centage of the diet (in their case, 20%) overlooks the potential difference between using resources and the fitness return on their use (Fig. 4.8 b).

In conclusion, omnivory clearly is a striking example of dietary versatility. The capacity of some organisms to combine routinely diverse food types together in their diets clearly makes them more generalized than organisms restricted to feeding on a single trophic stratum. However, the combination of herbivory with carnivory provides some operational difficulties. For example, the use of phytoplankton in aquatic systems by an organism like the opossum shrimp probably does not involve much additional processing capability beyond that required for consuming herbivorous zooplankton. The same probably holds for terrestrial birds and mammals utilizing fruit or grain in conjunction with animal prey. The problems are most severe for omnivores that have to contend with plant materials such as the leaves of trees or grasses, or marine algae, which are generally either low in nutritional value (per unit volume) or dosed with chemical or physical deterrents, or both. Thus, omnivory at the plant–herbivore interface is not a homogeneous process because of the variety of plant materials. These differences may have important ramifications at all scales up to overall ecosystem function (Hairston and Hairston 1993).

4.5 Mutualism

Symbiosis originally was a term coined to denote the 'living together of dissimilarly named organisms' (de Bary 1879 cited in Starr 1975), which clearly makes no distinction between the respective effects on the cohabitants of the association (the 'symbionts'). Since then, the term 'symbiosis' has acquired a connotation of mutual benefit in the vernacular and in much of the scientific literature, although many specialists in the area have returned to de Bary's usage (e.g., Starr 1975, Smith and Douglas 1987, Werner 1992). In this section, symbiosis will be regarded in the original sense, although I will concentrate on the mutualistic aspects given that the versatility of parasites has already been discussed above.

Mutualisms are interspecific interactions in which there is a net fitness gain to each party (i.e., +/+). They have been connected intimately with coevolutionary processes and there are several extensive reviews covering this topic (e.g., Thompson 1982, Boucher *et al.* 1983, Schemske 1983, Addicott 1984, 1986). There are many forms of mutualism: pollination, plant-protection (e.g., ants), mycorrhiza and seed and fruit dispersal (Howe 1984).

Several writers have emphasized the difference between *mutualism* and *mutualistic association* (e.g., Starr 1975). Mutualisms are regarded as relatively ephemeral interactions (with respect to the duration of the lifetimes of the symbionts) between different organisms such as in pollination or fruit/seed transport. Mutualistic associations on the other hand are more prolonged interactions often involving a more or less complex train of recognition events, as typified by mycorrhizal or rhizobial formation (Harley and Smith 1983, Smith and Douglas 1987, Young and Johnston 1989, Brewin 1991). Addicott (1984) provided a more detailed classification of the types of mutualism based on: (1) degree of association (symbiotic or not); (2) obligatoriness of the association (obligatory or facultative); (3) nature of the benefits (dispersal, defence); (4) extent of reciprocal specialization by one partner in a mutualism on another (i.e., extent of coevolution); (5) temporal patterns in the association (continuous or intermittent) and the 'strength' of the interaction (variable or consistent); (6) proximity of the interaction (direct or indirect); and (7) numbers of participants (explicit bilateral or 'diffuse' mutualisms). For the purposes of this section however, I will use the simpler dichotomy of mutualisms and mutualistic associations because many of Addicott's distinctions are not fully independent of one another. The permanence of relationships as indicated by mutualisms and mutualistic associations perhaps emphasizes the most potent force in the set of factors referred to by Addicott. Addicott (1984: 443) himself remarked: 'We need to consider not only highly coevolved, symbiotic or obligate mutualisms, but also nonsymbiotic, facultative, intermittent, or diffuse mutualisms', which reflects the distinction adopted here. Of course, the ecological versatility of partners in mutualisms and mutualistic associations is expressed in terms of the specificity of hosts and symbionts for one-another, which forms the basis of the following discussions.

Mutualisms

The degree of versatility evident in mutualisms appears to be highly variable. Some workers have reported extreme specialization in some seed-parasite–pollination relationships (i.e., obligatory bilateral mutualisms), in which insects may be quite specific to a particular species of plant and vice versa (e.g., Pellmyr 1989, James *et al.* 1993). Bilateral mutualisms are relationships in which each member of the pair is the reciprocal resource for the other. However, there may be asymmetry in the relationship such that the pollinator for instance, may be relatively more generalized than the plant by gathering pollen from several local species, while the plant may rely on the one pollinator, or vice versa (Schemske 1983, Howe 1984). A similar asymmetry occurs in the ant-tending mutualisms between the lycaenid butterfly *Plebejus argus* and ants of the genus *Lasius* (Jordano and Thomas 1992, Jordano *et al.* 1992).

Extreme specificity does not seem to be the norm for most mutualisms (Schemske 1983, Pierce 1989, Sanders 1993). In some situations, one population may play the mutualistic partner in a sequential manner to several other heterospecific populations (Waser and Real 1979). It is important to remember that many mutualisms are restricted to just one life history stage of a symbiont (e.g., the seed in myrmecochorous plants; Addicott 1986) so that there may be selective pressures affecting the population dynamics of a mutualist apart from the mutualism itself (e.g., intensity of predation on the adults of such a plant).

Although there appear to be some cases in which specificities are high, Howe (1984: 767) provided a list of cases in which nonspecificity characterizes pollination and dispersal mutualisms. He also believed that history, dynamic changes in community and climate in relatively short geological time-spans and pleiotropic genetic effects probably preclude the effective coevolution of mutualists (at least for pollinator/disperser systems). On this basis, specialized mutualisms between plants and animals may be only local and ephemeral.

Addicott (1984: 446–447) cited several situations with supporting evidence to show that mutualisms are spatially and temporally variable in terms of the probable benefits to the participants. There may also be density-dependent or frequency-dependent effects such that interactions that are demonstrably mutualistic at low densities may become transformed into commensal or parasitic interactions at higher densities. In other words, the benefit to one party at low densities becomes a cost at higher densities (see Addicott 1986). Thus, there may be many ecological circumstances having the potential to override the coevolution of specific, obligatory mutualisms, several of which are discussed eloquently by Addicott (1986: 431–435).

The potential for extremely tight, specialized mutualisms is probably diluted by the 'accumulation' of visitor species over evolutionary time, which reduces the strength of selection between the dominant mutualistic pair of species (Herrera 1988, Thompson 1989, Eckhardt 1992). For example, empirical studies of the interaction between a lepidopteran, *Greya politella*, and a species of small herb, *Lithophragma parviflorum*, at sites in Washington state, USA, convinced Thompson and Pellmyr (1992) that extreme specialization in mutualistic relationships often will be governed by the community setting in which a putative mutualism is set. They pointed out that detailed analysis of cost–benefit relationships of mutualist pairs may count for little if the background community is not concurrently evaluated. Abundant co-pollinating species appeared to dilute the possibility of evolution of an extreme specialist relationship between *G. politella* and *L. parviflorum* at the studied locations (Thompson and Pellmyr 1992). Herrera (1988) came to much the same conclusions in his study of the pollinators of *Lavandula latifolia* at sites in the Sierra de Cazorla in south-eastern Spain. These observations appear to sit comfortably with ideas expressed by Howe (1984), who doubted whether bilateral mutualisms are of much importance in natural communities. He argued that the characteristically turbulent nature of most environments should facilitate generalized rather than specialized use of resources. Some restricted conditions may favour obligatorily specialized mutualisms, perhaps in pairs of species isolated from the interference of other taxa.

Mutualistic associations

There are many mutualistic associations involving fungi, bacteria, algae, higher plants and prokaryotes, which have been reviewed extensively by Harley and Smith (1983), Werner (1992) and others. Most associations

Table 4.5. *A nonexhaustive list of fungal species (symbionts) that are capable of forming ectomycorrhiza with two species of Pinus (hosts) in culture*

Pinus silvestris	*Amanita muscaria, A. pantharina, Cennococcum graniforme, Clitopilus prunulus, Cortinarius glaucopus, Co. mucosus, Lactarius deliciosus, La. helvus, Lyophyllum immundum, Rhizopogon roseolus, Rhi. luteolus, Rhodophyllus rhodopolius, Russula emetica, Scleroderma aurantium, Suillus bovinus. Su. flavidus, Su. flavovirens, Su. luteus variegatus, Tricholoma flavobrunneum, T. flavovirens, T. imbricatum, T. pessundatum, T. saponaceum, T. vaccinium*
Pinus strobus	*Amanita muscaria, Boletinus pictus, Boletus rubellus, Cantharellus cibarius, Cennococcum geophilum, Endogone lactiflua, Gyrodon meruloides, Gyroporus castaneus, Lactarius chrysorrheus, L. deliciosus, Russula lepida, Scleroderma aurantium, Suillus granulatus, Su. luteus, Tuber maculatum, T. albidum*

Source: After Harley and Smith (1983).

involve one physically large partner, the *host* or *macrosymbiont*, and one small partner, the *microsymbiont* (Werner 1992).

Generally speaking, specificity in mutualistic associations appears to be relatively low. For example, species of trees may play host to dozens of fungal species when forming ectomycorrhiza. Laboratory inoculations have shown that at least 24 species of fungus are capable of initiating ectomycorrhiza with *Pinus sylvestris*, while the figure is more than 16 for *Pinus strobus* (Table 4.5). Similarly, microsymbiotic fungi may be nonspecific in relation to hosts, with the fungus *Amanita muscaria* being capable of forming ectomycorrhiza with at least 26 species of trees. Other fungi show similar levels of nonspecificity (Table 4.6). Not only are some hosts comparatively nonspecific in relation to species of fungi giving rise to ectomycorrhiza, they also can concurrently form associations yielding up to four kinds of mycorrhiza (e.g., Ericales, Largent *et al.* 1980; see also Harley and Harley 1987). In an especially important form of mutualism from the human agricultural perspective, nodulation in legumes, Young and Johnston (1989) reported a huge variation in the specificities of both hosts and bacteria. They attributed at least part of the limited bacterial specificity to the fact that the bacteria probably are not obligatory symbionts, ranging freely in soils and, therefore, coming into contact with many different species of plants.

Table 4.6. *A nonexhaustive list of host plants with which* various *species of fungus can form mycorrhiza in culture*

Fungus	Host plants
Amanita muscaria[a]	*Eucalyptus camaldulensis, E. calophylla, E. dalrympleana, E. diversicola, E. maculata, E. marginata, E. obliqua, E. regnans, E. st johnii, E. sierberi, Betula pendula, Larix decidula, L. occidentalis, Picea abies, P. sichensis, Pinus contorta, Pin. echinata, Pin. monticola, Pin. mugo, Pin. ponderosa, Pin. radiata, Pin. strobus, Pin. sylvestris, Pin. taeda, Pin. virginiana, Pseudosuga menziesii*
Glomus mossae[b]	≥20 species (Mosse 1973)
G. monosporus	*Bellis perennis, Lycopersicum esculentum, Maiathemum dilatatum, Trillium ovatum, Zea mays*
G. microcarpus	*Fragaria* spp., *Geum* sp., *Phleum pratense, Rubus spectabilis, Taxus brevifolius, Thuja plicata, Zea mays, Juniperus communis* var *siberica*
G. fasciculatus	≥20 species
G. macrocarpus var. *macrocarpus*	*Allium cepa, Epibolium glandulosum, Fragaria chiloensis, Galium aparine, Stachys mexicana, Trifolium repens, Triticum aestivum, Zea mays*
G. macrocarpus var. *geosporus*	*Lycopersicum esculentum, Zea mays, Fragaria* sp.
G. caledonius	*Zea mays, Triticum* sp.

Notes: [a]*ectomycorrhiza*; [b]*Glomus* form vesicular–arbuscular mycorrhiza.
Source: After Harley and Smith (1983).

Laboratory cultures do not necessarily demonstrate the realized degree of host or microsymbiont specificity found in the field (Harley and Smith 1983: 358). Apart from the availability of potential hosts, which will limit the realized ecological versatility of fungal species at any one location (e.g., *Eucalyptus* and *Pinus* are not naturally sympatric, see Table 4.6), Smith and Douglas (1987: 255) also noted that the effectiveness of an association in the laboratory may not reflect that in the field because of more rigorous conditions encountered in the latter. In addition, there are questions concerning the difference between the capacity of a given mutualistic association to form a successful mycorrhiza and the capacity to produce fruiting bodies by the fungus, as noted by Harley and Smith (1983). However, *in vitro* capabilities do show that there is often a relatively great potential for mutualistic associations and that close coevolution between symbionts probably is unusual.

Are ectomycorrhiza representative of mutualistic associations in terms of specificity? Smith and Douglas (1987) surveyed the specificity of hosts and microsymbionts in eight forms of association (Table 4.7). They quantified specificity in terms of the available information on the taxonomic diversity of known hosts (for microsymbionts) and microsymbionts (for hosts). Thus, restriction to a single strain or subspecies of host or microsymbiont was regarded as 'very high' specificity, to a single species as 'high,' and so on (see footnote Table 4.7). The results suggested that microsymbionts show very little host specificity in any form of mutualistic association, whereas hosts usually display rather greater specificity for their partners. Ectomycorrhizal mutualists appear to be the least specific.

Smith and Douglas (1987) attributed this greater specificity of hosts to a need for them to avoid the infiltration of unsuitable parasitic or pathogenic microsymbionts. To this end the processes by which mutualistic associations are formed normally involve several stages of recognition and establishment; within each the host may terminate the process (see Young and Johnston 1989). Given that mutualisms often greatly increase the fitness of hosts, one expects the occurrence of mechanisms of active microsymbiont recognition and assistance by hosts, although this clearly does not occur at a very restricted level. The mechanisms involved in recognition and establishment vary greatly between and within association types, but cell-to-cell biochemical mechanisms would seem to be important, although not the only factors involved.

Ecological versatility in mutualistic associations therefore appears to be relatively great. As with many of the interspecific interactions considered in this chapter, the diversity of resources used (here the host and the microsymbiont reciprocally) seems to be largely or partly limited by the local availability of potential symbionts rather than by an intrinsic species-for-species specificity.

Mutualisms – an overview

Mutualisms and mutualistic associations are forms of symbiosis in which individuals of pairs of species 'combine' to produce a mutually beneficial outcome (i.e., +/+). The former are comparatively brief encounters such as in pollination and fruit or seed dispersal while mutualistic associations are longer-term commitments often involving physiologi-

Table 4.7. *A general overview of the specificity displayed by hosts and symbionts in a variety of forms of symbioses. 'host' means macrosymbiont, while 'symb' denotes microsymbiont (see text)*

symbiosis	Freshwater hosts and *Chlorella*		Marine hosts and *Symbiodinum*		Rhizobia and legumes		Cyano-bacteria		Lichens		Ecto-mycorrhizas		Vesicular-arbuscular mycorrhizas		Bio-luminescence symbiosis	
specificity	host	symb	host	symb	host	symb	host	symb	host	symb	host	symb	host	symb	host	symb
Very high[1]					●											
High[2]			●		●				●						●	
Moderate[3]	●					●	●		●	●			●			
Low[4]										●	●	●				●
Very low[5]		●		●				●				●		●		

Notes: Range of suitable partners – within: [1]one strain or other subspecific taxon; [2]one species; [3]one genus; [4]one family or order; [5]any higher taxonomic grouping.
Source: After Smith and Douglas (1987).

cal coupling but not always (e.g., ant-protection in plants). Although mutualisms have been regarded as interactions in which high specificity between partners might be expected to occur, several factors appear to mitigate against the development of generally high levels of specialization. Howe's (1984) critique detailed a suite of ecological factors that should prevent the widespread occurrence of highly specific mutualisms. These include variable and often unpredictable environments and especially the community background, which provides a ready source of alternative mutualists (i.e., the dilution effect). Much evidence suggests that mutualisms generally are not very specific (Briand and Yodzis 1982), indicating that the ecological versatility of mutualists is not limited.

There are many forms of mutualistic association, which are longer-term interdependencies than mutualisms. The available information indicates that the specificity of hosts and microsymbionts also is comparatively low. This probably is to be expected for microsymbionts because of the need to keep their options open, and also because close coevolutionary biochemical coupling may be unnecessary in situations in which hosts are not strongly favoured to reject the microsymbiont (compared with parasites, Vanderplank 1978, Harley and Smith 1983). Host specificity seems to be somewhat greater, most probably because they have to be wary of potentially injurious partners, especially parasites and cheats in general.

4.6 Summary and conclusions

The interactions between species are a major focus of community ecology. For this reason, Chapter 4 is a key part of this book because it involves a review of many of the ideas that have been raised to account for the specificity or otherwise of relationships between species. In other words, the topic of this book largely relates to this chapter.

This review necessarily is a comparatively brief one given that every one of the interactions considered has had several or dozens of specialized volumes devoted to it. The main purpose has been to look for common elements among the welter of different viewpoints that have arisen in each of the fields. In terms of the sorts of explanations for versatility of resource use that characterize each interaction, it seems as though one can broadly group predation, herbivory by larger animals and omnivory into one set, and parasitism, mutualism and phytophagy by small herbivores into another.

In the first set of interactions the common threads are issues involved in finding, selecting, 'catching' and processing of prey (animal or plant). In general, the consumer is either absolutely large or relatively large compared with the prey (call them 'macropredators'). There appear to be some promising developments in generalizing the theoretical treatment of macropredators. For example, Taghon *et al.* (1990) and others have commented on the potential for applying the DRM and related models to any macropredator in which alternative prey types differ substantially in digestibility, an idea that arose originally from models for large herbivores. Some ecologists working on aquatic systems in which the physical distinction between plants and animals (e.g., phytoplankton and zooplankton) is comparatively little, environments relatively unstructured and feeding methods simple, have argued that trophic levels are indistinct, which may offer some promise for linking together processes involving macropredators in these systems.

However, these comments should not be taken to mean that a general method for accounting for or predicting dietary versatility will emerge for macropredators. No general model is likely because of the diverse biologies of species and the differences in the strengths of factors influencing dietary selection (e.g., predation risk, variation in food availability and quality, processing time versus search time, learning, imprinting, the degree of omnivory, social behaviour and physiological status, Courtney 1988, Taghon *et al.* 1990, Thompson and Pellmyr 1991; see Fig. 4.10).

In the second group of interactions, parasitism, mutualism and insect phytophagy ('micropredation'), there tends to be a sizeable difference between the participants – with the victims or macrosymbionts (hosts) usually being much larger than the exploiters or microsymbionts. It is in this group of interactions that most attention has been drawn to high levels of specificity. The main reason for this is some empirical evidence (and theoretical assumptions) that these interactions often are dominated by coevolution or 'arms races' (if antagonistic). Although there is evidence for high specificity in some cases (e.g., marine trematodes) extreme specificity is probably exceptional rather than the rule and is very often restricted to only one phase of the life cycle. Coevolution alone does not seem capable of accounting for the different degrees of host specificity seen in parasites and especially in phytophagous insects, and few mutualisms or mutualistic associations appear to be very specific.

A common theme in these interactions is that host specificity often is

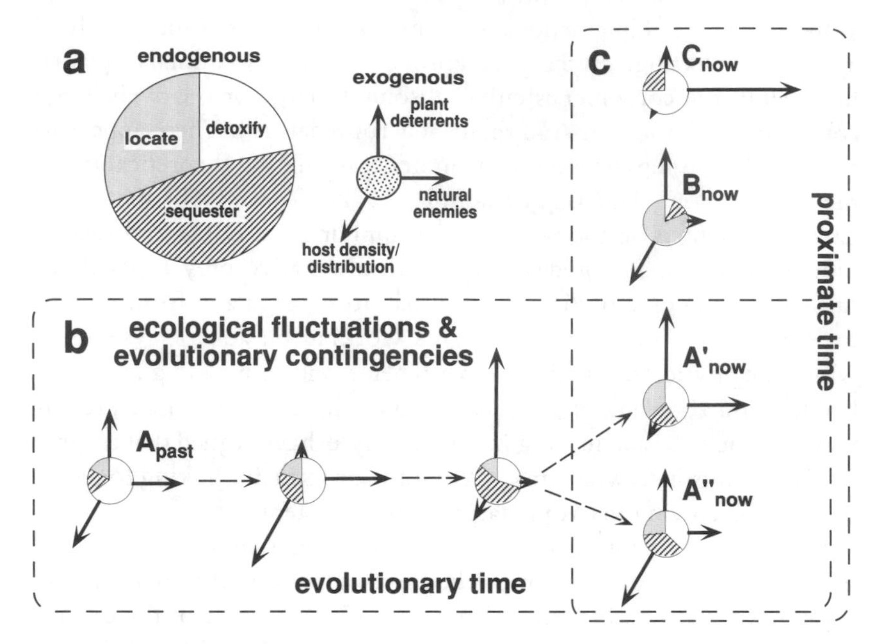

Fig. 4.10 (a) there are several to many factors controlling resource use, which is illustrated here by a hypothetical example involving three exogenous and three endogenous factors affecting phytophagous insects. The respective 'strengths' of each exogenous factor are shown by the lengths of the vectors – the longer the vector the more important is that factor. Also shown within the disc are the relative magnitudes of endogenous factors on versatility. (b) relative effects will vary during the evolutionary history of a species due to ecological fluctuations. (c) similarly, relative effects will differ between sets of contemporary, related species and between isolated populations of the one species.

an ecological rather than a physiological phenomenon – limited host range then is a reflection of the lack of availability of alternative hosts either in space or time. However, as argued in previous chapters, ecological versatility is defined as the degree to which individuals can fully exploit the resources that are naturally *available* to them in their environments. Even though an organism *can* use other resources if they are artificially brought into contact with those resources, the fact remains that they *do* use a restricted set of resources under natural conditions. The ectomycorrhizal fungi are a good example of a high capacity to use a wide variety of hosts but a limited realized versatility in the field. Specificity, such as it is, in this set of interactions also seems to be a function of many competing processes and, as with macropredators, no general model will suffice – explanations will need to be system-specific and they may have a strong element of historical contingency as well (Fig. 4.10). This is especially true of the evolution of idiosyncratic adaptations in some lineages that allow species greater versatility – leaf-trenching in phytophagous insects is one example.

Few ideas span the gulf between macropredation and micropredation. One that does is the capacity of exploiters to detect and divine the characteristics of prey or microsymbionts, namely behavioural and biophysical capability. In most interactions the versatility of resource use has been seen to be a function of the capabilities of the macropredator or microsymbiont. Thus, large insectivorous bats appear at first sight to be relatively selective but in reality may be unable to detect the full range of items that occur in their environments so that they are opportunistic on prey that they can detect (Barclay and Brigham 1991). Limited neural capacity has been suggested as an important factor in limiting host use in phytophagous insects (Jermy 1988). Latitudinal gradients in the diversity of parasitoids have been attributed to the difficulty of locating particular host insects in the diverse floristic communities of the tropics compared with the more depauperate temperate regions (Janzen 1981). Learning may lead to restricted resource use as seen in some birds and invertebrates (e.g., Greenberg 1983). I suspect that differences in behavioural and biophysical abilities may be one of the major contributors to variability in versatility in natural populations and that we currently have only a very rudimentary understanding of the reasons for these differences. Why individuals of the Cocos finch (Werner and Sherry 1987) and the sea slug *Placida dendritica* (Trowbridge 1991) should have idiosyncratic, specialized patterns of resource use when other species, such as the stonefly *Megarcys signata* (Peckarsky *et al.* 1994), do not is not well understood. Thus, much greater attention needs to be paid to the behavioural controls on ecological versatility, such as learning, early imprinting and social interactions.

Of course, hosts and victims cannot be regarded as evolutionarily passive players. Species (or subsets within species) display a variety of behavioural or physical adaptations to reduce or avert their susceptibility to macropredators and microsymbionts, so that the versatility of the latter depends upon the characteristics of their victims as well.

Intermediate-sized herbivores would seem to be a group of organisms to which much more work and thought needs to be directed. There must be a point or range of sizes at which the processes pertinent to large herbivores give way to those that are more significant for smaller herbivores. The factors controlling resource use in these 'mesoherbivores' must be extraordinarily complex and may provide an important testing ground for many of the ideas that have been suggested for herbivores.

In this chapter, we have seen that specificity between participants in interactions may vary dramatically between stages of organisms, which is particularly obvious in parasites. In many or perhaps most parasites, some life history stages are catholic in their use of hosts, with host restriction being mainly ecological in character. Other stages of the same populations, however, show extreme specificity, being unable to infect any but a single species (and possibly only a particular race of a species). This clearly is physiological specificity. The differentiation in versatility expressed at different stages is indicative of the need to consider the ecological versatility of a population in terms of its constituents, whether these are age classes, stages, or gender groups. Some of the processes leading to the differentiation of resource use within populations are considered next.

5 · *The influence of population structure on versatility*

We have already seen numerous examples of the increase in ecological versatility associated with differentiation within populations. Such differentiation comes in many forms, including gender-based, age- and stage-specific, and polymorphic differences in resource use. There also are instances in which phenotypically indistinguishable individuals develop particular patterns of resource use due to an environmental interaction early in their lifetimes. Thus, the algal species to which an individual of the sea slug *Placida dendritica* is exposed first largely determines the diet of that individual, and this appears to be unrelated to underlying genotypes (Trowbridge 1991). Similarly, individuals of the Cocos finch, *Pinaroloxias inornata*, develop idiosyncratic foraging behaviour apparently by watching older, experienced birds (i.e., cultural transmission, Werner and Sherry 1987; see also Wcislo 1989).

However, ecological differentiation within populations can also arise from the occurrence of morphologically distinguishable 'types' within populations, which increases the range of resources used (e.g., Decho and Fleeger 1988, Malmquist 1992). For example, in many species of anuran amphibians, juveniles are obligatorily herbivorous or detritivorous while adults are strictly carnivorous (Toft 1985), while the reverse occurs in some fishes (Braband 1985). Gender-based (e.g., Freeman *et al.* 1976, Dawson and Ehleringer 1993, Le V. dit Durell *et al.* 1993) or morphological (e.g., Ehlinger 1990) differentiation of resource utilization is common. Even parthenogenetic clones have different ecological characteristics from one-another (e.g., Jaenike *et al.* 1980, Weider and Hebert 1987), as have strains of some parasites (e.g., the parasitic nematode *Howardula aoronymphium* of *Drosophila*, Jaenike 1993). Stages of many species of parasites utilize very distinct sets of hosts, often displaying a large difference in specificity (Jones 1967).

Apart from its effect on versatility, the differentiation of resource use among the components of the population has profound implications for population dynamics because of the potential for spatial and temporal variability in the availabilities of different resources (Wilson 1992). Potentially independent variation in resource availability may mean that some components of a population enjoy bloom conditions while others suffer shortages. Clearly, these differences can perturb population dynamics (e.g., Kadmon 1993).

In this chapter, I examine some of the ideas on heterogeneity within local populations and the effects that this heterogeneity can have on patterns of resource exploitation. The topics are divided into four main areas: facultative exploitation and behavioural plasticity, developmental plasticity (adaptive polyphenism), polymorphism, and differences due to age- or stage-structure within populations. The first topic, facultative exploitation and behavioural plasticity, refers to differentiation within populations that need not be morphologically based. Facultative exploitation is when individuals within a population expand or contract their use of resources in a plastic way (Glasser 1982). In some populations, individuals specialize on resources in a manner that is not related directly to genotype because of behavioural plasticity, such as in the Cocos finch and sea slug examples (West-Eberhard 1989). This is referred to here as *ecological polymorphism*.

Developmental plasticity, or adaptive polyphenism, is the capacity of the one genotype to develop into distinct alternative phenotypes depending upon environmental conditions during development (Scheiner 1993). It would seem to depend upon the correlation between environmental cues during development and the conditions that will be experienced during adulthood (West-Eberhard 1989: 253–254, Moran 1992). While plasticity and polyphenism permit some degree of phenotypic variability with respect to a given genotype, polymorphism is the dynamic coexistence of conspecific individuals displaying different phenotypes on the basis of their genotypes (Getz and Kaitala 1989). That is, polymorphism is more 'hard coded' than either ecological or phenotypic plasticity or polyphenism (West-Eberhard 1989). Some of the issues involved in how age- and stage-structure within populations affect patterns of resource exploitation, and hence ecological versatility, also are considered.

It will prove useful to outline a logical scheme for characterizing the components of variation of resource use within populations that has been developed over the past couple of decades. Roughgarden (1972) noted that generalized populations may consist of either monomorphic generalists, or polymorphic populations of pure specialists, at least as the extremes of a range (Fig. 5.1). Thus, he distinguished two distinct

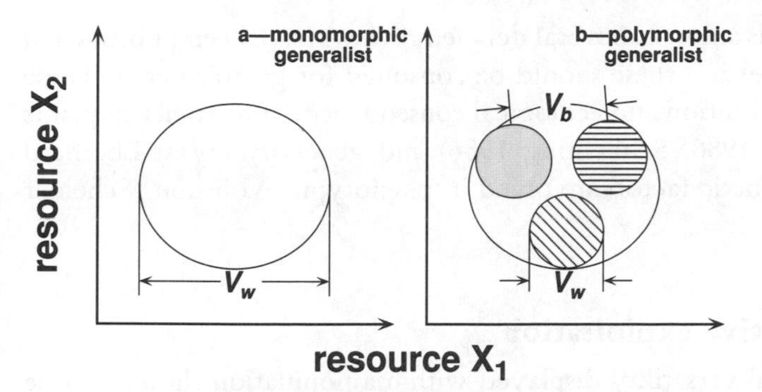

Fig. 5.1 Extreme patterns of resource use in generalized populations. (a) a monomorphic population in which all individuals have the same, broadly based use of resources. (b) a polymorphic population with three morphs utilizing only restricted sections of each resource axis, making them each relatively specialized; the total resource use when summed over the entire population is as variable as in the monomorphic population. Thus, two sources of variation in resource use are recognized: within-phenotype (V_w) and between-phenotypes (V_b).

categories of generalization: (1) within-phenotype, where a population essentially consists of a single, generalized phenotype; and (2) between-phenotype, where generalization is a property of the summed effects of a set of specialized phenotypes exploiting the range of resource availability.

These ideas lead Van Valen and Grant (1970) and Roughgarden (1972) to partition variance in niche breadth into two components, the within-phenotype (V_w) and the between-phenotype contributions. The latter was further divided by Christiansen and Fenchel (1977), who distinguished between variance due to differences between phenotypes of the same 'age' (V_b) and variance due to differences between age classes (V_a; also known as the ontogenetic component, Ebenman 1987). In some models, a temporal variance component also has been recognized (see Siegismund *et al.* 1990). Based on work by Soulé and Stewart (1970), Roughgarden (1972) suggested that the within-phenotype and between-phenotype components may vary independently so that a generalized strategy may consist of any combination of contributions from these sources. This logical structure is used later in this chapter as a basis for defining four 'idealized' exploitation strategies, which are used in the models of Chapters 6 and 7.

The material covered here addresses only some of the issues involved in the occurrence and evolution of phenotypic plasticity, and certainly is biased towards animals. Several detailed reviews have been published in the last decade, and these should be consulted for greater detail. These include the evolution and ecological consequences of plasticity in plants (Grime *et al.* 1986, Schlichting 1986) and generally (West-Eberhard 1989), and genetic factors involved in phenotypic evolution (Scheiner 1993).

5.1 Facultative exploitation

The ecological versatility displayed within a population should not be regarded as being 'fixed', as shown by time-dependent variation in the conformance statistics of some field studies referred to in Chapter 3. One might expect that some populations or morphological types within populations alter their patterns of resource use as conditions change, leading to variation in their versatilities. These conditions could be predictable (e.g., seasonal) or unpredictable (e.g., storms) changes in climate, the appearance of immigrant competitors or enemies, perturbations in the density or stage-structure of conspecifics, or fluctuations in the availability of resources.

By using arguments similar to those of Svärdson (1949), Glasser (1982) suggested that the degree of versatility is a direct function of relative resource scarcity, as measured by the ratio of the current population size, N, to the carrying capacity of the population, K (i.e., N/K). He suggested that populations might have narrow niches (i.e., low versatility) when resources are superabundant compared with utilization, but that at densities close to carrying capacity, versatility necessarily must be high. Evolutionary effects were expected to lead to different exploitation efficiencies for each resource between syntopic populations, and consequently, the lowest possible versatilities (Glasser 1982: 252).

This analysis was based on the partial niche model of Vandermeer (1972). The partial niche uses the idea that there is a certain gradation of suitability or utility of different resource types (or habitats, in the appropriate context; Fig. 5.2). As population density increases, the most favoured resource becomes fully utilized, leading to resource categories that were initially less favoured being used, and so on (Vandermeer 1972). This process leads to increased versatility as the population grows.

As we have seen already (§4.3), Glasser (1982) simulated alternative exploitation strategies by using algorithms having different rules for the order in which resources are selected. The pattern of niche expansion as populations grow towards their carrying capacities depended on the

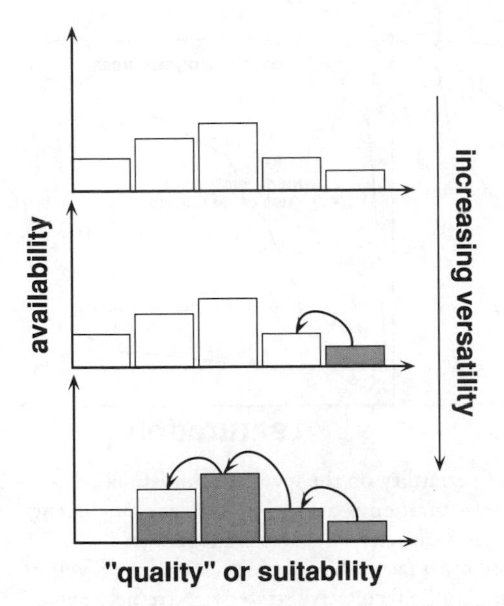

Fig. 5.2 The concept of the partial niche (Vandermeer 1972). As the population grows (top to bottom panels), the highest quality or most suitability resources are utilized in succession, leading to increased versatility. Shading indicates resource use.

algorithm for selecting the next resource (Fig. 5.3). For example, niche breadth grew slowly as a population expanded if individuals within the population select resources in efficiency-order of exploiting the resources. This form of versatility was termed *obligatory specialization*. If resources are added to repertoires based on the greatest unused capacity of individuals of that population, then the population exhibits *facultative specialization*. This strategy led to an intermediate rate of expansion of niche breadth as population size increased towards the carrying capacity. The fastest rate of increase in niche breadth occurred when resources were used in proportion to their overall free availability, a mechanism referred to as *obligatory generalization*.

What are the conditions under which these three exploitation strategies might be expected to occur? Obligatory specialization was predicted to occur when the population density is held to small values relative to the potential carrying capacity (Glasser 1982). Conversely, obligatory generalization is likely to be associated with near-saturation population densities. Under constant conditions, the capacity of populations to expand or contract their resource utilization might never, or

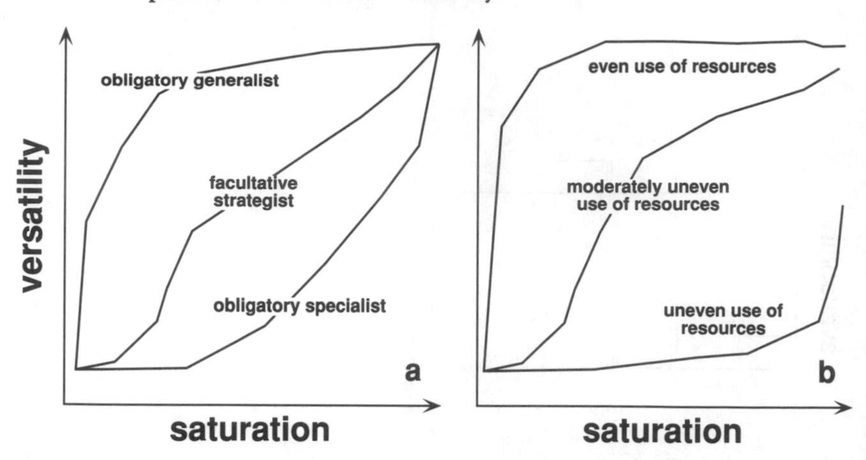

Fig. 5.3 Functional dependence of versatility on the level of population saturation of an environment, where saturation is gauged by the ratio of current population density to nominal carrying capacity. (a) three strategies: an obligatory specialist and a generalist and a facultative exploiter. (b) three levels of evenness of efficiency of resource use in the facultative strategy: extremely even, intermediate and uneven efficiencies (adapted from Glasser and Price 1982).

rarely, be used, which could lead to 'excess baggage' in terms of genetic diversity or phenotypic potential (West-Eberhard 1989). The latter might adversely affect the efficiency of resource use or lead to reduced population sizes, thereby making such populations vulnerable to invasion and/or local extinction.

By assuming that specialization should occur if possible, Glasser (1982) deduced that within-phenotype generalization is likely to arise only when the population size usually is close to carrying capacity (i.e., $N/K \rightarrow 1$), but with a large variance (Fig. 5.4). On the other hand, if the variance is small, between-phenotype generalization is more likely.

The picture for facultative specialization is more complicated, being a function of both the environment in which the strategy evolved and the nature of the environment in which a population currently occurs. Recall that a facultative specialist concentrates on preferred resources when $N/K \rightarrow 0$, but expands its repertoire to include other resources in decreasing preference order as $N/K \rightarrow 1$. Thus, facultative behaviour may originate in environments in which the variability of N/K is high. A spectrum of facultative specialization exists. The extremes of this range are populations utilizing resources in either (1) a frequency-dependent fashion, a strategy evolved under conditions in which average N/K

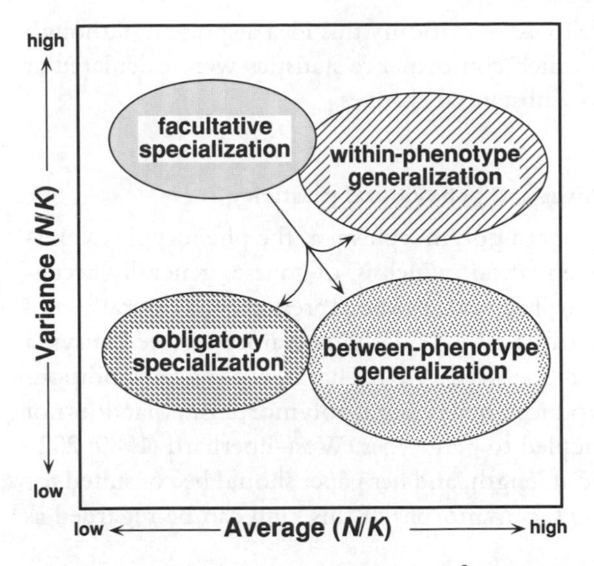

Fig. 5.4 Expected strategies or patterns of resource use given contrasting means and variances of population saturation in evolutionary time (after Glasser 1982).

values are near unity; or (2) a frequency-independent selection of resource states, which is characteristic of origins in situations in which mean N/K is close to zero (Fig. 5.4).

In summary, Glasser (1982) developed a model of facultative resource utilization linking strategies with current fluctuations in resource availability and resource variation in evolutionary time. Strategies were deemed to depend upon both the 'saturation' of communities, in the sense of the relative amounts of unutilized resources, and on trophic level. Obligatory specialists are expected in constant, unsaturated environments and in lower trophic strata under constant, saturated conditions. Obligatory generalists should occur in saturated communities, particularly at higher trophic levels. Facultative specialists ought to prevail in fluctuating environments, resembling obligatory generalists at higher trophic levels and obligatory specialists at lower levels. The efficiency with which such facultative specialists exploit resources possibly may be less than that of corresponding obligatory strategies, although the support for this assertion is not particularly convincing (see Glasser 1982: 256). Because resource availabilities vary in nature and tracking of these variations is imperfect (Boyce and Daley 1980, Stephens 1987), Glasser (1982) predicted that the majority of populations should display facultative use of resources. For reasons outlined in

Chapter 3, it is difficult to assess critically this idea at present, although most of the studies in which conformance statistics were calculated at different times did show substantial changes.

Ecological polymorphism and behavioural plasticity

There is a tendency to regard polymorphism as the phenotypic expression of genotypic differentiation, which is, of course, generally acceptable. However, polymorphism can also be expressed behaviourally, and in behaviour not necessarily tied to underlying genotypes (see Futuyma and Moreno 1988, West-Eberhard 1989). That is, ecological conditions and/or learning may produce a behavioural polymorphism that does not appear to be closely coupled to genotypes. West-Eberhard (1989: 252–260) discussed this topic at length, and her paper should be consulted for more detail. In any event, polymorphs of this kind can be regarded as *ecological phenotypes*.

Behavioural mechanisms can affect the patterns of ecological polymorphism, as we have seen already with studies such as those of Werner and Sherry (1987) and Trowbridge (1991). Another example is the work of Holbrook and Schmitt (1992) on dietary differentiation in striped surfperch (*Embiotoca lateralis*) in waters off Santa Cruz Island, California. They found dietary specialization and generalization associated with a social hierarchy. The hierarchy involved sets of aggressive individuals within each cohort that were able to dominate isolated patches of the red alga *Gelidium robustum*. The capacity to exclude other individuals allowed access to *Gelidium* patches that harbour high availabilities of caprellid amphipods (Fig. 5.5). Diets consisting mainly of these crustaceans provided the highest energetic returns. Other individuals foraged on alternative patches, exhibiting either generalized diets or diets rich in gammarid amphipods. In this example, dietary versatility is a passive reflection of microhabitat versatility because individual surfperch appear to feed opportunistically within microhabitats (Fig. 5.5). Holbrook and Schmitt estimated that the lifetime fecundity of caprellid specialists was about 10% higher than that of the generalists and approximately 20% higher than that of the gammarid specialists. Thus, versatility did not relate directly to fitness because the fitness of generalists was intermediate between those of the two specialists. This example shows how social behaviour can determine or affect versatility in some populations.

An ingenious way for populations to expand their ecological versatili-

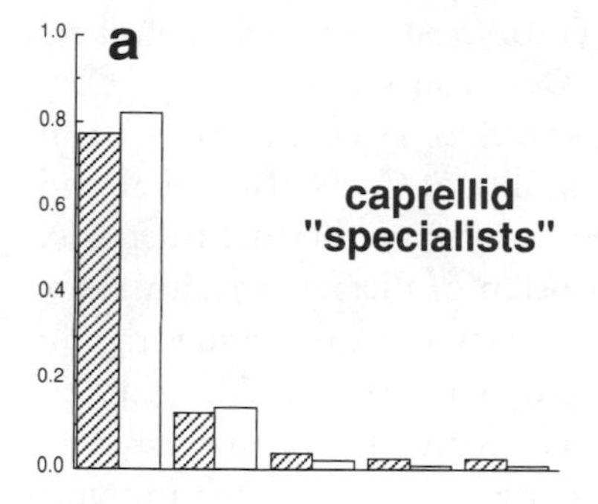

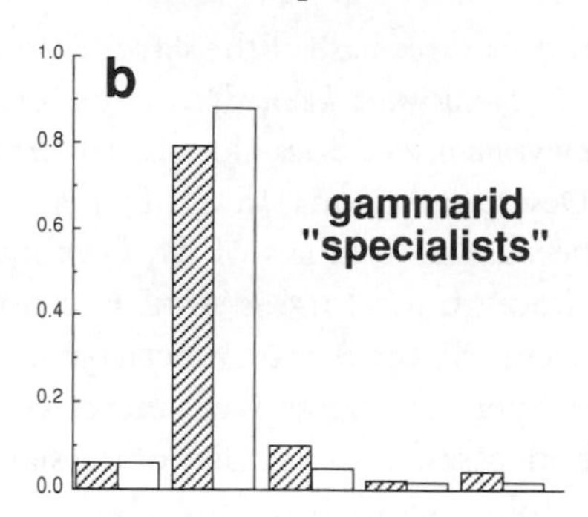

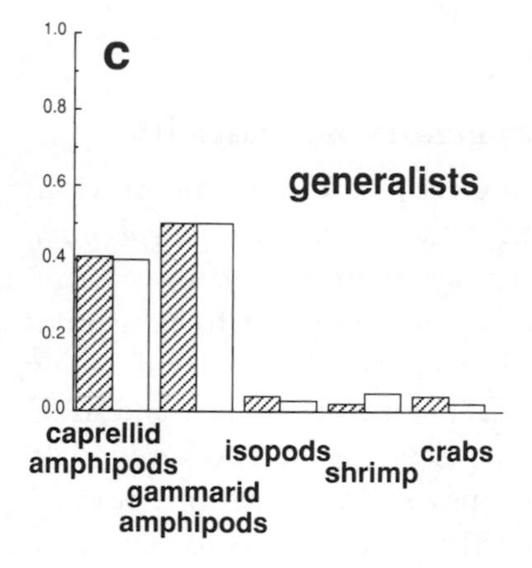

Fig. 5.5 Diet as a passive reflection of food availability in the striped surfperch (*Embiotoca lateralis*), with diets indicated by open bars and food availability shown by hatched bars. Populations consist of three ecological phenotypes: (a) individuals concentrating on patches of the red alga *Gelidium robustum*, which is rich in caprellid amphipods; (b) individuals using other microhabitats, which house high densities of gammarid amphipods; and (c) individuals feeding at random from all microhabitats (generalists). Diets closely match relative food availability, which indicates opportunistic feeding within microhabitats but selective use of microhabitants in (a) and (b). After Holbrook and Schmitt (1992).

ties is to enslave individuals of other species (Bernstein 1979), which can be likened to an ecological polymorphism. An example is that of species of ants in the south-western United States, in which workers from slaver colonies kidnap pupae of other species to act as workers for them (Bernstein 1978). The enslaved individuals, by having different thermal tolerances and ways of collecting food to the slavers, effectively expand the resource base of the slaver colonies. Thus, *Conomyrma bicolor* enslaves *Myrmecocystus kennedyi* in the Mojave Desert in California and *Conomyrma insana* does likewise to *Crematogaster emeryana* in the Great Basin Desert of Arizona. In California, the versatility of *Co. bicolor* is increased because individuals of *M. kennedyi* have a different thermal tolerance, although food use is similar so no expansion of dietary breadth *per se* occurs. In the Arizona species-pair, both dietary versatility and foraging temperature range were increased markedly (Bernstein 1978). Thus, in both cases, the versatility of the slaver species is substantially increased by employing this behavioural adaptation (Bernstein 1979), which amounts to the annexation of another ecological phenotype.

5.2 Adaptive polyphenism and phenotypic plasticity

A concept bridging the ideas on ecological plasticity in the previous section and polymorphism in the next is that of *adaptive polyphenism*. This is the switching between alternative developmental pathways and corresponding phenotypes in one population in response to exogenous stimuli (Shapiro 1976). Adaptive polyphenism has been reported widely in insects, marine invertebrates, terrestrial plants, amphibians and fishes (e.g., Meyer 1987, Via 1987, Stearns 1989, West-Eberhard 1989, Wheeler 1991). Plasticity of this form appears to be controlled both by pleiotropic and epistatic effects (Scheiner 1993), where the former is the degree of expression of a genotype in different environments, and the latter is the modulation of the expression of some genes by other genes ('switches' or 'controllers').

Phenotypic responsiveness to environmental cues and selection regimes clearly has a bearing on the capacity of populations to display different degrees of ecological versatility. For example, Moran (1992) cited several cases in which the loss of polyphenic capacity has led to a contracted pattern of resource use and thus, to reduced versatility.

Populations generating a mix of alternative phenotypes in response to different environmental conditions may be more versatile than phenotypically fixed populations. However, there must be a consistent mapping between the physical differentiation of phenotypes and the corresponding patterns of resource exploitation that they display (i.e., that 'morphological' phenotypes are reasonably well-defined 'ecological' phenotypes too). Otherwise, polyphenism would not be relevant to ecological versatility *per se*.

An example of such a mapping between morphological and ecological phenotypes based on adaptive polyphenism is that of the tadpoles of the southern spadefoot toad (*Scaphiopus multiplicatus*) of North America (Pfennig 1992). These larvae occur in one of three forms in ponds and pools: (1) carnivores, which are larger, developmentally more advanced, solitary predators on fairy shrimps (Anostraca) and other tadpoles; (2) omnivores, which are small, developmentally retarded, gregarious feeders on algae and detritus; and (3) an intermediate form that is midway in size and development between the others, but has dietary habits more akin to the omnivores than to the carnivores. Pfennig (1992) showed experimentally that morph determination is related to the impact of exogenous thyroxine – the addition of this thyroid hormone causes a larva to develop into the predatory morph, while its absence leads to the omnivorous form. Thus, this is a true polyphenism (Fig. 5.6).

Both fairy shrimps and tadpoles contain thyroxine itself or a precursor, so that the consumption of either appears to trigger the accelerated development characteristic of the predatory morph. Pfennig believed that the polyphenism is adaptive because the density of fairy shrimps is negatively correlated with pond longevity, and that the developmental advancement of the predatory morph produces a higher likelihood of maturation before ponds dry. Thus, the density of shrimps seems to be a cue upon which the tadpoles can 'predict' the permanence of the pond. High densities of shrimp signal conditions that would favour the predatory morph, while lower densities indicate a longer-lived pond in which fast maturation is not so critical.

Environmental predictability has been one of the main foci in the development of ideas on polyphenism. Moran (1992) explored the effects of environmental predictability (in space and time) and of developmental constraints and the costs of the maintenance of adaptive polyphenism. Although her arguments generally referred to phenotypic responses to different environments (habitats), there seems little to prevent one discussing her results in terms of sets of local resources such as microhabitats or hosts (for instance, see page 975 and the Discussion in her paper).

A key issue in the model was the distinction between environmental cues that influence the choice of developmental pathway (the developmental effect) and environmental impacts causing differential effects on the relative fitnesses of alternative phenotypes (the selection effect). The degree to which the latter can be gauged from the former appears to

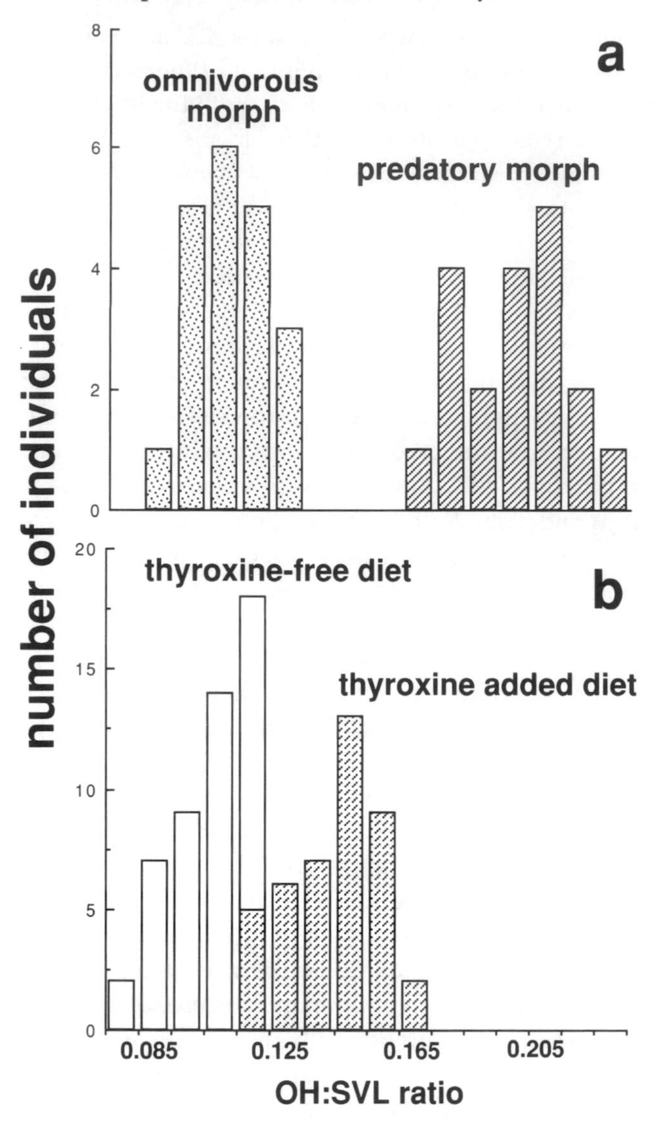

Fig. 5.6 The effect on a morphological indicator of thyroxine in the diet of tadpoles of the southern spadefoot toad (*Scaphiopus multiplicatus*). (a) the distributions for 8-day-old tadpoles collected from a single pond, with omnivorous and intermediate morphs grouped together. (b) the distributions for tadpoles kept on diets with and without thyroxine, administered for four days to 4-day-old tadpoles. OH:SVL is the ratio of the width of the primary buccal cavity adductor muscle (OH) to the snout-vent length (SVL) of the individual. After Pfennig (1992).

govern the likelihood that polyphenism can be maintained against monophenism (i.e., phenotypic fixation).

The maintenance of the polyphenic strategy against sets of specialized monophenic strategies depended on the accuracy of phenotype–environment mapping. If this were imperfect (which seems likely), then polyphenism would be lost if the level of matching were only moderate or if the relative mix of (local) environments changed. The levels at which the comparative advantages of polyphenism and monophenism prevail were affected also by: (1) the costs of switching involved in the polyphenic strategy, such as the implementation of alternative developmental pathways and antagonistic pleiotropy (i.e., opposing selective pressures on a given allele in different phenotypes; see Scheiner 1993); and (2) the comparative fitness advantages of specialists in their preferred environments and disadvantages in their nonpreferred ones.

A polyphenic strategy also may suffer a threat from a single generalized strategist. The stability of a polyphenic strategy against a generalized strategist required that the weighted, average fitness of the polyphenic phenotypes in various environments exceeds the mean fitness of the generalist. The weighting factor in this case was the accuracy of the phenotype–environment matching in each environment. The 'correct' phenotype for each environment was assumed to show higher levels of fitness than the generalized strategist (i.e., there is a cost to generalization). In essence, polyphenism would prevail if the matching were sufficiently good to produce almost always the appropriate phenotype for the conditions.

A feature of Moran's (1992: 980) approach is that the conditions under which several strategies might coexist locally are extremely unlikely. Either polyphenic, multiple monophenic, or generalized monophenic strategies might prevail under particular conditions, but not more than one. Whether these conditions differ sufficiently over large spatial scales to sustain several alternative strategies is an important issue. Are locally entrenched strategies continually under threat from alternative immigrant strategies? If they are, then the large-scale spatial dynamics of the ecological versatility *of the species* must be sensitive to local conditions and immigrant pools. However, Moran (1992) noted a prevalence of temporal over spatial polyphenism, which she thought may be linked to the greater reliability of temporal cues (e.g., photoperiod) over spatial cues as predictors of the ultimate selection environment. This suggests that the balance between alternative strategies, if they exist, may depend upon spatial factors having strongly associated temporal components, such as latitude.

5.3 Polymorphism

Models of polymorphism have a long history, owing much to the interest in this topic of population geneticists (e.g., Levene 1953, Dempster 1955, Levins and MacArthur 1966). Many models comprise analyses of diallelic, single-locus systems, mostly of diploids but some haploids too (e.g., Czochor and Leonard 1982). In these models, alternative morphs arise due to the presence of one or another allele at a locus – the 'allelic-switch' concept. The aim is to identify the conditions that permit a stable polymorphism rather than a monomorphism (e.g., Gillespie 1974, Campbell 1981). Spatial heterogeneity has played a key role in many of these models (e.g., Templeton and Rothman 1974, Yokoyama and Schaal 1985, Holsinger and Pacala 1990). Of course, we are interested in polymorphisms of this form only insofar as they relate to different patterns of resource use, so that, for example, colour morphs that are not ecologically differentiated are not of interest here.

Many genetic issues have been discussed in the context of the maintenance of polymorphisms, including the relationships between gene flow and selection pressures in heterogeneous space (Speith 1979), differential fitness, dispersal, assortative mating, habitat selection (Scott and McClelland 1977), and dominance relationships between homozygotes and heterozygotes (Wilson and Turelli 1986). The conditions leading to polymorphism usually are determined by equilibrium methods, although some workers have cautioned against an over-reliance on them (Speith 1979).

Some of the more recent models have included a greater number of ecological features in the models. For example, Getz and Kaitala (1989) proposed a model organism in which there were elements of natal host resources (with density-dependent interactions and survivorship) and the capacity of adults to select alternative host resources to use for oviposition. They found that protected polymorphisms, in which each allele increases when rare, were more likely if either density dependence or intraspecific competition were comparatively weak. On the other hand, habitat selection by adults did not appear to promote protected polymorphisms. To the extent that morphs differ in resource use, the factors that influence the likelihood of protected polymorphisms would correspond to increased (higher likelihoods) or decreased (lower likelihoods) ecological versatility.

An important distinction was raised by Getz and Kaitala (1989), namely the need to distinguish between 'protection' and 'stability' in polymorphism. They described many conditions under which the polymorphism was protected but the population dynamics were unstable. Some situations permitted periodic, even chaotic, dynamics of the densities of morphs. Getz and Kaitala did not specify densities below which morphs might go extinct locally, either stochastically (Shaffer 1981) or deterministically (Hopf and Hopf 1985), but they did recognize that even protected polymorphisms might become fixed due to density fluctuations. Thus, ecological conditions that produce unstable but protected polymorphisms ultimately may lead to reduced versatility because some morphs become extinct. The rather stringent requirements that seem to be necessary for the stable coexistence of nonplastic (allelic-switch) polymorphisms may be responsible for their comparative rarity in nature (West-Eberhard 1989: 258).

The way in which populations might expand the range of resources used (i.e., generalize), specifically to exploit unutilized resources, has been studied by Wilson and Turelli (1986), among others. They considered the effect of the appearance of a novel mutant in a population consisting of individuals that are specialized for a resource that is completely utilized. In addition to this resource, the mutant also can use an unutilized resource, albeit in an inefficient way. The model was restricted to the effects of exploitation competition only (i.e., not to interference competition).

Wilson and Turelli (1986) found that the fitness of the mutant may exceed greatly that of the original type because it can use an untapped resource. Several alternative outcomes emerged from this model, including a stable polymorphism, or the elimination of the finely specialized original type by the mutant (under some conditions). Either scenario would expand the versatility of the population from its specialized beginning. From this analysis, Wilson and Turelli suggested that well-adapted specialists may be eliminated by poorly adapted generalists under some conditions. The applicability of this analysis depends upon the coincidence of the appearance a suitable mutant with the availability of an unused or underutilized resource, which seems unlikely in general.

Polymorphism can be manifested in a temporal sequence too, so that one morph appears at one time and others subsequently. A useful illustration of this point was reviewed by Moran (1988), who looked at host-plant alternation in morphs of aphids. Moran suggested that a seasonal shift between distinct sets of host plants is common in aphids, but not in many other groups of insects. She attributed this to evolutionary constraints on one morph, the fundatrix, which is primitively specialized for the utilization of woody plants. Fundatrices may be limited by a short window of moderate suitability of their hosts in spring, while the other morph, the summer female, is less constrained because it typically utilizes herbaceous angiosperms. Many species of aphids appear to have lost the fundatrix morph so that resource use now is restricted to the hosts of the summer females and versatility therefore is reduced. Species of aphids with both morphs perhaps should be regarded as serially specialized morphs.

The effects of intraspecific competition on versatility were considered in some detail by Roughgarden (1972). His approach was to consider resource utilization patterns of phenotypic classes within a population, where each phenotype was indexed by its position on a one-dimensional resource axis (e.g., food size, moisture content). He also compared the rates and patterns of evolutionary change in versatility between asexually and sexually reproducing populations.

As noted earlier, Roughgarden distinguished between within-phenotype (the range of resources along the axis utilized by each phenotype) and between-phenotype (the distribution of means of phenotypes along the axis) variances in utilization. Another variance arises when sexually reproducing populations are considered, namely the variance in the distribution of phenotypes in the offspring. This is a measure of the range of phenotypes produced by recombination from one type of mating (Fig. 5.7). Roughgarden (1972: 707) argued that this component in sexual populations is as important in determining the equilibrium distribution of densities along the resource axis as within-phenotype and between-phenotype variances. Thus, this model implies that the versatility of a sexual population may depend significantly on the spread of recombinant products.

All studies mentioned so far concern polymorphism in populations of conspecific individuals. However, the existing patterns of ecological versatility in natural populations may reflect something of the stability or resistance of such populations to the impact of other species. Some models have considered the interactions between polymorphic and monomorphic populations of different species. For example, Milligan

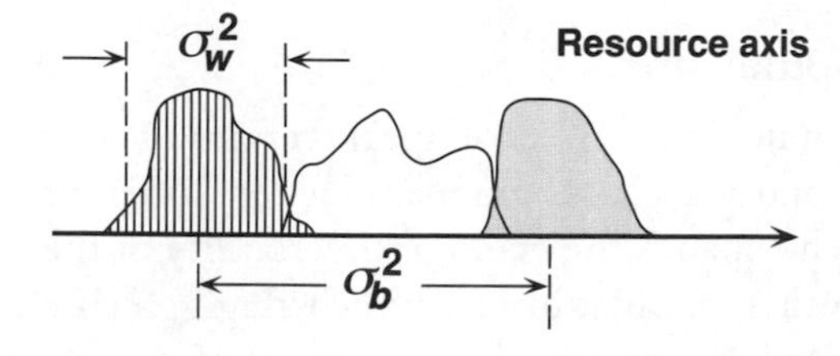

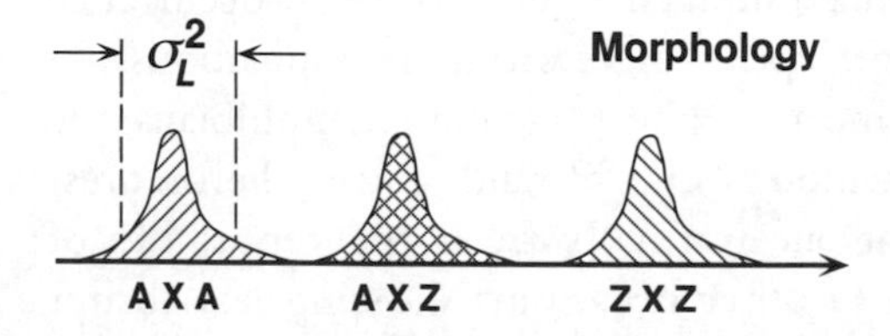

Fig. 5.7 Components of variation in resource use (top) and variances in recombinant products (bottom) in sexual populations envisaged by Roughgarden (1972). Variances are within-phenotypes (σ_w^2), between-phenotypes (σ_b^2), and in offspring phenotypic distribution (σ_L^2). Shadings indicate different phenotypes based on different genotypes. A × A, etc. indicate different mating combinations.

(1986) investigated the ability of populations (either monomorphic or polymorphic) to invade habitats occupied by other species, which often has been regarded as a criterion for competitive coexistence (e.g., Strobeck 1973, Turelli 1981, Haefner and Edson 1984, Loeschcke 1985, Case 1990).

Milligan's model was comparatively limited because the resource set consisted of a one-dimensional resource axis (i.e., food particle size), phenotypes used resources along the axis in accordance with a morphological index, and resident populations failed to exploit all of the resources (cf. Brew 1982). Nevertheless, Milligan's (1986) work suggested that the influence of polymorphism on coexistence could be ordered in terms of the 'ease of invasion.' For species pairs, Milligan suggested the order for ease of invasion is:

{a monomorphic invader and a monomorphic resident} is harder than
{a monomorphic invader and a polymorphic resident} which is harder than
{a polymorphic invader and a polymorphic resident}.

There appears to be much scope for continuing to model the impact of populations displaying different polymorphic structure on one another.

5.4 Stage-structured populations

The impact of stages in populations on ecological versatility was outlined briefly above. Stages and age-classes can manifest very different patterns of resource use, which expands the ecological versatility of the population compared with other populations in which stage-specific resource use is not so pronounced. It turns out that the degree of somatic reconstruction (and related behavioural changes) and the co-occurrence of stages have influenced the perception of versatility in populations. To use the common example of metamorphosis in anuran amphibians, the nature of morphological transition is so profound (aquatic herbivores/detritivores → carnivores) that one normally treats the versatilities of adults and larvae separately. In other organisms showing less abrupt changes or with direct development, such a division would seem to be less desirable. Also, if stages do not occur together at the same time, it is often more useful to regard the use of resources by each component separately (e.g., Moran 1988). Thus, significant differentiation in resource use by various stages of populations is an age-specific or stage-specific component of versatility analogous to the between-phenotype component recognized by Roughgarden (1972).

Werner and Gilliam (1984) reviewed size-dependent shifts in resource use, and concluded that the majority of species show marked ontogenetic changes in resource use, particularly in diet and habitat use. They coined the term *ontogenetic niche shift* for this phenomenon – the change in patterns of resource exploitation as organisms increase in age and usually in size as well. The importance of stage-dependence to resource use, usually overlooked in early theoretical studies (Werner and Gilliam 1984), now forms a key area in understanding the dynamics of both single and interacting populations (e.g., Auger 1985, Ebenman 1988, Bishir and Namkoong 1992, De Roos *et al.* 1992, McCauley *et al.* 1993).

The effect of stages and age-classes on versatility depends upon the amount of differentiation of morphology and behaviour involved. Differences between stages of a population may be relatively continuous (direct development), or discrete (metamorphism) (Werner 1988). There often will be only a limited size-range over which a particular somatic design is effective at exploiting a given type of resource, making a profound transformation necessary. On the other hand, some forms of major niche shift do not necessarily require substantial reorganization (Werner and Gilliam 1984, Braband 1985, Werner 1988).

Versatility and continuous development

Werner and Gilliam (1984) described pronounced size-dependent dietary shifts in many taxa that grow in a continuous fashion. Many piscivorous and planktivorous fishes, water snakes, turtles, lizards, and gastropods fit this prescription. For example, experimental work on a species of planktivorous fish, the bluegill sunfish (*Lepomis macrochirus*), has shown that there is size-dependent diet selection. Perhaps not surprisingly, the size of prey attacked by smaller individuals generally is small compared with that of larger individuals (Fig. 5.8). The predictions of two theoretical models in relation to changes in diet selection as individuals grow larger have been studied by using this species, namely the apparent size model (ASM, O'Brien *et al.* 1976, Li *et al.* 1985) and the contingency or optimal foraging theory that was described in the previous chapter (OFT, Walton *et al.* 1992). Neither model provided a reasonable prediction of diet selection for the smallest size-classes of fish (*c.* 11 mm in length; Fig. 5.8 a). However, the ASM performed well for fish of intermediate size (*c.* 29 mm in length; Fig. 5.8 b), but large sunfish (≥60 mm; Fig. 5.8 c) preferred large prey to a greater extent than expected on the basis of the ASM predictions (Fig. 5.8 c). The OFT clearly was an inappropriate description of diet selection in this species, underestimating dietary breadth for each size-class of fish (Fig. 5.8 a–c). The poor performance of the ASM for small fish may be related to their capacity to handle and digest prey, which may limit them to the smallest-sized prey (Walton *et al.* 1992). Dietary versatility is increased in this species because of a change from diets consisting mostly of small-sized prey in small individuals to larger prey in moderate to large-sized individuals (Fig. 5.8 d). Once fish reach about 30 mm in length, dietary versatility changes relatively little.

Versatility and metamorphosis

Metamorphosis is common in animals, with perhaps as many as 80% of species having metamorphic life cycles (Werner 1988). The general evolutionary trend is towards metamorphism from an ancestral pattern of relatively continuous development (Ebenman 1992; Fig. 5.9 a). It is important to distinguish between species in which there are extremely pronounced changes (called here *disjunctive metamorphosis*, or, equivalently, complex life cycles, Wilbur 1980) in somatic organization and those in which differentiation is much less marked (*nondisjunctive*

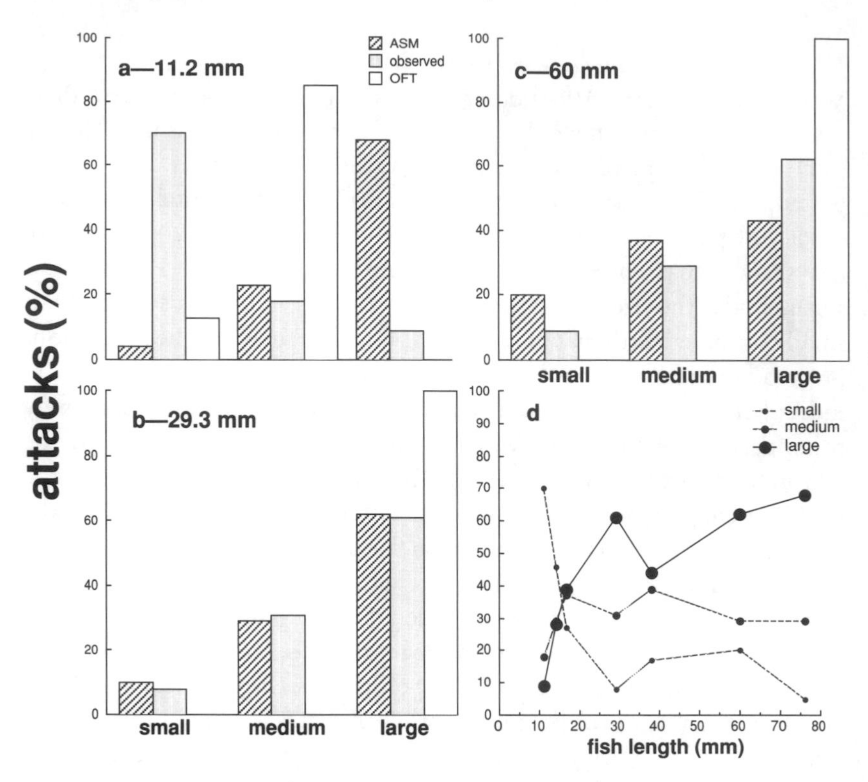

Fig. 5.8 Experimental results of diet selection in bluegill sunfish (*Lepomis macrochirus*) (after Li *et al.* 1985 and Walton *et al.* 1992). Data are distributions of attacks on three size classes of *Daphnia pulex* by individuals of three sizes: (a) 11.2 mm; (b) 29.3 mm; and (c) 60 mm, compared with the predicted distributions of two theoretical models, the apparent size model (ASM) and optimal foraging theory (OFT). Percentages of attacks by all seven sizes of sunfish reported by Walton *et al.* (1992) on the three size-classes of *Daphnia* are shown in (d).

metamorphosis, Fig. 5.9 b). Disjunctive metamorphosis commonly is associated with parasites, anuran amphibians, holometabolous insects, and perhaps most species of marine invertebrates (Jones 1967, Kuijt 1969, Wilbur and Collins 1973, Wilbur 1980). Larvae of the latter commonly are planktonic, while adults are sedentary on or in benthic or intertidal substrates. Instances of nondisjunctive metamorphosis include multiple-instar development in many arthropods, in which the incremental changes in morphology are comparatively small.

Niche shifts accompanying metamorphic progression may be relatively small in populations showing nondisjunctive metamorphosis, but

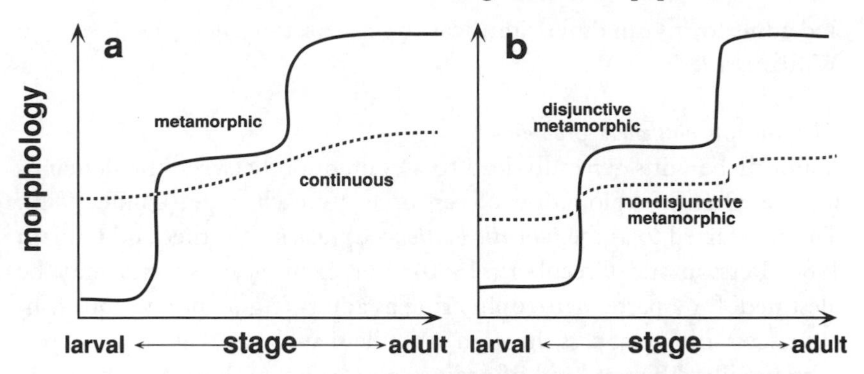

Fig. 5.9 Schematic representation of changes in morphology with respect to stage. (a) patterns in metamorphic and continuously developing organisms (after Ebenman 1992); (b) contrast between metamorphoses involving dramatic somatic reorganization (disjunctive) and less pronounced changes (nondisjunctive).

differentiation can be pronounced in some species. For example, in a study of the scorpion *Paruroctonus mesaensis*, Polis (1984) showed that the ecological differentiation between instars exceeded that between comparable species of scorpions, leading him to conclude that the stages of populations in this species function as ecological species.

The versatility of populations in which nondisjunctive metamorphosis occurs may be coloured by the degree to which various components of the population are contempories. If all stages appear together, as is typical of many anoline lizards (e.g., Schoener and Gorman 1968, Roughgarden 1974a), then the population may be characterized as being relatively versatile overall but with a substantial within-population component of versatility. If the stages appear sequentially through time, however, the versatility of the population at any time may appear to be low. Perhaps such populations should be thought of as serially specialized, as in the case of polymorphic aphids (§5.3).

Species with disjunct metamorphosis normally would not be regarded as extraordinarily generalized just because of the different resources exploited by different stages (i.e., resource use integrated over all components of the population may span a wide range, e.g., anuran amphibians). Populations may be deemed to consist of sets of nonoverlapping, relatively specialized subpopulations. Comparisons of versatilities usually would be restricted to between adults of different taxa, or between their larvae, but not with respect to entire populations. Juvenile and adult forms can differ significantly in versatility (e.g., Delbeek and Williams 1987).

The juvenile bottleneck problem

Major niche shifts generally lead to a contention between the demands for the efficient exploitation of resources by each stage (Lande 1982). This is referred to as the *juvenile bottleneck problem* (Werner and Gilliam 1984, Persson and Greenberg 1990). For example, a species may be 'destined' for a particular trophic role as adults (e.g., a piscivorous fish) but necessarily must pass through juvenile size-classes that have to use a substantially different kind of prey (e.g., plankton). Juveniles therefore may be burdened with restrictions that are required by the ultimate adult morphology. Werner and Gilliam (1984) believed that this burden may place such juveniles at a competitive disadvantage to species in which both larvae and adults specialize on plankton. To this effect we might add that juvenile piscivores may not be able to focus well on those plankton species offering the best returns. These juveniles might then appear more generalized (perhaps less optimized would be a better term) than the juveniles of the planktivorous species due to the restrictions that must be met as adults. This suggests that less specific utilization of resources might be anticipated for populations in which stages differ significantly in resource use (e.g., diet) and must accommodate these changes with the one basic somatic design.

Ebenman (1992) extended the analysis of contention between stages to interacting populations with both simple and complex life cycles. His studies suggested that if niche differences were pronounced in simple cycles, then compromises between stages generally lead to greater specialization in each stage, with efficiencies in juveniles being maximized at the expense of adult fitness. For less niche differentiation, the compromise may be more symmetrical and the efficiency in neither adults nor juveniles would be maximized (hence little stage-specific specialization). Greater juvenile efficiencies were to be expected if the availability of resources were more restricted in the juvenile niche, but this would be at the expense of reduced adult efficiencies. The reverse was predicted if adults faced more stringency in resource availability. If the nexus were broken between the developmental programs of each stage (i.e., disjunctive metamorphosis), then the efficiencies of exploitation by stages can be maximized almost independently of one another.

What then do these ideas on stage-structured populations tell us about versatility? The most important point is that the existence of stages (or other subsets such as age-classes) within the vast majority of populations expands the resource base of the population. The ecologist's perception of the versatility of a population depends in some measure on the degree of morphological abruptness between stages. In disjunctive metamorphosis (e.g., tadpoles → frogs), the level of somatic and behavioural reorganization involved makes speaking of such populations as 'very versatile' of dubious value. Versatility is better specified with respect to each stage separately. Indeed, this provokes questions as to whether corresponding stages of conspecific populations with disjunctive metamorphosis display significantly different levels of ecological versatility (e.g., Delbeek and Williams 1987), and if so, why?

Where either continuous development or nondisjunctive metamorphosis occurs, the way in which to view versatility seems to be less clear, but it does depend upon the simultaneous occurrence of different stages in the same place. If stages are sequential (i.e., each stage matures abruptly into the following one) or allotopic, then the existence of significant stage differentiation would not lead us to think of the local population as being excessively versatile. Stage-differentiated populations in which stages are contemporaries and syntopic must be viewed as exhibiting high ecological versatility with a significant stage-specific component (e.g., Polis 1984).

5.5 An idealized set of exploitation strategies

The forms of heterogeneity of resource use attributable to structural differentiation in populations have been described in this chapter. It is a convenient point to use these ideas to introduce the concept of *exploitation strategies*, which are employed in Chapters 6 and 7. It is rather obvious that the variety of exploitation strategies displayed within natural populations is potentially limitless. But for the purposes of modelling, a limited set of exploitation strategies can be considered in a relatively straightforward framework. This framework is based on the degree of similarity of resource exploitation shown between individuals within a population (i.e., do all individuals exploit resources in similar ways?), and how resource exploitation conforms with availability (i.e., does exploitation match availability or not?). Together, these two aspects circumscribe four 'idealized' or extreme kinds of versatility.

To begin with, the most general description of resource exploitation within a population at a particular location can be described completely by these quantities:

- within-individual temporal variation in exploitation of a given resource;
- between-individual temporal variation in exploitation of a given resource;
- within-individual temporal variation in exploitation of all resources; and
- between-individual temporal variation in exploitation of all resources.

This description emphasizes several points in characterizing versatility. First, an estimate of resource exploitation is dependent on how, when and where a sampling program is conducted and whether data are collected at various locations and through time. If information were collected at only one location and at only one time, then major sources of variability will not be addressed and only a limited amount of interpretation is possible. Second, exploitation patterns of individuals may vary and it usually is important to try to gauge whether individual variation is of the same magnitude as variation between individuals. Third, it also is necessary to diagnose whether versatility arises from all individuals in a population being equally versatile or whether there is substantial between-individual variability but little within-individual variability (Jaenike *et al.* 1980, Johnson 1986, Milligan 1986, West 1988). And last, versatility is not merely a measure of between-resource variability in exploitation, which has been a common view (e.g., Petraitis 1979, 1985, Ricklefs and Lau 1980, Feinsinger *et al.* 1981), but also depends upon differences in exploitation patterns displayed by individuals or segments within populations (Wissinger 1992).

Four idealized exploitation strategies

If between-individual variation generally exceeds within-individual variation by a significant amount, then the four main kinds of versatility that can be displayed within a population are (see Fig. 5.10):

(1) *coherent* versatility: there are no significant differences between individuals in exploiting resources in the proportions in which they are 'available';
(2) *resource-like* versatility: relative exploitation of a set of resources, when summed over the population, is in the same proportions as the availability of the resources but individuals differ significantly in the

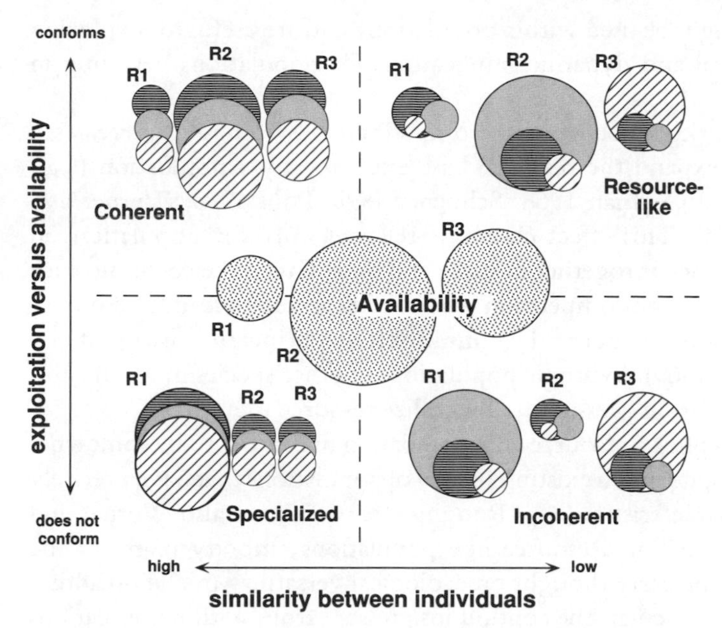

Fig. 5.10 A representation of the four idealized types of versatility in populations. Availabilities of resources R1, R2, and R3 are proportional to the areas of the circles in the centre of the figure. For simplicity, the resource exploitation patterns of just three conspecific individuals (distinguished by type of shading) within each type of population are shown. In *coherent* versatility, the total exploitation of all individuals in the population of each resource is proportional to availability, and individuals do not differ from one another. The total exploitation of all individuals in the population of each resource also is proportional to availability in *resource-like* versatility, but individuals differ significantly in their patterns of resource use. Total exploitation does not conform to availability in either *specialized* or *incoherent* versatility – individuals do not differ in patterns of exploitation in specialized populations but do so in incoherently versatile populations.

resources that they exploit (thus integrated resource use over the whole population resembles, or is 'like,' relative resource availability hence the term 'resource-like');

(3) *specialized* versatility: all individuals exploit resources in the same way but not in direct proportion to relative resource availability; and
(4) *incoherent* versatility: there are significant differences between individuals and exploitation differs significantly from availability.

These patterns, or resource *exploitation strategies* as they are called here, are idealized ones. They represent some of the possible ways in which resources might be used within populations, and are useful for exploring the ecological and dynamic implications for populations behaving in these ways.

Many workers believe that competition between stages creates a pressure to expand the resource base exploited by a population (e.g., Schoener and Gorman 1968, Schoener 1970, Polis 1984, Werner and Gilliam 1984). This effect clearly is relevant only for populations in which stages occur together in space and time. Given the co-occurrence of stages, interstage competition would be expected to lead to resource-like populations under the minimization principle discussed by MacArthur (1969). Syntopic populations of other species may curb this within-population pressure to generalize resource use.

The concept of a resource-like population may seem to be somewhat artificial, despite much existing theory of populations that have precisely the same characteristics (e.g., Roughgarden 1972; see also Werner and Sherry 1987: 5506). Resource-like populations embody many of the concepts of the early thought on ecological versatility (niche breadth – see §7.1), in particular, the continuous pressure from within populations to utilize all available resources (e.g., MacArthur 1969, Real 1975, Gatto 1990). Resource-like strategies therefore seem to be central to discussions of the origin, maintenance and dynamics of local versatility in populations.

It is important to reiterate the basis for these strategies given that they are used extensively in subsequent chapters. The four kinds of versatility are distinguished by using two simple criteria. First, in populations displaying either coherent and resource-like versatility, utilization of each resource closely matches availability. In coherency, all individuals within the population utilize resources in the same way. Exploitation matches availability only when integrated over the entire population in resource-like versatility – there are significant differences between individuals. Exploitation does not match availability in either specialized or incoherent versatility. The second criterion is the degree of similarity in resource use between different individuals within populations, the subject of the current chapter. There are no significant differences between individuals in both coherent and specialized versatility. Individuals differ substantially from one another in both resource-like and incoherent versatility.

5.6 Summary

The differentiation of resource use between individuals or components of populations appears to be one of the main ways in which the diversity of resource use is increased. The dynamic plasticity model of Glasser (1982) addresses quite general issues related to patterns of versatility evident in contemporary communities. However, he envisaged resource differentiation within populations arising only under special conditions, namely, when mean saturation is high (populations almost always near their carrying capacities). Ecological plasticity need not necessarily be an expression of underlying genetic differences because ecological polymorphism can occur in which behaviour (e.g., learning and experience, dominance hierarchies) causes individuals to adopt different patterns of resource use without a strong coupling to genotype (e.g., striped surfperch, *Embiotoca lateralis,* Holbrook and Schmitt 1992).

Polyphenism, the capacity of organisms to follow alternative developmental pathways according to ecological cues experienced during development, also contributes to an expanded versatility providing there is a consistent mapping between phenotypic differentiation and dissimilar resource use, which is exemplified by larvae of the southern spadefoot toad of North America (*Scaphiopus multiplicatus*, Pfennig 1992). Polyphenism seems more likely to be a temporal effect given the greater reliability of temporal cues over spatial ones, so that polyphenism probably expands versatility most often in a sequential manner than in a simultaneous way. Under these circumstances, the population at any time may appear to be relatively specialized but the nature of the specialization changes through time (e.g., Moran 1988).

Genotypically related polymorphism also can affect versatility if there are ecological differences between morphs. Differences may be expressed secularly, when morphs have dissimilar phenologies hinging on particular resources. Morphs may be more or less specialized for different sets of resources, which increases the spread of resources used by the population because of the existence of the polymorphism.

Perhaps the most pervasive effect in differentiated populations is that due to ontogeny. It seems as though most organisms exhibit ontogenetic shifts in resource use, which is associated with increased versatility. Ontogenetic changes in resource use are exemplified by diet selection in the bluegill sunfish, *Lepomis macrochirus* (Walton *et al.* 1992).

The focus on ecological differentiation within populations in this chapter is used to introduce the concept of idealized, resource exploitation strategies, in which individuals may or may not differ in patterns of resource use, and resource utilization when summed over the population may or may not conform with the relative availability of resources. Four idealized strategies are defined and these are used in the next chapter as a basis for modelling the dynamic effects of alternative patterns of resource use within populations.

6 · *Ecological versatility and population dynamics*

This book up to here has largely been a commentary on the specific, that is, on the variety and complexity of patterns of ecological versatility evident in nature. It is desirable to pass on from this particularity to produce or look for a more general picture of the reasons for, and consequences of, different patterns of versatility in natural populations. How might this be accomplished? The usual approach has been to develop models that link patterns of resource use with population dynamics in either a contemporary or evolutionary context (e.g., Roughgarden 1972, Eldredge 1989, Ebenman 1992, Holt and Gaines 1992). It is worth emphasizing that population dynamics have been intimately connected with the breadth of resource use in the development of ecological theory, as we have seen previously with the models of Glasser. That there should be a connexion between densities and versatility is a much older idea than this and was formally developed by Svärdson (1949) and others many years ago. Thus, I believe that the modelling of the relationship between versatility and population dynamics is a necessary component of a monograph devoted to the topic of ecological versatility.

An avenue to modelling that seems to have been the first (or perhaps preferred) mode of attack has been to use mathematical models (e.g., Roughgarden 1979, Mueller and Ayala 1981, Nisbet and Gurney 1982, Ginzburg 1986, Hofbauer and Sigmund 1988). Population dynamics in these models often are represented in a 'phenomenological' way, which simply means that the equations describe the numerical dynamics of populations without necessarily enunciating the mechanisms underlying those dynamics (León and Tumpson 1975). Thus, these models describe *how* populations behave rather than *why* they do so.

The *logistic* model of population dynamics is a widely used phenomenological model (Hall 1988) in which the density-dependent inhibition of the growth in density increases as the density approaches the *carrying capacity* of the population in that environment. The nature of the density dependence usually is not made explicit (hence the phenomenological nature of the model) but resource limitation, increased rates and opportunities for disease and parasite transmission ('crowding') are some of the factors that are thought to underlie the dependence (see also, Roughgarden 1979: 302). Density dependence is implemented by using the carrying capacity – as densities approach the carrying capacity, the rate of growth decelerates. This reliance on a specified maximum density, the carrying capacity, is onerous because it is a parameter that does not readily allow for the variation in resource availability and other effects. The logistic model often underpins mathematical models of the relationship between versatility and population dynamics (e.g., Glasser 1984).

However, there are many problems with using mathematical models, such as the logistic, to relate population dynamics and resource use. Turelli (1977) wrote that models of population dynamics incorporating realistic structure and limitations may be beyond analytical interpretation. Hall (1988: Table 1) went further by saying that the number of problems that can be potentially addressed by using mathematical techniques is relatively small. There also are excessively restrictive mathematical assumptions required for tractability (e.g., Gurney and Nisbet 1980: 331; Tuljapurkar and Orzack 1980: 320) in addition to the host of ecological and biological approximations that one has to make. As Hall (1991: 512) pointed out: 'The analytical models most often used ... are selected for mathematically [sic] solvability' rather than for ecological relevance *per se*. It seems that some of the assumptions made for mathematical convenience become uncritically ingrained into the modelling 'culture' and even may be confused with model predictions (Armstrong and McGehee 1980). Hall also queried whether we should expect nature to be representable by simple mathematical models (at least, the ones that can be analysed; Hall 1988).

These difficulties have led some workers to shun mathematical modelling in favour of (computer) simulation to explore population dynamics (e.g., Haefner and Edson 1984, Shaffer and Samson 1985, Janovy *et al.* 1990). It seems that simulation offers several advantages over conventional mathematical models in relation to ecological modelling. Some of the more obvious benefits include the capacity to represent conditional behaviour and the potential for finding a range of outcomes depending upon initial values. Conditional behaviour is not easy to implement in mathematical models. For example, trajectory switching under certain conditions (e.g., *if–then–else* logic) is difficult to incorporate into mathematical models but clearly is an important characteristic of living organisms. Organisms often show the capacity to switch behaviours or strategies in response to different conditions. Switching obviously is easy to effect in the algorithmic approach. Simulation also allows an estimate to be made of the range of the variation of results by performing large numbers of simulations (i.e., *realizations*, Shaffer and Samson 1985). This is particularly important for the stochastic situations that seem to be most the appropriate ones for natural populations. Most useful of all, computer simulations can be designed to be more mechanistic than phenomenological and, from the preceding comments, this clearly is a desirable feature.

My approach is encapsulated well by the comment of Hall and DeAngelis (1985), who wrote: 'field-oriented ecologists, who, if they ever use modeling at all, confine their efforts to computer simulation models that differ radically in the spirit and approach from abstract analytical models'. Thus, as such an ecologist it seems appropriate that the modelling I use in this chapter be based on a comparatively simple algorithmic description of the relationships between the dynamics and the resource exploitation strategy of a population. This is undertaken by using computer simulations, the algorithms for which are described in Appendix B. In the process, it will become evident why there may have been such a poor correspondence between the results of mathematical models (of any complexity) and the ecological behaviour of natural populations.

6.1 The principles underlying the models

The models developed here are based on the simple premise of providing a set of resources for a population and studying the behaviour of the population through time, which is, of course, a common way of studying the dynamics of model populations. The use of the algorithmic approach allows arbitrary numbers and types of model population with arbitrary initial densities and patterns of resource use.

Before investigating the dynamics associated with alternative exploitation strategies, it is necessary to describe the fundamentals of the modelling. The key aspect of the method is that organisms within populations engage in *exploitation competition* for the resources. Eventually, the resources become fully utilized by the population and, at this stage, a variety of responses are envisioned as ways to cope with the limited availability of resources relative to the demand for them. By using just a few simple distinctions, a rich assortment of responses can be pictured. The nature and importance of each of these distinctions are described in this section.

Utilization pressure

The model dynamics are built on the concept of *utilization pressure*. This is a generalization of Solomon's (1949) ideas on the rates of resource use by predators. Solomon noted that the predation pressure exerted on prey is a function of two components: (1) the number of items consumed per unit time by each predator (corresponding to the average utilization rate per individual), and (2) the number of individuals comprising the predator population (the population density). The change in the rate of consumption (i.e., (1), sometimes called *voracity* for food resources, Matessi and Gatto 1984, Gatto 1990) in relation to different prey availabilities is usually called the *functional response*, while the corresponding change in density (i.e., (2)) is the *numerical response* of the predator to changes in prey availability (Holling 1959).

Armed with these ideas, it now is possible to define the *utilization pressure* on a resource. This simply is the product of the density of individuals using a resource and the average rate of use of the resource by those individuals, that is, population density × utilization rate. Clearly, the utilization pressure exerted on a resource by a population will vary in accordance with both the functional and numerical responses of the population.

The main assumption that I use is that there always is an exponential growth rate of utilization pressure on a resource until the total pressure exerted on that resource equals the availability (of nondegradable resources, such as nesting holes) or production (of degradable resources, such as food) of the resource. This principle differs from the usual way in which population dynamics are modelled in which the growth of population density, when there are unlimited resources, is exponential (see Cooper 1984: part I, Getz 1984).

This emphasis on the exponential growth of utilization pressure rather than of population density arises because the latter usually assumes, at least implicitly, that organisms do not adjust their utilization rates in response to changes in resource availability. Given the wealth of evidence that consumers as different as vertebrate predators and plants alter their rates of resource utilization as the relative availability of those resources changes, it seems unreasonable to focus on population density alone.

Notwithstanding these comments, the distinction often (but not always) may not be particularly important during the early stages of population growth when, given a superabundance of resources, organisms usually utilize resources at the fastest rate of which they are capable. During this stage, the exponential increase in the rate of utilization pressure often will be manifested as an exponential rate of increase in density.

I introduced, without much fanfare, a rather significant point in the last paragraph, namely, that utilization rates are bounded. This means that there exists a satiation point beyond which a consumer cannot further utilize resources irrespective of their availabilities, in other words, there is a *maximum* utilization rate. Some of the biological reasons for this are finite gut volumes, maximum biochemical absorption rates, limited hunting times, and so on. Many models of population dynamics do not impose a maximum rate of resource utilization (Łomnicki and Ombach 1984), which makes it difficult to judge the applicability of their results because of this unrealistic condition.

To this point, we have considered what happens during the early phases of the growth of a population in which there is a superabundance of resources. Generally, individuals within such a population are expected to utilize resources at the maximum rate of which they are capable, produce offspring, and show a concomitant exponential rate of growth in population density. This, of course, differs little from the situation envisaged for most other ways of modelling population dynamics.

However, the growth rate of the population is eventually stemmed because of 'environmental resistance', or some form of density dependence. The method for implementing density-dependence used here is an explicit limitation based on the availability of resources. One reasonable way of handling overexploitation in models is to reduce the utilization pressure exerted by each population (or component of a population) in direct proportion to the degree to which that population (or component) contributes to the excess pressure (Cooper 1984: model I, Rogers 1986). Populations that exert the greatest pressure are made to reduce their pressure proportionately more than populations exerting less pressure on a fully utilized resource.

This form of response to overexploitation is characteristic of *exploitation* competition, but not of *interference* competition (Rogers 1986: 364). Densities and utilization pressure in interference competition are held lower than is potentially possible, given the availability of resources, by behavioural mechanisms such as territoriality (Murray 1981). For

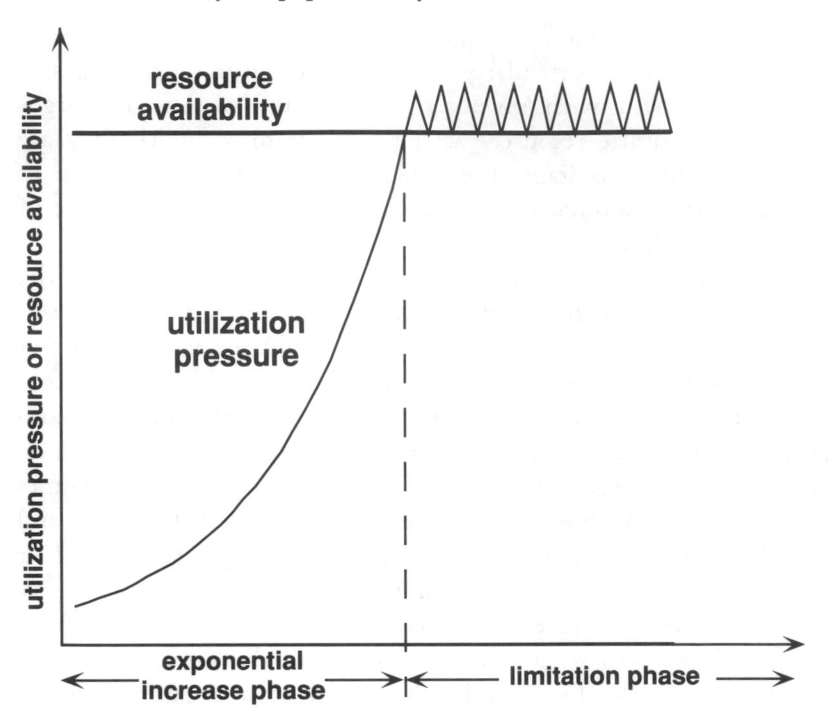

Fig. 6.1 Utilization pressure, the product of utilization rate and density, grows exponentially when resource availability exceeds demand (exponential increase phase). When demand exceeds availability, utilization pressure must decrease but will oscillate for reasons noted in the text.

example, several species of colonial honeyeaters (Meliphagidae) in Australia (bell miner, *Manorina melanophrys*, and noisy miner, *Manorina melanocephala*) aggressively defend areas within eucalypt forests and in the process exclude other small species of foliage-gleaning birds. That this interference results in underutilized resources is shown by the outbreak of folivorous insects (the food of most of the excluded bird species) within these areas, leading to die-back in the trees (Loyn *et al.* 1983). Van Horne (1983) cited a number of similar examples. Densities under these conditions depend in part on density-dependent mechanisms that are not directly related to resource availability, but these are not considered here.

The upshot of these ideas are the dynamics illustrated in Fig. 6.1. In this particular example, resource availability is fixed at a constant level (for example, nesting hollows in old trees for parrots) and the utilization pressure grows initially at an exponential rate. At some point, demand exceeds availability and the utilization pressure has to be reduced to 'match' availability. However, there will be an ongoing tendency for the pressure to grow again. This can happen in several ways: (1) the progeny produced as a result of the use of the resource now enter into competition for it; (2) existing individuals within the population that were forced to forego the resource so that utilization pressure matched the sustainable level (for example, floaters or satellites in bird or frog populations, Smith 1978) resume attempts to acquire the resource; and (3) individuals immigrate into the area and also seek to use the resource. A characteristic saw-tooth pattern results (a two-point limit cycle, Fig. 6.1), with alternating increases and decreases in utilization pressure. These oscillations resemble overcompensating, density-dependent models based on simple survival functions but that do not explicitly refer to resource utilization and/or availability (e.g., Grenfell *et al.* 1992).

Functional and numerical responses

We have seen that a generalization of Solomon's (1949) ideas on predation leads to the concepts of the functional and numerical responses. That is, the demand for a resource by a population is a function of the utilization rate of that resource by different individuals, and also the number of individuals, the population density.

If the total demand for a resource exceeds its availability, then that demand (utilization pressure) must be reduced in one of two ways, or both. First, the rate of utilization might be reduced with either little or no effect on population density. Alternatively, the rate of utilization by each individual may remain fixed, the consequence of which is that the population density must decrease (i.e., by death and/or emigration, or possibly abstention, Glasser 1983) to accommodate the necessary reduction in demand. A third possibility is a mixed response in which both the utilization rate and density change to reduce utilization pressure.

These ideas are shown graphically in Fig. 6.2. Here we consider a cyclically varying resource (Fig. 6.2 a). A purely numerical response involves the maintenance of the utilization rate at the one level, but with changes in the density of individuals that reflect the pattern of changes in resource availability (at least in the limitation phase, Fig. 6.2 b). A purely functional response on the other hand, involves a constant density in the limitation phase that is offset by changes in the rate of utilization, which then accommodates the fluctuations in resource availability (Fig. 6.2 c). A more complicated scenario occurs when there is a mixed response,

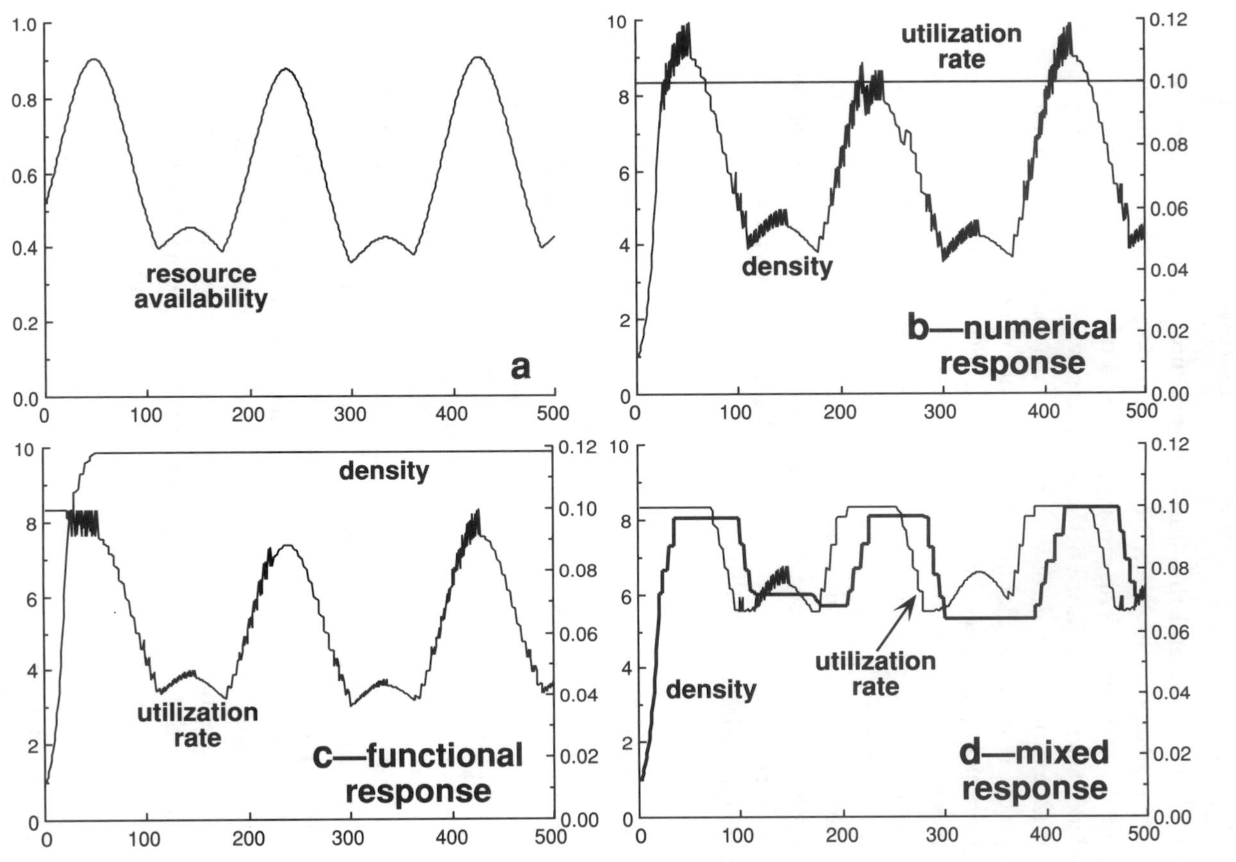

Fig. 6.2 The possible responses to overutilization of resources: (a) time-course of the change in resource availability; (b) pure numerical response; (c) pure functional response; and (d) mixed response. Ordinate scales on the left in (b)–(d) refer to densities, while those on the right refer to utilization rates.

with variation in both utilization rate and density to deal with the variation in resource availability (Fig. 6.2 d). There is, of course, an infinite number of possible patterns for mixed responses.

The marginal utilization rate

The capacity of organisms to vary their utilization rates raises the question of how far the rate can be reduced. At some point, the rate will be so low that the organism is unable to persist on that rate of resource acquisition – it then becomes uneconomic to use the resource (MacArthur 1970). Chesson's (1988) term 'maintenance requirement' referred to a similar idea, a basal level below which rates cannot fall. This minimum rate provides a marginal existence in ecological terms so it is called here the *marginal utilization rate*. Without incorporating this constraint, model populations might reach astronomically high densities based on minuscule utilization rates, which is ecologically and thermodynamically unacceptable.

We are now able to draw together these ideas on utilization rates, resource availabilities, and the economics of resource use (Fig. 6.3). At low levels of availability, an organism does not use the resource (the uneconomic zone). At some point however, the resource becomes sufficiently abundant to be economically useable and the resource is utilized at the marginal rate. At higher levels of availability again, the utilization rate also in greater, although the exact form of the change in utilization rate cannot be specified in a general way, hence the '?' in Fig. 6.3. In some organisms (I), the response to an increased level of resource availability ($A_{marginal}$) might be an immediate jump to the maximum utilization rate of which they are capable, which effectively bypasses the marginal rate. In other organisms (II), the utilization rate may remain at the marginal level until a certain level of resource availability is reached (A_{max}), and then increases immediately to the maximum rate. Other organisms might show linear (III), convex (IV) or concave (V) functional responses in the interval between $A_{marginal}$ and A_{max}. At superabundant availabilities, biological constraints such as handling or hunting time, gut capacity, or biochemical diffusion rates, enforce a maximum utilization rate.

The flexibility of utilization rates – hard and soft exploitation

The dynamic implications of the degree to which organisms can adjust their rates of resource utilization are profound, particularly in terms of

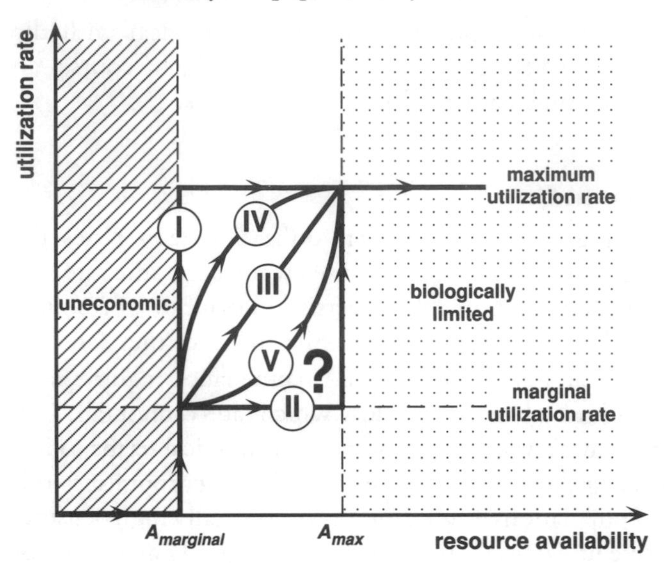

Fig. 6.3 Diagram relating rates of resource utilization (arrowed lines) to resource availability. At low levels of availability ($< A_{marginal}$), an organism does not use the resource (cross-hatched: the 'uneconomic zone'), but at higher levels, the resource is used at an increasing rate although the exact form of the rate-change is not specified (hence the '?' – see text for explanation of I–V). Once the resource becomes superabundantly available ($\geq A_{max}$, stippled), biological constraints such as handling or search time or gut volume (if food) enforce a maximum utilization rate.

the effects on densities. For example, if we were to monitor the changes in the density of a population, we might find little difference over a long period, suggesting that the population was stable, and that resource availability was sufficient to ensure the continuation of that stability. This might well be a misapprehension if the population were showing purely functional responses to changes in the availability of resources, such as in Fig. 6.2 c. If the utilization rate during the period of stability were marginal, then even a comparatively small proportional change in the availability of the resource may force a sharp decline in the density of the population. This may be difficult to relate directly to the magnitude of the change in availability under field conditions.

It seems reasonable to expect that there will be a wide range in the capacity of different organisms to adjust their rates of resource utilization. For example, the energy requirements of some small endothermic vertebrates (e.g., shrews, hummingbirds) are so high that they can show little flexibility in the rate at which they need to gather food. Of course,

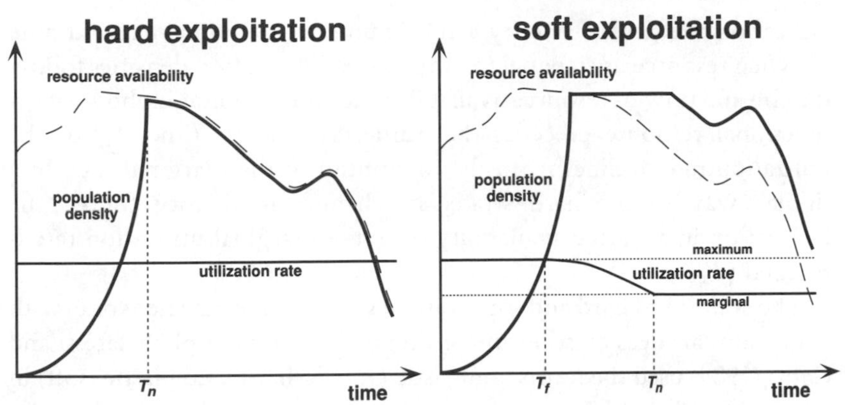

Fig. 6.4 Time courses of changes in population density (thick line) and resource utilization rates (thin line) for hard and soft strategies. In both cases, T_n indicates the time at which a numerical response to declining resource levels (dashed vertical line) occurs. For soft strategies, a functional response precedes the onset of the numerical response (at time T_f). This is accomplished by a reduction in the utilization rate from the maximum to the marginal rate (horizontal dotted lines).

many animals and plants have pronounced abilities to vary their utilization rates in response to changes in availability – the fasting of snakes and dormancy in a host of plants, vertebrates, and invertebrates are examples.

I distinguish between utilization strategies on the basis of the degree of flexibility shown. At one extreme are organisms that have resource requirements so stringent that they are completely inflexible in their utilization rates. These can be thought of as having an obligatory maximum rate. They show only numerical responses to the overutilization of resources, which are manifested as death or emigration depending upon the mobility of the organism in question. Organisms at this extreme are called *hard strategists*.

Any organism showing some degree of flexibility is referred to as a *soft strategist*. By definition, any strategy that is not hard is soft, but it is the degree of flexibility that is important. The degree of flexibility is a function of the difference between the magnitudes of the maximum and marginal rates. If these differ to only a small degree, then the dynamic behaviour will resemble that of a hard strategist. If the rates are extremely dissimilar, then the dynamics will be different from those of a hard strategist.

The relationship between the flexibility of resource use and population dynamics is shown in Fig. 6.4. In hard exploitation, with a fixed utilization rate (scaled to unity here), a numerical response occurs at time T_n when resource use is equal to availability. Thereafter, densities follow the ebb and flow of resource availability. In soft exploitation however, a functional response precedes the numerical one (at time T_f) as the utilization rate declines from the maximum to the marginal rate. In a similar way to the hard strategist, changes in density follow the fluctuation in resource availability once the marginal utilization rate is reached (T_n).

The concept of hard and soft strategies is by no means a new one, with many similar ideas existing in the literature. For example, Matessi and Gatto (1984) used the terms nonplastic (akin to hard) and plastic (soft) in relation to the ability of *r*-selected strategists to alter dynamically utilization rates to contend with resource variation. Łomnicki and Ombach (1984) modelled resource 'share' as variable, with a maximum value when resource availability exceeded demand, and falling off as resources became overutilized (i.e., a soft strategy). Botsford (1981) showed that growth rates, which can plausibly be linked to utilization rates, could be held at high or low levels depending upon population densities. Ginzburg and Akçakaya (1992) also noted that several studies have shown that consumption rates decrease as the density of consumers increases. Abrams (1984) and Chesson (1988) have closely linked variation in resource utilization rates with probabilities of coexistence of interacting populations. Variability in the uptake of nutrients (i.e., utilization rates) appears to characterize most terrestrial plants (Osborne and Whittington 1981, Bloom *et al.* 1985, Lambers and Poorter 1992). For example, seasonal variation in phosphorus absorption was reported by Chapin and Bloom (1976) for tundra graminoids, while similar seasonal fluctuations were found for nitrogen uptake in *Deschampsia flexuosa* on acid soils and *Zerna erecta* on calcareous soils in Britain (Taylor *et al.* 1982). And finally, Abrams (1987b) has entreated workers to incorporate variable utilization rates into mathematical models of consumer-resource dynamics for greater realism. Thus, contrasts between the dynamics of soft and hard strategists is a crucial component in any modelling scheme purporting to link population dynamics and resource utilization.

The model strategies

The main purpose of Chapter 6 is to begin to explore the relationships between different patterns of resource use, or ecological versatility, and population dynamics. To do so, I use the idealized strategies introduced in §5.5. These strategies represent the following kinds of versatility: (1) coherent; (2) resource-like; (3) specialized; and (4) incoherent. Relative rates of resource use conform with the relative availability of resources in (1) and (2) ('generalized' in the usual sense), but do not in (3) and (4). The conformance constraint in coherent and resource-like strategies means that these are regarded here as *obligatory* generalists (cf. *facultative* strategists, see Glasser 1982, Glasser and Price 1982). Phenotypes within populations differ significantly in their use of resources in (2) and (4), but populations of (1) and (3) effectively consist of a single ecological phenotype.

It is important to reflect briefly on the usefulness of each of these strategies for the purposes of modelling population dynamics, which amounts to the question: how well can the dynamics be constrained? The more non-arbitrary constraints that are associated with a strategy, the more general the results will be. It should be clear that coherent generalization, by definition, is completely constrained because there is no arbitrariness associated with the structure of the population (there is none – individuals are identical ecologically), and the relative rates of resource use conform with the respective availabilities. On the other hand, although conformance is constrained in resource-like exploitation, there may be an arbitrary number of phenotypes and an arbitrary number of ways for each phenotype to use the resources. Specialized exploitation involves a single phenotype like coherent versatility, but the respective use of each resource can be defined in any number of ways. For example, if there were three resources, relative utilization rates might be [1.0, 0.5, 0.1], or [1.0, 0.1, 0.0], or any other combination of numbers ≥ 0.0 and ≤ 1.0. For incoherent versatility, there are effectively no constraints at all, which makes that strategy of limited use in the present context. Thus, the results of this chapter and the next are restricted to a discussion of some of the effects of ecological versatility on population dynamics in the coherent, resource-like, and specialized strategies only. The algorithms for implementing these strategies require some additional constraints for resource-like and specialized exploitation, the bases for which are outlined in Appendix B.

The assumptions of many-resource models

Up to now, we have considered how population dynamics relate to the exploitation of a single resource. The extension of the modelling to environments consisting of many resources requires the imposition of several general assumptions. The purpose of noting them here is to emphasize the dependence of the results derived from general models on the particular assumptions of the models. Obviously, the dynamics of natural populations cannot be expected to resemble those of the model populations if these assumptions are seriously violated.

The first main assumption is that resources can be consistently distinguished by the model organisms, and hence, that no aliasing occurs. This assumption also implies that individuals differentiate between resource classes in the same way as one another. Second, resources are regarded as substitutable and not complementary (see León and Tumpson 1975; Chapter 2). Therefore, they are of the same general type (e.g., food types), and switching between them is possible where necessary. Although substitutable, the resources need not be perfectly substitutable for all populations. For the coherent strategy, the resources are regarded as perfectly substitutable so that switching between them occurs readily. For a specialized strategist, the capacity to perceive, intercept, handle, and process particular resources is known to be resource-dependent (e.g., Wainwright 1988). This means that there will be different maximum utilization rates for each resource (sometimes zero for unusable resources). This also is the case for each phenotype in a resource-like population. Third, constraints on resource use such as abiotic gradients or natural predators are not modelled here. And last, resources are assumed to be distributed randomly in space, which means that availabilities do not depend functionally on location. A more detailed discussion of the modelling assumptions is presented in §B.5.

The test environments

The form of variation in resource availability dictates the patterns of resource use and also the population dynamics of any exploitation strategy. To provide an indication of the sensitivity to different resource regimes, two alternative environments are used initially. The first is a *variable* environment in which there are significant changes in the absolute availabilities of some resources, while others remain at more or less constant values (Fig. 6.5 a). In the *constant* environment however, all resources remain at the same levels of availability through time. Note that these environments have no significance in their own right, but the use of several forms of resource variation will help to convey the complex dependence of population dynamics and exploitation strategies on the nature of fluctuations in resource availability.

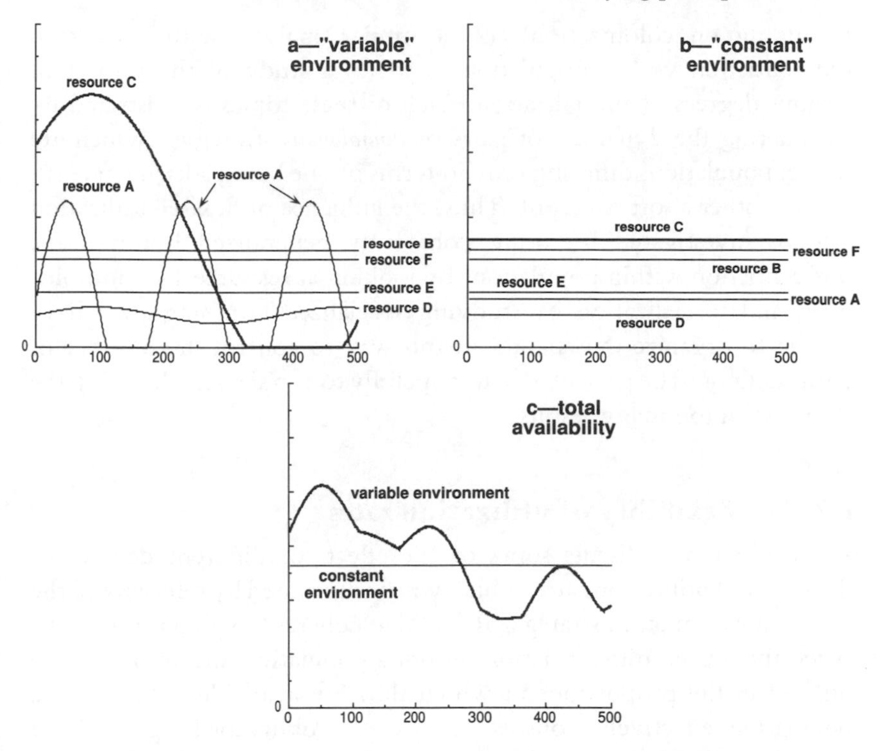

Fig. 6.5 Patterns of resource variation in two six-resource 'environments'. (a) the variable environment; (b) the constant environment; and (c) the total resource availability in each environment. Note that the ordinate in (c) is 1/3 the scale of (a) and (b).

The variable environment consists of three resources having a constant absolute availability (resources B, E, F) and three others showing sinusoidal fluctuations (A, C, D) (Fig. 6.5 a). Resources A and C are the most profusely available resources at some times, but each becomes unavailable (asynchronously with each other) at certain times. In the constant environment, all resources have constant absolute availabilities in the order C > F > B > E > A > D (Fig. 6.5 b). The total resource availabilities in each environment are similar on average (Fig. 6.5 c).

The remainder of Chapter 6

We now have reached the end of an admittedly lengthy preamble, but one that is necessary to provide the background for the models discussed here and in Chapter 7. The remainder of this chapter involves three main themes: (1) the effect of flexible utilization rates on population dynamics (including on colonization); (2) a similar analysis with respect to differentiation within populations; and (3) a study of the impact of various degrees of specialization. Each of these topics is illustrated by considering the dynamics of pairs of *homologous* strategies, which are model populations differing only in terms of one being a hard strategist and the other a soft strategist. Thus, the influence of flexible utilization rates is investigated by using coherently generalized homologues, differentiation within populations by looking at resource-like homologues, and specialization by studying specialized homologues. I have chosen to organize the chapter in this way to control the number of permutations to be presented, and hopefully to avoid overwhelming the reader with too many results.

6.2 The flexibility of utilization rates

In this section, I discuss some of the effects of different degrees of flexibility of utilization rates, which will be illustrated by reference to the coherent exploitation strategy. Recall that coherent exploitation occurs when there is no differentiation within a population and resources are utilized in the proportions in which they are available. Therefore, a population effectively consists of a single, obligatorily generalized phenotype. Furthermore, the resources are perfectly substitutable for coherent strategists. For convenience, I use the abbreviations **Ch** (i.e., **C**oherent–**h**ard) and **Cs** (**C**oherent–**s**oft) for the hard coherent and soft coherent strategies respectively.

The growth in the density of a **Ch** population in the constant environment, if initially rare, follows an exponential course, the rate being determined by the utilization rate. Recall that only numerical responses occur in **Ch** populations. Therefore, when the utilization pressure exceeds the availability of a resource, densities must decrease. If the availabilities of resources are constant, a characteristic saw-tooth oscillation in density or two-point limit cycle results from the alternation of overutilization and underutilization (Fig. 6.6 a). Note that oscillations are not damped (cf. Southwood *et al.* 1974) but continue at the same amplitude indefinitely.

As with **Ch** and indeed all other strategies, the density of a **Cs** population grows exponentially if initially rare (Fig. 6.6 b). When utilization surpasses availability, the utilization rate is reduced and the density remains constant (Fig. 6.6 b). A **Cs** population in a constant environment would maintain a uniform density for as long as the

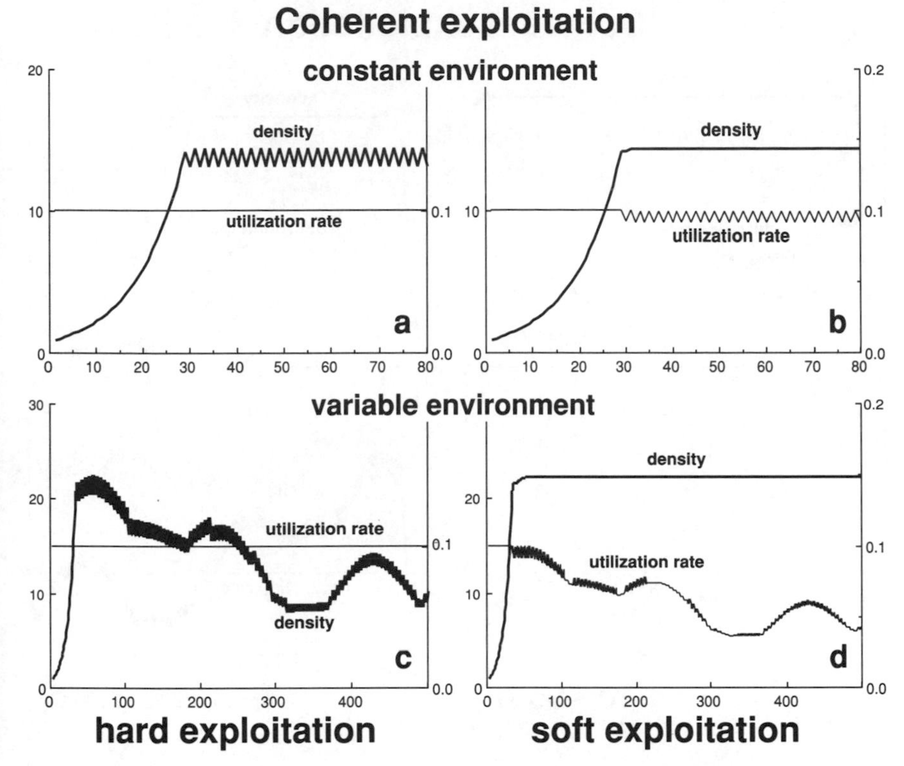

Fig. 6.6 Density and utilization rate trajectories of coherent strategists in the constant (a, b) and variable (c, d) environments. Trajectories are for both hard (a, c) and soft strategists (b, d). Ordinate scales on the left of each plot refer to densities while those on the right refer to utilization rates. Note the different scales on the density axes and abscissae between the top pair of plots and the bottom pair.

resources remained constant, with the uniformity being maintained by oscillations in the utilization rates of individuals (Fig. 6.6 b).

The changes in the density of a **Ch** population follow a quite different course if the resource availabilities vary greatly through time. Growth follows the usual exponential trajectory until saturation is reached (Fig. 6.6 c). At this point, densities are subject to the fluctuations in resource availability because individuals are incapable, by definition, of adjusting their utilization rates. This causes densities to crash, either by a net emigration flux or by an appropriate level of local mortality (Fig. 6.6 c). Densities more or less track total resource availability in **Ch** populations because of the generalized exploitation strategy.

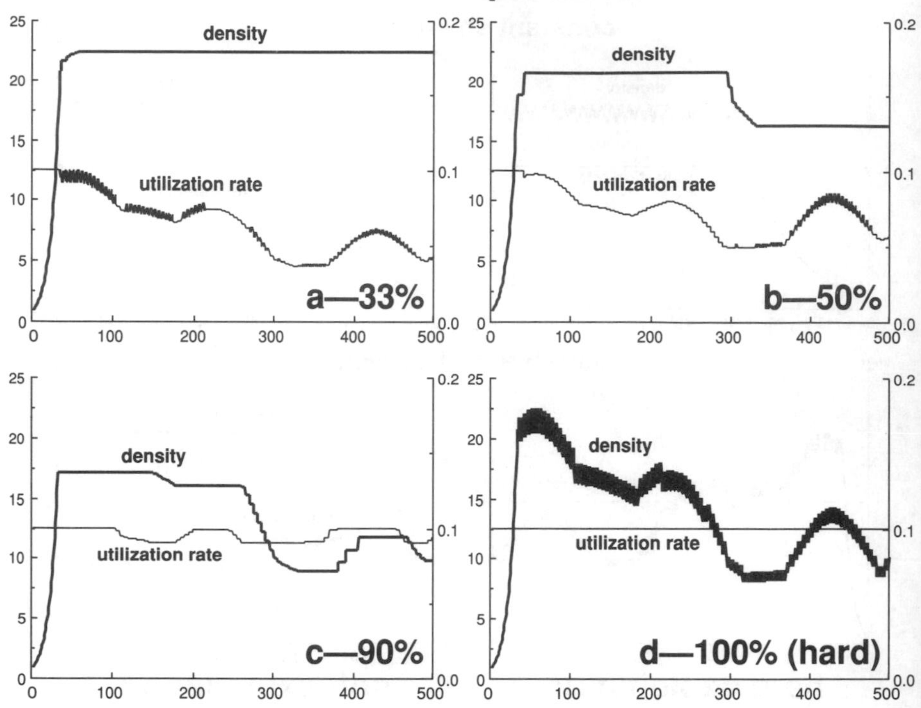

Fig. 6.7 Expression of the soft to hard continuum in relation to density (scale on left) and utilization rate (scale on right) trajectories: (a) with the marginal utilization rate one-third of the maximum rate; (b) one-half; (c) 90%; and (d) equal (hard strategy).

As one might expect, the situation for a **Cs** population in a varying environment involves an initial exponential growth in density (if rare) followed by a levelling out at the time at which utilization matches availability (Fig. 6.6 d). At that same time, individuals alter their utilization rate in such a way as to mirror the overall fluctuation in resource availability (Fig. 6.6 d).

This hard and soft dichotomy really is a reflection of two extremes of a continuum running from individuals having no capacity to alter utilization rates (hard) to those that have a large capacity to do so (soft). The critical aspect is the 'room to move' between the maximum and marginal utilization rates. As these rates converge, the trajectories evident for the **Cs** strategy gradually are transformed into those for the **Ch** strategy (Fig. 6.7). For example, in the variable environment, the density trajectory changes from being an insensitive indicator of fluctuations in resource availability (Fig. 6.7 a) to reflecting those fluctuations directly (Fig. 6.7 d) as strategies go from soft to hard. Correspondingly, the trajectory of the utilization rate switches from being an indicator of resource variation (Fig. 6.7 a) through to one that is insensitive of that variation (Fig. 6.7 d). Between the extremes, mixed numerical and functional responses are evident (Fig. 6.7 b, c).

High density influxes

Most models of population dynamics, including the logistic model, begin with model populations growing from small initial densities in resource-rich or unconstrained environments. Under these conditions, model populations generally expand exponentially at first before density-dependent restrictions (including resource limitation) rein in the growth rate. Rarely are situations considered in which the initial densities are very high compared with the sustainable levels of resource availability. Fluxes of this form may be quite common, particularly in the seasonal migrations of many vertebrates and some invertebrates. Thus, the purpose of this subsection is to look at whether high and low initial densities affect subsequent population dynamics in different ways.

In turns out that the effects of high-density influxes on the population dynamics of soft strategists are substantially greater than on the hard. The main reason for the difference is the compressibility of the utilization rates in soft strategies. This permits comparatively high initial densities to be sustained for relatively long periods at utilization rates near to the marginal rates. Hard strategists quickly equilibrate to sustainable densities and subsequently follow trajectories similar to those when model populations grow from low initial densities.

To make these points clearer, consider **Ch** and **Cs** populations invading the variable environment. The initial densities were much greater than can be sustained, even at the marginal rate (i.e., for **Cs**). The effects on the population of **Ch** are minimal compared with when the population grows from small initial densities (Fig. 6.8 a). The density immediately falls to a 'sustainable' level. Thereafter, densities follow similar time-trajectories to the case of low initial densities (Fig. 6.6 c, Fig. 6.8 a). Overall, it seems likely that few if any consistent changes can be attributed to differences in initial densities in **Ch** populations.

However, there are pronounced effects of high influx densities on a soft strategy such as **Cs**. When expanding from low densities, utilization

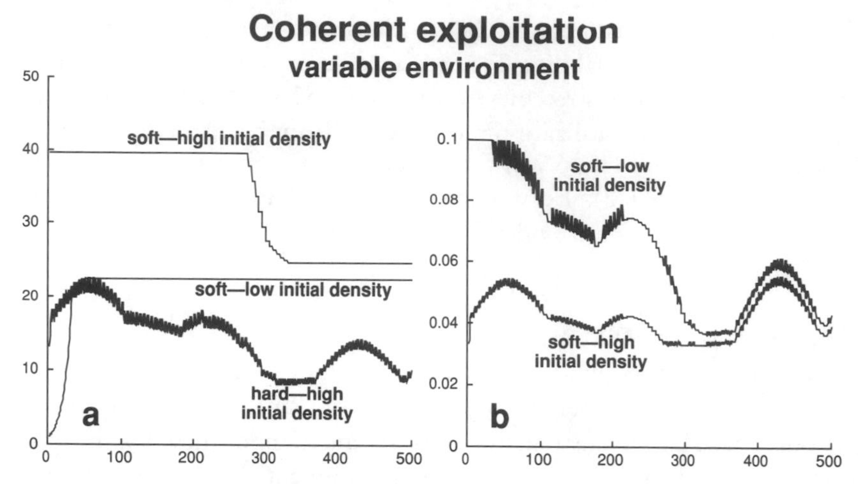

Fig. 6.8 Comparative effect of the influx of high densities into the variable environment. Initial influx densities far exceed carrying capacities for the 'high' initial density trajectories. Plots are of (a) the densities of **Ch** (high initial density) and of **Cs** (both high and low initial densities), and (b) the utilization rates of **Cs** (both high and low initial densities).

rates are at the maximum level until overutilization occurs (Fig. 6.8 b). Functional responses thereafter are needed to relieve overutilization, which, in the current case allows the maintenance of a constant density. However, model populations starting from a large influx density, that is, from an immigration of many individuals into an unoccupied environment, show two effects. If the influx density is sufficiently high, utilization rates are pushed as low as possible (i.e., to the marginal rate, Fig. 6.8 b) yielding extremely high initial densities (Fig. 6.8 a). Density trajectories after this initial phase differ markedly from the low initial density case for long periods (Fig. 6.8 a). The trajectories do eventually converge (Fig. 6.8 a). Utilization rates also show significant differences for long periods – the convergence of rates occurs at the same time as densities (Fig. 6.8 b). Therefore, the general effect of high influx densities on the population dynamics of a **Cs** strategist is to establish a high-density–low-utilization rate regime that may persist for long periods, but that eventually stabilizes to a pattern similar to that originating from an initially low-density situation.

Overview

The variation in utilization rates that is evident in soft strategies may explain why population densities rarely appear to be directly linked to fluctuations in resource availability (see Wiens 1984: 416–419). The greater the capacity of individuals to vary their consumption rates (i.e., the greater the disparity between the maximum and marginal utilization rates), the less clear will be the correspondence between changes in densities and changes in resource availability (Fig. 6.7). Unless individuals have obligatorily fixed utilization rates (hard strategists), there probably is no reason to expect that densities will closely follow variation in resource availability.

The dynamics and resource exploitation patterns of hard strategists may hardly depend upon the initial density of a population, except when the time taken to reach overutilization is extremely long (i.e., a relatively extended exponential growth phase). The picture for soft strategies is significantly different. The effects of various initial densities can be propagated for long periods, depending upon several factors but mainly on the degree to which the utilization rates are forced down by high influx densities. The different behaviour of hard and soft strategies in relation to high-density influxes suggests that the likelihood of extended colonization may depend sensitively on the amount of flexibility in utilization rates. A hard colonist may encounter sufficient resources to support only a few individuals, making the extinction probability relatively high in the short term. A soft strategist, with similar characteristics to the hard strategist apart from its flexibility, may be able to occur in relatively higher densities for longer periods. Even if the soft colony does not eventually develop into a self-sustaining population, it nevertheless might be recorded for a lengthy period following the initial influx. This effect may seriously affect estimates of the rates of colonization and extinction that are central to several disciplines (e.g., biogeography, habitat fragmentation).

6.3 Differentiation within populations

Studies of natural populations have uncovered ecological differences between the components of populations as large or larger than those between closely-related species (e.g., Polis 1984, Werner and Sherry 1987). The widespread occurrence of substantial within-population variation has stimulated an increase in the number and diversity of models of resource use in populations consisting of distinct stages (e.g., Tschumy 1982, Werner and Gilliam 1984, Ebenman 1987, 1992, Siegismund *et al.* 1990, De Roos *et al.* 1992, McCauley *et al.* 1993).

Most of these models involve modelling with explicit schedules of natality and mortality rates and transitions between stages. As the focus of this book is on resource use rather than on stage-structured populations *per se*, the models considered here are limited to the dynamics of populations consisting of ecological phenotypes, reminiscent of the surfperch populations that have been studied by Holbrook and Schmitt (1992). Only the resource-like and incoherent strategies consist of phenotypically differentiated populations. However, as discussed above, there are virtually no (general) constraints on models for incoherently versatile populations, so that this section involves a discussion of resource-like exploitation alone.

The utilization of resources in resource-like exploitation, when summed over the entire population, is in the same proportions as the relative availability of the resources. Unlike coherent versatility, phenotypes differ significantly in the respective proportions of each resource that they utilize. These are the only constraints, which means that utilization pressures and resources used by various phenotypes may be quite volatile, especially in variable environments.

The modelling of resource-like behaviour is complicated because of the phenotypic structure within the population. The total utilization pressure on each resource must reflect the respective availability of the resource, which intimately connects all of the phenotype-specific densities and utilization pressures of a resource-like population. A model of resource-like exploitation must include explicit mechanisms for forcing phenotypes to differ substantially yet leave the integrated utilization of each resource proportional to its relative availability. One way in which these conditions can be met is described in Appendix B, and this method underpins the algorithms used here.

The resource-like algorithms are dynamic to the extent that once a phenotype becomes locally extinct, the remaining phenotypes expand their rates of utilization to exploit the resources freed by the extinction. The algorithms are structured in such a way that phenotypes compete with one another, which causes phenotypic extinctions. Thus, the dynamics include an ongoing competitive pressure within populations that is consistent with many of the existing models for resource use within populations (e.g., MacArthur 1969, Roughgarden 1972).

It will be useful to be able to display the conformance between the use and availability of resources, which I will refer to as B_{pop}. This is easy to do by using the proportional similarity measure encountered in Chapter 3 (Feinsinger *et al.* 1981):

$$B_{pop} = 1 - \frac{1}{2} \sum_{r=1}^{R} |U_r - A_r|. \tag{6.1}$$

In equation (6.1), there are R resources, and U_r and A_r are the utilization rates and availabilities of the rth resource respectively. Also, as we are dealing with models of differentiated populations, we also require a summary statistic to capture the between-phenotype variation in resource use. The Czekanowski similarity measure is used here for this purpose. It ranges from 0.0 for complete dissimilarity (i.e., no overlap) to 1.0 for complete overlap, and is described more fully in Appendix B.1.

In this first part of the analysis of the behaviour of resource-like populations, I consider populations comprising three ecological phenotypes in the variable environment. The initial densities of each phenotype are the same, as are the total utilization rates. In the hard strategist (**Rh**), the total rates remain constant for each phenotype individually while in the soft strategist (**Rs**) the total utilization rates of each phenotype, although initially the same, vary independently of one another in accordance with the demands, availabilities, and utilization rates of the resources that each uses.

The early stages of the growth in the density of **Rh** populations is exponential, as expected (Fig. 6.9 a). This exponential rate of increase is a reflection of the exponential growth rate in each phenotype separately. The densities eventually are limited by the availability of resources and follow trajectories through time that are broadly reminiscent of the patterns shown by **Ch** populations (cf. Fig. 6.9 a and Fig. 6.6 c). That the algorithm performs successfully is shown by: (1) conformance generally fluctuates close to unity, as is required for the resource-like criterion; and (2) the between-phenotype similarity remains substantially different from unity throughout the time-course of the simulation (Fig. 6.9 b). The results presented here are representative ones for **Rh** populations in the variable environment.

There is much more variability in the dynamics of **Rs** populations because of the flexibility in utilization rates shown by each phenotype. Two realizations are provided to give an idea of some of this variability. In a three-phenotype **Rs** population, one phenotype usually becomes (locally) extinct in a short period of time (by timestep 300 or so), and only one persists in the longer term. All phenotypes show an exponential growth in the beginning (Fig. 6.10 a, c). The complex interaction between utilization and availability makes it difficult to predict just which phenotype eventually persists at the expense of the others. For example, at about timestep 200 in both examples, the eventual 'winner' lagged behind the phenotype that became extinct second (Fig. 6.10 a, c). Conformance remained high (≥ 0.9) in both examples (Fig. 6.10 b, d),

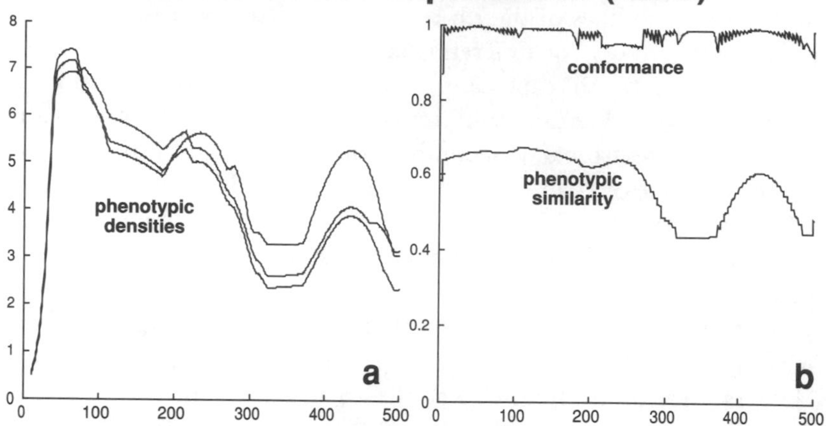

Fig. 6.9 (a) Density trajectories of the three ecological phenotypes of a **Rh** population in the variable environment. (b) The trajectories for conformance and phenotypic similarities for the same population.

despite the demise of two of the phenotypes. This shows the capacity of the algorithm to permit the surviving phenotypes to expand their use of resources in the desired way. The adaptive expansion of resource use leads to an increase in phenotypic similarity as phenotypes become extinct, which is particularly evident in Fig. 6.10 d.

The stability of ecological polymorphisms

Some of the thought on polymorphisms and their relationships to ecological versatility was reviewed in §5.3. The observation that phenotypes become extinct in resource-like model populations, especially **Rs** ones, suggests that an analysis of the stability of ecological polymorphisms should be considered in more detail. By stability, I do not intend to look for the mathematical conditions that might be associated with a stable ecological polymorphism, but rather, large numbers of simulations are performed with an aim to make some comments on when ecological polymorphisms might be expected to occur.

To do so, **Rh** and **Rs** populations were modelled in both the variable and the constant environments. The three-phenotype populations do not give much scope for finding differences between strategies or with respect to environments, so eight-phenotype populations were considered instead. Again, in all cases, the total utilization rates and initial

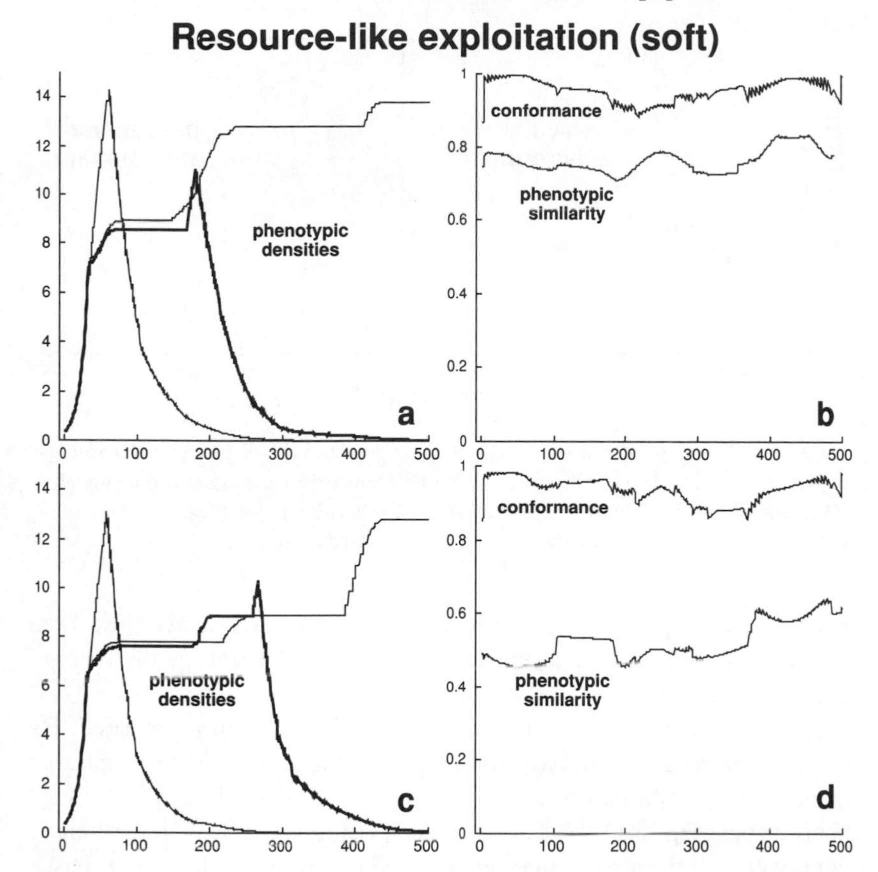

Fig. 6.10 Results of two realizations of **Rs** populations in the variable environment. (a) and (c) density trajectories; (b) and (d) trajectories for conformance and phenotypic similarities.

densities were the same for each phenotype. Simulations were run for 500, 1000, 2000, 5000, and 10000 timesteps to gauge the effect of duration on stability.

In both environments, ecological polymorphism was significantly higher in **Rh** populations than in **Rs** populations, irrespective of duration (Fig. 6.11). In the variable environment, most phenotypes were lost from **Rs** populations within 500 timesteps, and only one persisted for durations of ≥ 1000 timesteps (Fig. 6.11 a). For progressively greater durations, the number of surviving phenotypes declined dramatically in **Rh** populations, from over five after 500 timesteps, to an asymptotic mean number of two surviving phenotypes (Fig. 6.11 a). Thus, the

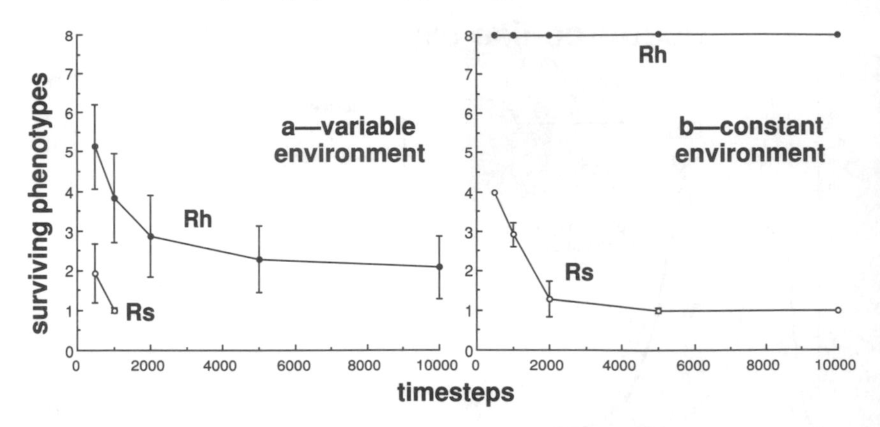

Fig. 6.11 The numbers of surviving phenotypes of eight-phenotype **Rh** and **Rs** populations in the variable (a) and constant (b) environments after durations of 500, 1000, 2000, 5000, and 10,000 timesteps. Each value is the mean of 500 separate realizations and errors bars are ±1 standard deviation.

initially diverse population of eight phenotypes was reduced to just three within 2000 timesteps, although only one more phenotype on average was lost after that.

In the constant environment a rather different picture emerged. **Rs** populations remained more polymorphic for longer than in the variable environment, although by 2000 timesteps, the average had fallen to 1.27 phenotypes (Fig. 6.11 b). The more surprising result was that all eight phenotypes of the **Rh** population persisted regardless of the duration of a simulation, which indicates an absolute stability of more phenotypes than there were resources (i.e., eight and six respectively). Admittedly, an absolutely constant environment is an abstraction but nevertheless it seems as though there may be circumstances in which a rich and varied ecological polymorphism might be stable for long durations providing that there are no disturbances or external influences.

The results of this section indicate that ecological polymorphism in **Rh** populations probably will be maintained for long periods providing phenotypes have similar total utilization rates. On the basis of the limited analyses considered here, more diverse polymorphisms might be expected if resources are relatively constant in availability (Fig. 6.11). However, if utilization rates are flexible, as in **Rs** strategists, polymorphism is likely to be lost rather quickly because imbalances develop in total utilization rates. These imbalances lead to most phenotypes being outcompeted by their conspecifics, and, eventually, to just a single

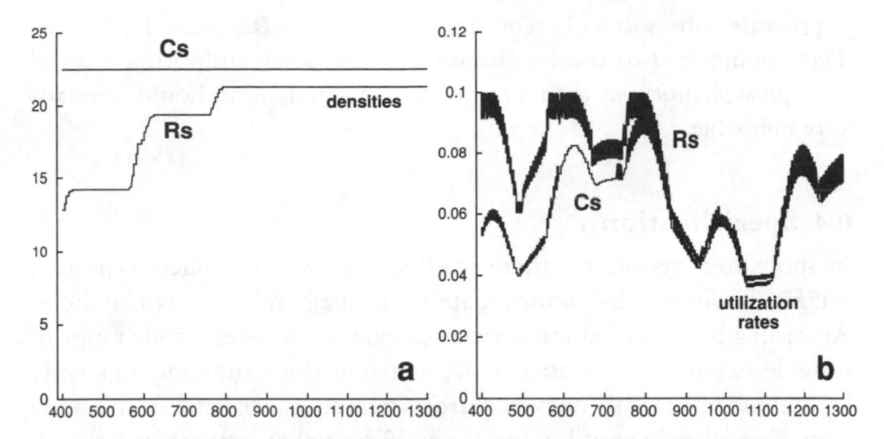

Fig. 6.12 The degeneration of a **Rs** population to a **Cs** one. The surviving phenotype in the **Rs** population (after 400 timesteps in this example) converges in both density (a) and utilization rate (b) to the pattern displayed by a **Cs** population in the variable environment.

phenotype persisting. This process would appear to proceed more rapidly in situations in which resource availability varies to an appreciable extent (Fig. 6.11).

What becomes of an **Rs** population following the loss of phenotypic diversity? It turns out that **Rs** populations eventually converge to the same pattern as a **Cs** strategist. The density of the surviving phenotype ultimately increases to a level similar to that of a **Cs** population (Fig. 6.12 a) and, correspondingly, the utilization rate of that phenotype becomes similar to that of the **Cs** strategist (Fig. 6.12 b). Once all phenotypes except the sole surviving one become extinct in a **Rs** population there is no way to distinguish the former **Rs** population from a **Cs** population.

Overview

It would seem that ecological polymorphism can be maintained in a resource-like population only if it is a hard strategist (i.e., **Rh**). The diversity of polymorphisms within **Rh** populations is unlikely to be high, irrespective of the initial diversity, and is expected to decrease through time if there is any level of variability in resource availability. In soft resource-like populations on the other hand, those phenotypes that reduce their (total) utilization rates soonest are rapidly outcompeted by other phenotypes and driven to extinction. The surviving phenotypes expand their patterns of resource use, so that, ultimately, **Rs** populations degenerate into soft coherent populations (i.e., **Rs** → **Cs**, Fig. 6.12). These points lead to the conclusion that the (total) utilization rates of ecological phenotypes within resource-like populations should be relatively inflexible.

6.4 Specialization

In specialized versatility, there is effectively a single phenotype that utilizes resources disproportionately to their relative availabilities. According to this definition, specialization encompasses a wide range of possible exploitation strategies, from populations showing relatively high conformances to ones that are extremely specialized in the usual sense. Specialized versatility implies a differential capacity to utilize each resource. Different maximum utilization rates will arise because of the interaction between the capabilities of the consumer and the nature of the resources.

For example, the whelk *Morula marginalba* preys on a variety of invertebrates on rocky intertidal platforms in south-eastern Australia. Fairweather and Underwood (1983) showed that the handling times of the whelk for its main prey species vary greatly, with an average processing time for the tube-dwelling worm *Galeolaria caespitosa* and adults of the barnacle *Tesseropora rosea* being between two and eight times the averages for juvenile barnacles and the limpet *Patelloida latistrigata* respectively (Table 6.1). The characteristics of the prey influence the handling time – *Morula* is able to insert its proboscis between the shell margin and the substrate for small limpets, a technique that greatly accelerates the rate of processing. However, the same technique cannot be used on larger limpets. These observations mean that the maximum possible utilization rates for each prey species necessarily must differ greatly too. In an idealized situation with unlimited prey availabilities, insatiable appetites, no interference from predators or competitors, and with generally free rein, an individual whelk on average could not utilize *Galeolaria* at more than 3.8 individuals per week. The analogous figure is 34.3 for the limpet. Thus, maximum utilization rates are resource dependent in *Morula*.

In some circumstances we might envision a consumer that needs to maintain constant utilization rates of a range of resources. The nutritional requirements of a large herbivore might be one example. If all individuals in a population have this inflexibility (and the same relative utilization rates), then the only possible response when there is an

Table 6.1. *The handling times of the main prey species of the muricid whelk* Morula marginalba *on intertidal sandstone platforms*

Prey species	Type of organism	*N*	Handling time (hours) Mean	SD	Maximum
Galeolaria caespitosa	Tube-worm	68	44.3	21.0	97.0
Patelloida latistrigata	Limpet	75	4.9	4.5	22.0
Tesseropora rosea	Barnacle				
Juvenile (<3.5 mm)		98	21.3	7.5	43.5
Adult (>3.5 mm)		58	42.1	14.9	88.1

Source: From Fairweather and Underwood (1983).

overutilization of resources is to decrease the population density (i.e., a numerical response). By analogy to the coherent and resource-like cases, such a population would display hard specialization (**Sh**).

In other specialized populations, the resource-specific utilization rates might be more flexible, being adjusted to accommodate changes in resource availability and utilization pressure. If all individuals change their rates simultaneously and to the same degree, then the population would continue to be regarded as a specialized one, albeit a soft-specialized population (hence **Ss**).

For a **Sh** population to persist, individuals cannot be obligatorily specialized on resources that become unavailable at some times because those individuals have fixed requirements for the resources that they use. That is, if individuals in the population *must* use a specific resource at a particular rate, then the population must become locally extinct if the availability of that resource drops to zero. This may involve either complete mortality, total emigration, or perhaps dormancy. However, the model structure developed here does not include a mixed strategy of utilization and dormancy, as discussed by Brown (1989). For this reason, I discuss the properties of specialized populations in relation to the constant rather than the variable environment because all resources remain at the same availabilities through time in the constant environment.

Two of the most likely consequences of specialization are: (1) reduced

Table 6.2. *Relative utilization rates of increasingly specialized populations. Q_s is a simple index of specialization – it is equal to the maximum utilization rate divided by the sum of all six utilization rates*

	Highest rate ← Resources → Lowest rate						
Strategy	1	2	3	4	5	6	Q_s
Generalized	1.0	1.0	1.0	1.0	1.0	1.0	0.17
.	1.0	0.75	0.75	0.75	0.75	0.75	0.21
.	1.0	0.5	0.5	0.5	0.5	0.5	0.29
.	1.0	0.25	0.25	0.25	0.25	0.25	0.44
.	1.0	0.1	0.1	0.1	0.1	0.1	0.67
.	1.0	0.04	0.04	0.04	0.04	0.04	0.83
Specialized	1.0	0	0	0	0	0	1.0

densities; and (2) a decline in the conformance between resource use and availability. Although these effects are to be expected, it is less clear how they are related to the nature of the specialization displayed by a population. For example, what are the differences in the effects on densities and conformance if individuals specialize on the most abundant resource, or on the least abundant?

These questions are addressed by examining the effect on the densities and conformances of populations having successively greater degrees of specialization on: (1) the most abundant resource in the constant environment, C; (2) a resource of intermediate abundance, B; and (3) the least abundant resource, D (see Fig. 6.5 b).

Seven levels of *potential* specialization were used, ranging from the most generalized strategy (all rates identical) through to a strategy completely specific for a single resource (Table 6.2). Potential specialization (denoted by Q_s) is defined as the proportion of the largest maximum utilization rate for any one resource to the sum of maximum utilization rates of all resources. Thus, if all six resources have the same maximum utilization rate, $Q_s = 1/(1+1+1+1+1+1) = 1/6 = 1.167$. On the other hand, if the maximum utilization rate of one resource is ten times that of all the others, in arbitrary units, $Q_s = 10/(10+1+1+1+1+1) = 10/15 = 0.667$. If only one resource is used, Q_s clearly is 1.0. Note that

Q_s is defined only in relation to the maximum potential utilization rates for each resource, and does not explicitly refer to the availabilities of those resources. Hence, it is only a measure of the potential specialization and not of ecological versatility *per se*.

A convenient measure of the effect on population density due to specialization is the ratio of the density of the specialized population to the corresponding density of a comparable **Ch** population in the same environment. Thus, if the density of the specialized population stabilizes at 10 individuals per unit area, say, and the analogous density of a **Ch** population is 25, then the relative density (reflecting the effect on density due to specialization) is 10/25 or 0.4.

By having these concepts, we can now look at the effects of specialization on some aspects of population dynamics. It is clear that increasing specialization leads to a depression of densities in **Sh** populations, but the effect is substantially more pronounced when that specialization is on the least abundant resource (resource D, Fig. 6.13 a). The densities are highest for specialization on resources B and C (the intermediate and most abundant resources respectively) when there is a marked imbalance in the maximum utilization rates ($0.3 \leq Q_s \leq 0.6$). In other words, densities were maximized when the utilization rate of the preferred resource was between two and ten times the rate of other resources. This effect did not occur for specialization on the least abundant resource (D) – density was maximized when all utilization rates were equal (Fig. 6.13 a). Conformance, like density, fell sharply as specialization increased, but again more sharply for specialization on the rarest resource (Fig. 6.13 b).

There were certain similarities in the trends for soft specialized (**Ss**) populations (Fig. 6.13 c, d). Recall that soft specialists can not only adjust utilization rates of particular resources, but they also can alter their overall or total utilization rate. That rate can be reduced to the marginal utilization rate, as we have seen previously in this chapter for the **Cs** and **Rs** strategies. The result is that **Ss** populations can develop higher densities than **Ch** populations in the same environments because the latter have an obligatorily high utilization rate, which necessitates a lower density. Thus, for $Q_s < 0.3$, **Ss** populations maintained densities in excess of the **Ch** figure and conformances exceeded 0.9 for specialization on any of the three resources B, C, and D (Fig. 6.13 c, d). Significantly depressed densities for specialization on resources B and C occurred only when Q_s exceeded 0.8.

In some respects, the most perplexing aspect of these results seems to

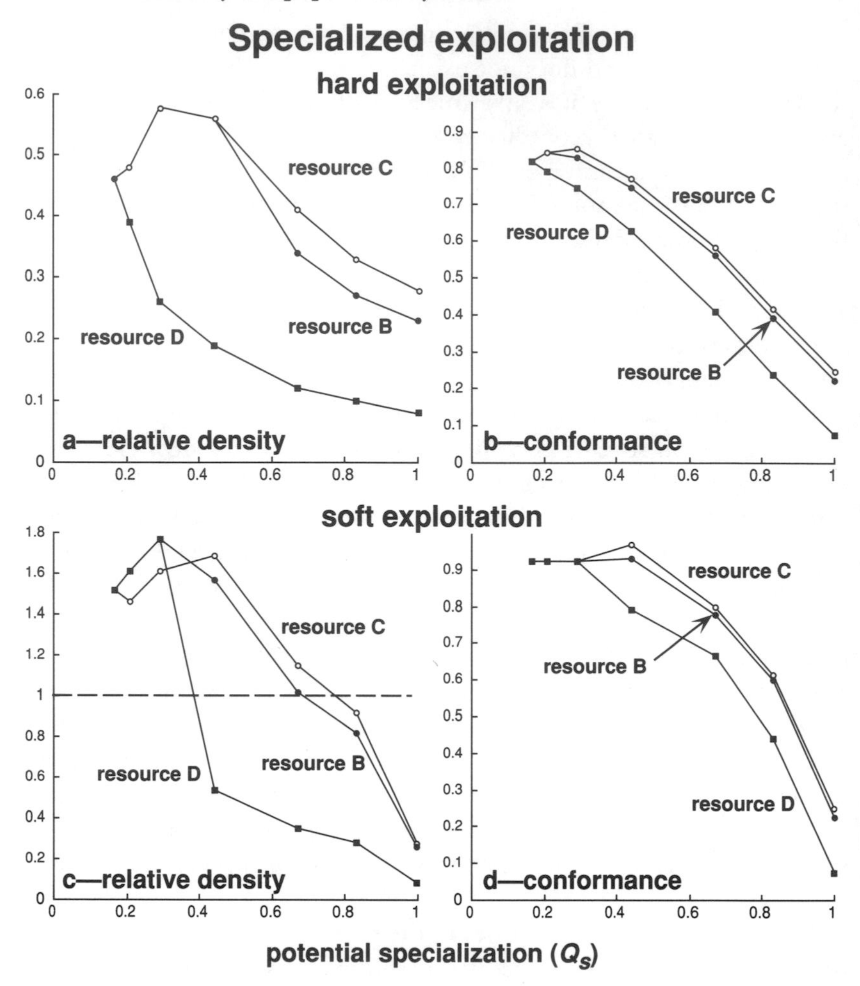

Fig. 6.13 The effects on relative density and conformance of increasing degrees of specialization (constant environment). The densities and conformance values are those at equilibration. Dashed line indicates density equal to that of a **Ch** population. Trajectories are shown for increasing specialization on the most abundant resource (C), on one of intermediate abundance (B) and on the least abundant resource (D). Relative abundance and the degree of specialization are defined in the text. (a) and (b) hard specialization, (c) and (d) soft specialization.

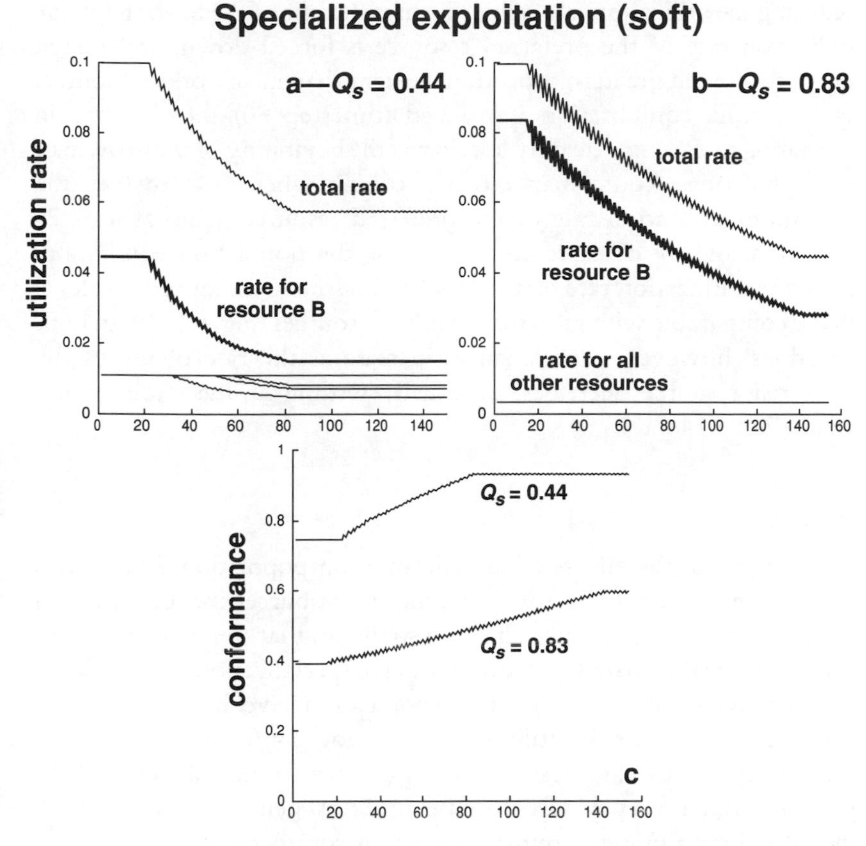

Fig. 6.14 The mechanism by which high levels of conformance are achieved in soft specialization (constant environment). (a) trajectories of utilization rates for all resources of a soft specialist capable of utilizing resource B at the highest rates, with an initial $Q_s = 0.44$. (b) As for (a) except that Q_s initially is 0.83. (c) trajectories of conformance for both initial Q_s values.

be the extraordinarily high conformances that arise for **Ss** strategists on any of the three resources for $Q_s < 0.5$. Note that this corresponds to up to a fourfold higher maximum rate on the preferred resource than on any other (Table 6.2). What is the mechanism responsible for this effect?

The reason for these high levels of conformance is that the utilization rates of different resources used by individuals of **Ss** populations with $Q_s < 0.5$ become more equal (Fig. 6.14 a). Naturally, the preferred resource is the one in most demand, which causes it to become overexploited sooner than any other resource. Given the option of reducing the utilization rate, which is characteristic of the **Ss** strategy, the utilization rate of the preferred resource is forced down, and forced down to a much greater proportional extent than in any other resource. By the time equilibration is reached (timestep 80) the variance in utilization rates is significantly less than at the beginning of the growth of the population. Conformance is not as high when $Q_s > 0.8$ (i.e., the maximum utilization rate of the preferred resource being at least 25 times that of any other resource) because the population equilibrates before the utilization rate on the preferred resource is reduced to a level that is comparable with rates for the other resources (Fig. 6.14 b). In both situations, however, conformances increase as the rate of use of the preferred resource decreases, eventually settling at the equilibration value (Fig. 6.14 c).

Overview

The analysis of the effects of specialization on population dynamics is more complicated than for coherent or resource-like exploitation because of the range of specialization strategies that populations might display. This is due partly to the definition of specialization adopted here, namely, given that all individuals in a population have the same strategy, specialization occurs if utilization does not conform strictly with availability. Thus, in these analyses, the behaviour of **Sh** and **Ss** populations that vary widely from being almost generalized (low Q_s) to specialized for a single resource ($Q_s = 1.0$) is considered.

Increasing specialization leads to depressed densities, but the degree to which densities are affected depends upon the resource upon which the population specializes (Fig. 6.13). Thus, increasing specialization on the more abundant resources does not depress densities as much as specialization on less abundant resources. In **Ss** strategists, densities may not be depressed until the levels of specialization are relatively extreme (high Q_s). For example, specialization to the extent that the maximum utilization rate of the preferred resource is ten times greater than that of other resources ($Q_s = 0.67$) still permits densities similar to those of a **Ch** population (Fig. 6.13 c). However, the densities of hard specialists (**Sh**) are less than half those of **Ch** populations under the same conditions (6.14 a). This means that one has to specify the nature of specialization to make a prediction as to the likely impact on densities: (1) is it a hard or soft strategist (**Sh** or **Ss**)? (2) which resource forms the basis of the specialization (abundant or rare)? and (3) what is the degree of potential specialization (relative maximum utilization rates of alternative resources)?

One of the unexpected outcomes was that conformances may remain quite high in **Ss** populations despite specialization to the extent that the maximum utilization rate of the preferred resource is four times greater than that of other resources ($Q_s = 0.44$; Fig. 6.13 d). This is attributable to the equalization of utilization rates in the soft specialist because of the greater utilization pressure on the preferred resource (Fig. 6.14). This phenomenon may explain, at least in part, the common observation that field rates of resource use (relative electivities) do not correspond to electivities measured in the laboratory. Resource selection by organisms in the field is undertaken in the context of the overall demand for the resources, whereas laboratory studies normally do not, or cannot, take this into account.

6.5 Additional remarks

The models considered in this chapter are contingent on many factors, but I think that two in particular deserve to be discussed in greater detail because they have the potential to influence greatly the conclusions drawn here. First, the efficiency with which harvested resources are converted into gains in fitness clearly is an important characteristic and has been included in models of resource use for many years. It is, of course, one of the key elements of optimal foraging theory. Thus, I make some observations below on the issue of efficiency in relation to models of population dynamics. And second, the majority of existing models of population dynamics and resource use treat populations as being 'closed' such that there is little impact of external populations on the dynamics. In recent years, ecologists have begun to view populations as more extended entities, which has led to the development of the concept of the *metapopulation*. Metapopulations essentially consist of populations occupying spatially segregated patches set in a matrix of unsuitable habitat. The dynamics within metapopulations then must take into account local- to intermediate-scale migration and dispersal between patches in addition to the local population dynamics. Therefore, some comments are made on the ways in which the existence of metapopulations might affect the results presented here and also those of other models.

In addition, there has been extensive criticism in the past couple of decades as to the relevance and applicability of theoretical models in ecology (e.g., Hall 1991, Peters 1991). Given the substance of this chapter and the next, I wish to make some comments on how I see the results of Chapters 6 and 7 relating to these criticisms.

Efficiency

Different organisms may use a resource at the same rate. However, there need not be an equal gain in fitness because the organisms have different efficiencies in converting the utilized amount of resource into a realized gain in fitness (MacArthur 1969, 1970, Schoener 1974b, Wilson and Turelli 1986). This aspect reflects the distinction drawn in Chapter 2 between utilization and exploitation. Utilization refers only to the relative use of resources while exploitation depends also on the translation of resources into fitness gains and, therefore, must take into account the efficiency of conversion (Wiens 1989a: 323). Efficiencies can be taken to embrace such things as handling costs, risk exposure and the mechanical and chemical means to process the harvested resources (if they are foods or nutrients). Different organisms are expected to show and do show different efficiencies (Wilson and Turelli 1986).

The effect of different utilization rates has been an important aspect of this chapter, particularly in relation to the concepts of hard and soft strategists. However, ideally one would like to relate exploitation strategies to population dynamics by using the exploitation rate rather than the utilization rate, where the exploitation rate is the product of the utilization rate and the efficiency. I have not provided any examples of the effects of different efficiencies because of the complexity of aspects already introduced in this chapter.

Nevertheless, a more comprehensive treatment of the effects of alternative exploitation strategies on population dynamics would need to include efficiency as an important element. This can be done at two levels. First, one could explore the dynamics of populations that are efficient in a general sense and ones that are inefficient. This involves the use of a global efficiency value so that the utilization rates for all resources are modified by the same proportion, with high constants (say, about 1.0) representing efficient exploiters, and low constants (<1.0) indicating poor converters. An alternative approach is to model efficiency at the level of each individual resource, and then, as J. W. Glasser has done, to explore the relationship between the variance in the efficiencies of resource use and population dynamics (Glasser and Price 1982, Glasser 1982, 1984; see Fig. 4.6). Both the global and resource-specific approaches are likely to influence substantially the patterns described in previous sections of this chapter.

Metapopulations

The numerical response of a population to resource shortages can occur in three distinct ways: (1) *in situ* mortality; (2) emigration; and (3) abstention. The particular mix of responses that will occur depends upon the relative mobility of individuals and also on the exploitation strategy. A relatively immobile organism, such as a limpet on an isolated rocky promontory, will be unable to emigrate, so that death or abstention are the only possibilities. For obligatory resource use, even abstention is not an option so that the numerical response is restricted to the death of at least some individuals.

That individuals of at least some species leave areas in which resources are overtaxed (or are attracted to places where resource blooms are occurring) is closely connected with the concept of the metapopulation. There has been a vigorous development of ideas concerning the ecological effects of the spatial distributions of conspecific individuals that no longer focuses on the abstract notion of the closed population (e.g., Fretwell 1972, Addicott *et al.* 1987, Gilpin and Hanski 1991). Some of these ideas are reviewed in greater detail in Chapter 8 but it is of sufficient importance to the modelling discussed in this chapter to warrant comment at this stage.

The modelling of spatial substructure within an extended population or metapopulation frequently addresses the question of habitat selection, especially in relation to the existence of 'source' and 'sink' habitats (e.g., Pulliam and Danielson 1991, Holt and Gaines 1992, Morris 1992). Metapopulation ideas also are crucial to conservation biology (e.g., habitat fragmentation, Opdam 1991) and the mediation of the coexistence of competitors (e.g., Rosenzweig 1991, Danielson 1992, Dunning *et al.* 1992, Palmer 1992, Tilman 1994).

The metapopulation dynamics of many of these models involve little more than attaching degrees of suitability to different patches of habitat that the metapopulation occupies. These suitabilities yield different levels of fitness in relation to the local population occupying a patch. For example, in the source–sink–unusable habitat model (Danielson 1992), source habitats are sufficiently productive or suitable to allow the production of more individuals that can be supported within the patch, which causes excess individuals to emigrate. Sink habitats can be occupied but are too poor to sustain an ongoing population without supplementation from source habitats, while unusable habitats effectively form the matrix in which the other habitat patches are set. Other models impose suitability onto the landscape in the form of a fractal map (e.g., Milne *et al.* 1992, Palmer 1992).

These theoretical extensions to population dynamics clearly are important because they reflect the empirical fact that many if not most species are more or less characterized by a metapopulation spatial structure than by sets of isolated, noninteracting populations. That is, most species display dispersal and migration. However, once one tries to take into account variation in resource availability and its effect on local population dynamics, we encounter a situation in which the number of unconstrained variables grows very rapidly. This is the main reason why I have not used explicit metapopulation dynamics here.

To see this more clearly, consider a metapopulation in which individuals move between patches. Dispersal of individuals between patches occurs in response to fluctuations in resource availability and is characterized by the fluxes J in Fig. 6.15. Patch A happens to be centrally located but also has the highest availabilities and diversity of resources, followed by patches C and B. The metapopulation as a whole also may be subject to immigration and emigration fluxes, so that individuals flow into and out of the metapopulation as well as between populations. This sort of structure probably characterizes many species of continental vertebrates (e.g., birds, mammals, and amphibians). In situations in which the migration fluxes greatly exceed the dispersal dynamics, the metapopulation model breaks down and such systems properly should be regarded as 'open systems', which appear to characterize many marine organisms (Gaines *et al.* 1985, Gaines and Roughgarden 1985, Underwood and Fairweather 1989, although see Butler and Keough 1990).

The problem in trying to model the dynamics of metapopulations lies in the large number of assumptions one has to impose on the system. For example, what is a meaningful distribution of fluxes, both within and between metapopulations? What are the spatial and temporal correlations and autocorrelations of resource variation among patches occupied by populations? What are the correlation distances of resource variation in relation to the typical movements of the species? What is the timing of immigration fluxes relative to blooms and dearths of resources?

To make any sort of general statement in such an extended model

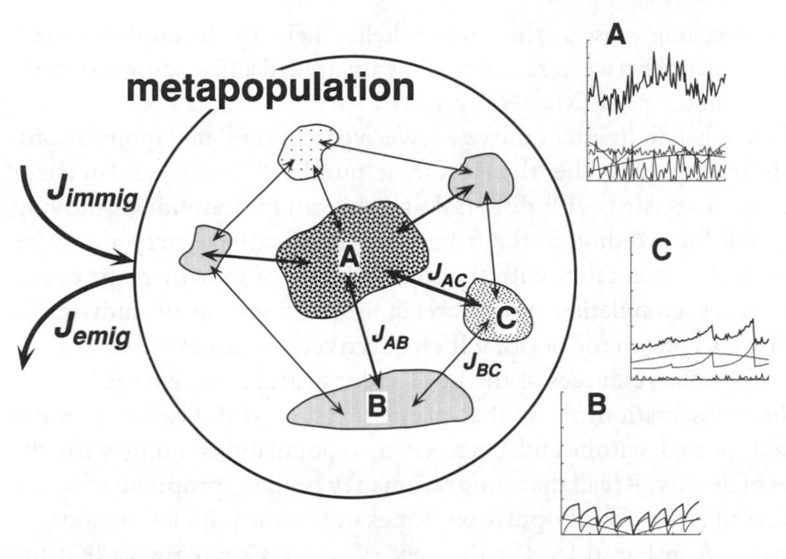

Fig. 6.15 Metapopulations consist of populations occupying spatially segregated patches (various stipplings) set in a matrix of unsuitable habitat, with communication between the patches by local dispersal (double-headed arrows, *J*s). Greater fluxes (e.g, J_{AB}) are shown by thicker arrows. The metapopulation as a whole is also affected by large-scale emigration (J_{emig}) and immigration fluxes (J_{immig}). Patch 'suitability' may be a function of a variety of factors, including constraints, but very often will be determined by the levels of resource availability and the diversity of resources. Resource fluctuations for three of the patches are illustrated (A, B, C), with total availability (bold line) and the diversity of resources (4 resources) each being higher in A than in C (3 resources), which in turn are higher than in B (2 resources).

seems to be difficult if not impossible. One of the aspects that needs to be considered in greater depth is the idea that fluxes between and within metapopulations are not merely numerical fluxes – we are not dealing just with the flow of individuals – but also with interchanges of the condition and resource utilization patterns of individuals. For example, many species of frogs migrate to pools to breed, forming classic metapopulations over potentially very large areas (e.g., Gulve 1994). In the process, both sexes import nutrients and energy into the pools in the form of reserves (e.g., fat-bodies and musculature in males) and eggs. This allochthonous input means that the population dynamics in any pool are financed to some extent by resources gathered and processed elsewhere and transported into the pool as somatic reserves and eggs (Mac Nally 1981). Although the reserves may be insufficient for the whole breeding season, they nevertheless help to decouple a direct correspondence between resources that are needed and resources that are used within the pool (Mac Nally 1987).

Thus, when individuals move between patches and metapopulations, they bear properties other than just their 'numbers'. Energy and nutrient fluxes are associated with dispersal and migration. Individuals moving into pools for breeding in the frog example effectively carry a positive energy and nutrient flux with them in the form of somatic reserves and eggs. After completing their breeding activities, spent individuals disperse away from the pools, which effectively decreases the utilization pressure on the resources of the pool and transfers it elsewhere.

These observations show that properties such as utilization pressure are transported within and between metapopulations along with the fluxes of density. Recall that emigration may be the appropriate response to overutilization for a population, for example, the population occupying patch A in Fig. 6.15. Let the species be a **Cs** strategist so that the utilization rate at the point at which emigration occurs is at the marginal rate. These individuals may immigrate into patch B, say, at a time when resources are overabundant there compared with demand. The result will be a mixture of individuals utilizing resources at the maximum rate (the residents) and the immigrants, whose rates of resource use may take some time to increase to that level. Clearly, this will perturb the population dynamics in patch B and the outcome will depend upon both the relative and absolute numbers of immigrants. In any event, in this example there is a flux of individuals from A, probably in relatively poor condition, into B (containing individuals in good condition), which effectively is a transfer from patch A to B of not only individuals but also of utilization pressure.

What does all this mean for the efficacy of modelling? The most significant aspect is that the modelling of metapopulation dynamics in relation to resource use cannot be one from which general conclusions can be derived. The timing of resource variation in different patches will be crucial, but equally, the absolute and relative numbers of immigrants and their exploitation characteristics also will have a major impact on the outcome. The emigrants from patch A may end up being distributed evenly among all other patches, but a more likely scenario will be an uneven distribution (e.g., possibly something like the ideal free distribution, Fretwell 1972), with some individuals being lost completely from the population.

Therefore, we return to the problem of having too many degrees of freedom, and the degrees increasing exponentially with the number of patches considered. Some of this freedom can be constrained by imposing correlations on resource variation among patches that are consistent with known natural systems. Migration fluxes also might be tied down in a similar way. But this approach leads to particular rather than general results. However, without these limitations the range of outcomes from metapopulation models is so great that inferences must necessarily be weak. This is the core justification for the analyses presented here and in Chapter 7, which mostly focus on the dynamics of effectively isolated populations. This does not mean that I think that metapopulations are not important, quite the contrary. It means that being able to ask *general* questions concerning the relationship between population dynamics and ecological versatility cannot be undertaken within the framework of the metapopulation.

Ecological modelling

I hope and believe that it is fair to paraphrase Hall's (1988, 1991) criticisms of the utility of 'theoretical ecology' under the following points.

(1) An envy of the physical sciences has led to an almost indiscriminant adoption of many of the modelling and analytical methods that have proved to be successful there. In a sense, the logical pathway of much of ecological theory has been: physical theory and associated methods → ecological theory → nature (perhaps), rather than nature → ecological theory.

(2) Closely related to (1), there has been a focus on the rigour of the solutions of equations that purport to represent ecological function rather than a searching examination of whether the forms of the equations do reflect ecological function (i.e., mathematical versus scientific rigour). Do mathematical solutions necessarily imply viable ecological solutions? Should ecological systems even be expected to conform with the implications of the behaviour of the models?

(3) Ecology as a discipline lacks the discriminatory tools to identify the theories or models that ultimately will be useful or accurate predictors of ecological behaviour from those that are merely intricate or interesting theoretically.

In essence, these criticisms amount to whether the theories have been sufficiently well-framed to be potentially falsifiable with respect to natural populations, and whether theoretical ecology has developed its own culture in which the models have gained a relevance of their own with little *need* for a relevance to nature. Having read some of the theoretical literature with a deliberate focus on the way in which assumptions flow from paper to paper, it seems inescapable that there is a self-serving, internal synergism, which supports Hall's contention that at least some parts of theoretical ecology only peripherally 'need' nature for their justification. Certainly, there are few data that unequivocally support the most modest of models (e.g., simple logistic growth) or the fundamental assumptions of those and derived models (Hall 1988). Few of Hall's criticisms seem unreasonable to me.

Why then do I devote the space and effort to modelling in this book if these criticisms are so valid? First, few workers who are interested in trying to come to grips with ecological complexity would eschew the value of modelling in any form. Another vocal critic of ecological theory as it stands, R. H. Peters, argued for an alternative methodology to ecological modelling rather than for no modelling at all (Peters 1991). Indeed, even Hall (1988, 1991) was more concerned with the existing form and paradigms of theoretical ecology than with the potential value of modelling as such. He advocates a broader approach, with a greater emphasis on the abiotic impacts on ecological behaviour and a theoretical development derived *for* ecology (not a transplanted methodology) with well-tested fundamental assumptions.

Modelling undoubtedly is a useful avenue for deriving insights and generating new predictions if it is anchored in ecologically meaningful terms. I have endeavoured to maintain ecological relevance in the algorithms developed here. Even so, one encounters problems such as just discussed in relation to metapopulations. I imposed the generally unrealistic condition that populations are isolated not because it is desirable, but because one can make so few comments on metapopulations consisting of arbitrary numbers of populations organized in arbitrary geometries, with arbitrary migratory and dispersal fluxes and arbitrary patterns of resource variation.

I see the value of the models developed here more in terms of their being alternatives to existing models of population dynamics and the relationship between dynamics and ecological versatility. The diversity of dynamics displayed by **Ch**, **Cs**, **Rh**, and **Rs** strategists in response to the same environment on the basis of few distinctions is evidence that one could expect only limited agreement in the field between models like the logistic and natural populations. One particular example is that of the carrying capacity, which is one of two critical parameters in the logistic growth model (Roughgarden 1979: 303). The carrying capacity implies some fixed amount of resource but if resource availability fluctuates, as is evident in most ecological systems, then the carrying capacity is variable for a population. Logistic behaviour is not to be expected for **Ch** or **Rh** populations in a varying environment. Conversely, **Cs** populations may show almost classical logistic behaviour even in the variable environment because individuals can vary their utilization rates (Fig. 6.6 d). But the degree to which variation in resource availability can be masked by this effect is controlled by the magnitude of the difference between the maximum and marginal utilization rates (Fig. 6.7).

In conclusion then, if we take into account these simple but biologically reasonable characteristics for individual consumers, then whether or not one expects to find logistic or any other particular form of population dynamics depends upon the characteristics of the organisms involved. There seems to be good evidence that organisms can vary their utilization rates, but the amount of variation, the 'compressibility' of the utilization rate, will vary dramatically between organisms. The sorts of numerical responses that are possible (e.g., immigration, death, abstention) also are specific to organisms and in large measure determine the nature of population dynamics. I believe that the fundamental assumption of my models – that utilization pressure tends to increase exponentially – is not unreasonable, but it needs to be evaluated for field populations. There is an urgent need for many more field measurcments of utilization pressure and utilization rates in relation to resource availability.

6.6 Summary

The relationship between population dynamics and the versatility of resource use is a crucial one, but one that has been difficult to characterize. This may have been due to the construction of models that are too bound by the demands for analytical tractability, and so, fail to represent the flexibilities that organisms are known to exhibit. Thus, in this chapter the influence of different patterns of versatility on population dynamics is explored by using computer simulations that are *not* underpinned by phenomenological, mathematical descriptions such as the logistic model. This amounts to allowing utilization rates to vary, dividing the populations into arbitrary numbers of ecologically differentiated phenotypes and permitting various degrees of specialization. These additional facilities are constrained in certain ways to limit the degrees of freedom. The single biggest limitation is that the models are restricted to a population-level analysis, notwithstanding the attractiveness of metapopulation-level modelling. The reasons for this restriction are explained in §6.5.

The models presented here are based on the concept of utilization pressure. This simply is the product of population density and the mean utilization rates of resources. I assume that utilization pressure grows exponentially when resources are unlimited. I also assume that utilization rates have an upper limit, the maximum utilization rate, beyond which greater resource availability is of no consequence. A variety of biological factors should enforce such an upper limit. These assumptions together imply that densities usually grow exponentially when resources are unlimited because individuals cannot use resources faster than at the maximum rate. In this respect, the models do not differ greatly from existing ideas such as the logistic model.

But what happens when resources become limited? There are two possible responses: a numerical response involving decreases in density and a functional response, in which utilization rates decline. The former is the conventional solution in most models of population dynamics. Although there has been some work done on the effects of functional responses on ecological dynamics (e.g., Holling 1965, Oaten and Murdoch 1975, Abrams 1980b, 1987a, 1989, Abrams and Allison 1982), the current analysis makes the interplay between functional and numerical responses the crux of the relationship between ecological versatility and population dynamics.

The assumption that at least some organisms can vary their utilization rates – show functional responses – has profound implications for population dynamics. Under resource limitation, when resource demands must be contained, much of the 'excess' utilization pressure can be alleviated by a reduction in the utilization rate. One of the key aspects of the models is that there also is a *minimum* rate, the marginal utilization rate below which an individual cannot sustain itself. Together, these ideas imply that densities may be stabilized at the expense of utilization rates up to a point after which densities must also decline.

I distinguish between hard and soft exploitation. In hard exploitation, individuals have an obligatory utilization rate so that their populations can exhibit numerical responses only. Soft exploitation allows both numerical and functional responses, which leads to rather more complex dynamics.

This dichotomy is linked to the strategies of ecological versatility introduced in §5.5, namely, undifferentiated populations of generalists (coherent strategists), differentiated populations of generalists (resource-like strategists), and undifferentiated populations of specialists (specialized strategists). The dynamics of populations showing distinct types of ecological versatility then are considered in relation to their having flexible or inflexible utilization rates. Thus, there are 'hard coherent' strategists (designated **Ch**) and 'soft coherent' strategists (**Cs**), with the corresponding strategies for resource-like (**Rh** and **Rs**) and specialized (**Sh** and **Ss**) populations respectively.

In this framework and given the limitations noted in the first paragraph of this summary section, the following conclusions were drawn. First, changes in the density of populations of hard strategists are tied to the levels of resource availability. For obligatory generalists such as **Ch** and **Rh**, densities reflect the fluctuations in the total availability of resources. For the obligatorily specialized strategy **Sh**, that relationship is less clear because the utilization rates are resource-specific.

Second, soft strategists (**Cs**, **Rs**, **Ss**) display more convoluted patterns than hard strategists because they exhibit both numerical and functional responses. There are frequent long periods during which densities do not change because the overutilization of resources is addressed by reducing utilization rates. These constant-density periods often precede sudden collapses in density because those high densities had been maintained at marginal utilization rates, and further declines in resource availability necessarily trigger mortality or emigration because utilization rates cannot be reduced any further (e.g., Fig. 6.2 d). The existence of flexible utilization rates implies that densities need not be expected to be closely related to resource availabilities (cf. Wiens 1984). Moreover, a soft strategist (e.g., **Cs**) may be able to colonize a new environment in greater densities than the homologous hard strategist (**Ch**) and persist there for long periods with depressed utilization rates. This observation may mean that measured immigration and extinction rates that are crucial to some models in biogeography (and elsewhere) may not accurately portray the critical rates, which really should include only self-sustaining populations.

Third, resource-like populations (generalized populations consisting of ecologically differentiated phenotypes) are expected to be **Rh** but not

Rs strategists. The nature of soft exploitation makes phenotypes vulnerable to extinction through intraspecific competition, so that, eventually, just a single, generalized phenotype persists. Thus, **Rs** populations should become transformed quickly into **Cs** populations. Similarly, very diverse **Rh** populations are not expected to occur for the same reasons – phenotypes are outcompeted by others that expand their use of resources once (or while) other phenotypes decline. However, it seems that several phenotypes might coexist for long periods in **Rh** populations even in strongly fluctuating resource environments (Fig. 6.11).

And last, soft specialized strategists (**Ss**) may exhibit an equalization of utilization rates for different resources even when there is a pronounced rank-order of maximum utilization rates for the resources (i.e., distinct preferences). Even when preferred resources increase in availability, these preferences may continue to be suppressed by the utilization pressure from within the population. This effect may contribute to the common observation of inconsistencies between the preference order for resources shown in laboratory experiments and those found by using field observations or experiments (e.g., Soberón 1986, Jaenike 1990, Crist and MacMahon 1992, Norbury and Sanson 1992). Individuals from field populations select resources from a natural and variable array and in the context of the relative demands for each resource from within the population. In general, these factors are not implemented appropriately in the laboratory so that such inconsistencies are to be expected.

7 · *Versatility and interspecific competition*

We can now move on to consider the relationship between interspecific competition and ecological versatility, after a rather lengthy postponement stretching back to Chapter 4. Having established a modelling framework suitable for investigating exploitation competition in the previous chapter, I now wish to explore whether the existing ideas on the effects of different species on each other's resource use (i.e., interspecific competition) provide much in the way of general prediction. In particular, I ask: to what extent do the more complex models of exploitation strategies and tactics developed here, and the more meaningful representation of the variability in resource availability, affect the expectations developed by community theorists over the past three or four decades?

Competition between species has been regarded as a major if not the predominant factor in controlling the range and diversity of resources used by populations. Clearly, if interspecific competition were such a pervasive influence on resource use, then much of the observed variation in ecological versatility must arise from the impact of this process. That interspecific competition is expected to affect the breadth of resource use is reflected by some of the most emotive jargon of ecology, such as 'competitive release', 'the compression hypothesis', 'resource partitioning [or segregation or differentiation]', and 'competitive exclusion'. The last term, competitive exclusion, seems to reflect an intuitive belief that species that are too similar to one another in ecological terms cannot coexist for long periods (Chesson 1991).

The possibility that interspecific competitive dynamics and resource use might be closely coupled was seized upon by theoretical ecologists in the late 1950s, and work in this vein continues to the present day (e.g., Abrams 1988b, Taylor 1988a, 1988b, Kishimoto 1990, Derrick and Metzgar 1991, Danielson and Stenseth 1992, Solé *et al.* 1992, Law and Blackford 1992, Law and Morton 1993). There can be little doubt that such a coupling, if substantiated, would have tremendous importance in advancing our understanding and predictive capability.

However, over the past 15 to 20 years, the idea that interspecific competition has such a significant effect on the patterns of resource use and hence on versatility has become less generally accepted. Several influential papers in the late 1970s and early 1980s (e.g., Peters 1976, Wiens 1977, Connell 1978, 1980) contested whether interspecific competition indeed leads to particular or specific patterns of resource use. Some parasitologists, for example Rohde (1979), refuted the idea that interspecific competition was the key agent in controlling parasite host range, while many botanists have expressed similar reservations for plants (e.g., Taylor and Aarssen 1990).

These observations lead me to raise a very important distinction – the difference between the *actuality* of the process of interspecific competition and the postulated *manifestations* of that process (or, equivalently, the *process* versus *pattern* dichotomy, Miller 1967, Wiens 1984). The actuality of interspecific competition refers to whether syntopic, synchronously active populations of different species adversely affect each other's population dynamics (density, age-structure) or affect aspects of the fitness of individuals (e.g., body mass, energy and nutrient stores or condition, fecundity, growth rate, mortality probability) (*depressive competition*, *sensu* Mac Nally 1983a: 1648–1649). The only convincing way in which the actuality of interspecific competition can be demonstrated is by using field experiments, notwithstanding the many limitations of this approach (Underwood and Denley 1984, Hairston 1985, Yodzis 1988, Tokeshi 1993). On balance, most general reviews of ecological field experiments indicate that interspecific competition frequently occurs in natural assemblages (Connell 1983, Schoener 1983, Hairston 1985, Underwood 1986, Gurevitch *et al.* 1992).

On the other hand, the postulated manifestations of interspecific competition (*interspecific interaction*, *sensu* Mac Nally 1983a: 1648) generally fall into two main classes: (1) direct resource segregation or partitioning; and (2) indirect resource segregation via morphological differentiation. In the former, morphologically undifferentiated species are thought to be able to coexist because they partition or segregate resources (Schoener 1974a, Toft 1985). MacArthur's (1958) work on the foraging microhabitat partitioning of migrant parulid wood warblers in North America is a famous example, but there is no shortage of other cases claiming to show a similar effect (e.g., Recher 1989).

Indirect resource differentiation through morphology is problematic because there is the additional inference that some degree of morphological differentiation corresponds to a certain level of resource differentiation. This probably is a rather weak inference and often will be violated. In any event, morphological differentiation due to interspecific competition has been held to be expressed sometimes as character displacement in which syntopic populations diverge in morphology compared with allotopic populations of the same species (e.g., Grant 1975).

Alternatively, morphological differentiation may be manifested as *Hutchinsonian series*, which are geometric series of species with the size of a morphological variable in the largest species being G times as large as in the second-largest species, which in turn is G times larger than the third-largest species, and so on. Typically, a value of $G \approx 2$ for a volumetric variable (body mass) or about 1.3 for linear variables such as jaw length, gape, culmen length or depth, has been advocated as being sufficient to permit coexistence (Hutchinson 1959, Horn and May 1977), at least under some circumstances (e.g., Pearson and Mury 1979). There have been many attempts to demonstrate or refute the occurrence of Hutchinsonian series in natural assemblages but when within-species variation in characters is taken into account, there seems to be little statistical support for the idea (Mac Nally 1988a). There are several reasons why Hutchinsonian series are not likely to be common in nature including the impact of environmental variation, a poor correspondence between morphological differences and resource segregation, other ways of 'partitioning' (e.g., diurnal activity) and whether the character that is selected for use in an analysis is an appropriate one (Simberloff and Boeklen 1981, Mac Nally 1988a).

As this book focuses on patterns of resource use, I will not comment much on whether morphological patterns are influenced significantly by the action of interspecific competition. This approach seems reasonable in any event because of the problems associated with the mapping of morphological differences onto differences in resource use that have just been discussed.

This chapter is divided into two main parts: (1) an overview of how theories of interspecific competition have been used to derive expected patterns of versatility in nature; and (2) an analysis of many of the predictions of existing competition theory in the light of the results of modelling based on the exploitation strategies defined in Chapter 6. Therefore, part (1) (§7.1) consists of short discussions of stability analysis and invasibility, the sorts of models used in the literature, and some of the theoretical results. This review is not meant to be exhaustive but should indicate the essence of the approach. Part (2) involves some comments on modelling issues (§7.2), and then analyses of persistence and invasibility, the effects of different exploitation characteristics on dynamics and interspecific interactions, environmental variation and processes in communities consisting of many populations (§7.3–§7.5). In §7.6, I discuss some of the more important findings of the analyses.

7.1 Versatility and interspecific competition – a short review

Equilibria, invasibility, and stability

One of the main methods by which interspecific competition and patterns of resource use have been linked is by using coevolutionary arguments. These rely on two main criteria for deriving the expected patterns of resource use: (1) *asymptotic, equilibrial stability* (Levins 1979, Law and Blackford 1992); and (2) *invasibility* (MacArthur and Wilson 1967, Roughgarden 1974b, Case 1990). In the former, patterns of resource use are sought that allow model populations to coexist at essentially fixed densities. On the other hand, invasibility is used to find particular combinations of populations whose joint resource use prevents the establishment of immigrant populations. In both cases the solutions are thought to be ones that should characterize natural communities because of the 'stability' aspect. Stability generally is gauged by reference to system behaviour following perturbations, which in many models amounts to analysing either the eigenvalues of the 'competition matrix' or the derivatives of the dynamic equations in the vicinity of equilibria.

The key assumption in this approach is that these equilibrial configurations will eventually be reached and that they should characterize existing natural systems (e.g., Nisbet *et al.* 1978, Pacala and Roughgarden 1982). It is assumed that sufficient time has elapsed for the systems to reach equilibrium (cf. Hubbell and Foster 1986). Whether natural systems are still in the process of moving toward these configurations, or whether they ever reach them, generally cannot be considered.

The features of populations (including resource use) associated with stable equilibrium solutions are assumed to maximize persistence and resilience (i.e., short return times to equilibria following disturbance, Pimm 1984). This idea usually is referred to as *feasibility*. Thus, stable organizations are thought to be feasible ones whether the organization is a characteristic of a population or of a community (Pimm and Lawton 1978, Briand and Cohen 1987).

Invasibility also employs coevolutionary criteria to identify particular sets of competitor populations that are 'resistant' to invasion by immigrants of other species. The main way in which invasibility is determined is based on the capacity of invading populations to increase when rare (i.e., from one or a few asexual propagules or a pregnant sexual female) (Armstrong and McGehee 1980, Turelli 1981, Hastings 1986). Configurations that might inhibit the successful establishment of immigrant populations are deemed to be more likely to persist for long periods, and so be more likely to characterize contemporary communities. Community assembly models (e.g., Drake 1990, Law and Blackford 1992) effectively are variants on the invasibility theme (see Hall and Raffaelli 1993: 226 ff.).

Many authors consider populations (or coevolved sets of competitors) failing to utilize fully available resources to be vulnerable to invasion (e.g., Case 1980, Brew 1982). Coevolution has been thought to lead to resource utilization rates that are proportional to the respective availabilities of resources (MacArthur 1969). This reflects the idea that the coevolution of competitors should lead to the minimization of the difference between the utilization and the availability of resources (MacArthur 1970, Matessi and Jayakar 1981, Gatto 1990, although see May 1974a: 33). Combinations of species that minimize this difference should be less susceptible to invasion because immigrant populations would find little free resources to sustain them.

Although superficially appealing, both stability and invasibility analysis have serious flaws (Hall 1991). Most stability analyses look for stable equilibria in model systems without recognizing the potential for long-term or short-term changes in environments, particularly in the availability of resources (e.g., Pimm and Lawton 1978, Ginzburg *et al.* 1988). The assumptions used in many stability analyses are overly restrictive, and apply only near to the equilibria (León and Tumpson 1975). It is now known that the coexistence of competitors in some mathematical models can occur if densities are in dynamic rather than static equilibrium (e.g., limit cycles, Armstrong and McGehee 1980). The implications for the versatility of resource use in such models are not clear. The mathematical concept of asymptotic convergence to stable configurations seems at odds with the ecological reality of ever-changing, patchy environments and substantial population fluxes (DeAngelis and Waterhouse 1987, Danielson 1991, Palmer 1992, Mac

Nally 1995). Moreover, resource availability, so often treated as a constant in models (either explicitly or implicitly) fluctuates markedly in nature at many time scales (i.e., daily, weekly, monthly, seasonally, and between years; see Levins 1979, Seastadt and Knapp 1993). Several models have shown the sensitivity of coexistence to the details of the time-course of resource fluctuations (e.g., Ebenhöh 1988). Time-courses clearly cannot be addressed by using conventional analyses that look only for asymptotically stable solutions.

Invasibility analysis shares many of these problems but also relies on the weak condition that invading populations must be able to penetrate existing systems from initially low densities (e.g., Case 1990). The possibility of high-density influxes is rarely considered, yet it may be an important ecological process in many mobile organisms such as birds, larger mammals, fishes, insects and marine invertebrates.

It is worth noting that many ecologists believe that equilibria do not characterize all or even most natural communities in any event (e.g., Andrewartha and Birch 1954, Wiens 1977, 1984, Connell 1978, Howe 1984). The rate of the process of 'exclusion' might be so slow (relative to human observation) that there may be no coherent manifestations of the process (e.g., character differentiation, niche segregation; see Hubbell and Foster 1986). Systems often may be prevented from reaching the point at which resources are fully utilized by forces unrelated to the competitive process. These include severe or catastrophic abiotic disturbances such as storms and various biological mechanisms such as predation, parasitism and disease (MacArthur 1970, Glasser 1983, Pimm and Redfearn 1988). Fluctuations in resource availability and concomitant switches in competitive dominance also may prevent equilibria from being realized (Abrams 1984, Rice and Menke 1985, Taylor and Aarssen 1990). The mobility of organisms within patchy landscapes involving frequent large- and small-scale migration and dispersal also should act to reduce the likelihood that equilibria will occur (DeAngelis and Waterhouse 1987, Kareiva 1990, Opdam 1991, Danielson 1991, 1992).

Some theorists have retreated from extreme positions on stability (e.g., Armstrong and McGehee 1980). Hastings (1988) conceded that the use of stability criteria now appears to be inappropriate and that frequent extinction and recolonization events are characteristic of most natural systems. Getz and Kaitala (1989) doubted whether the use of equilibrium and stability analyses for assessing the coexistence of groups of individuals (both conspecific and heterospecific) is justified. Therefore, the use of stability and invasibility criteria to predict patterns of resource use between local competitors seems increasingly unwarranted.

The dissatisfaction with stability/invasibility restrictions has led to the development of alternative approaches in recent years (Hall and Raffaelli 1993). One such possible alternative is the concept of *permanence* (Jansen 1987), which involves looking for community dynamics in which the population densities of all species remain positive (and finite) even though these dynamics may be chaotic or cyclical (i.e., non-equilibrial, Law and Blackford 1992). Whether permanence criteria say much about how resources might be partitioned among sets of competitors (and hence, versatility) has yet to be addressed in detail.

Models of interspecific competition

Interactions between syntopic populations, the vulnerability of resident populations to invasion by strategists of different kinds and the conditions for coexistence of populations have been woven together over the past 30 to 40 years to provide a picture of the patterns of resource use that might be expected to occur in natural communities governed by interspecific competition.

Underlying most of this work are assumptions concerning population and interactive dynamics of competitive situations. There have been disputes about the applicability of some of the main models (Mueller and Ayala 1981, Hall 1991, Peters 1991). Most theorists have opted for dynamics based on Lotka–Volterra models, which are referred to here as LVMs (e.g., MacArthur and Levins 1967, Gilpin and Case 1976, Lawlor and Maynard Smith 1976, Matessi and Jayakar 1981, Rummel and Roughgarden 1985, Hastings 1986, Chesson 1990, to name very few). Many variants or generalizations of LVMs have appeared (Roughgarden 1972, Milligan 1986, Taylor 1988a). May (1973) and Armstrong and McGehee (1980) provided general overviews of the properties of LVMs. The dominance of this way of representing community dynamics in theoretical terms led Hall and Raffaelli (1993) to refer to LVMs as the 'industry standard'. In many workers' opinions, it is doubtful that they deserve this accolade.

Although dominating much of the theory of interspecific competition, LVMs are by no means the only models used (e.g., Schoener 1976, Nunney 1980, Brew 1984, Vance 1985). Alternative theoretical bases have become more common in recent times (e.g., Taylor and Jonker 1978, Getz 1984, Sitaram and Varma 1984, Shipley 1987, Geritz *et al.*

1988, Hatfield and Chesson 1989, Cushing 1992). Turelli (1981) and others have stressed the importance of developing models of interspecific competition that are not so intimately connected with LVM dynamics, particularly as classical LVMs are not coupled tightly to underlying resources (León and Tumpson 1975, Cushing 1992) or the mechanisms involved in interactions (Vance 1985). Vance believed that LVMs are superficial descriptions of nature, while Abrams, a long-term advocate of explicit consumer-resource modelling, stated that LVMs generally are poor representations of the dynamics of competing species (Abrams 1987b). Several developments have avoided LVMs altogether (see §7.5). So it does seem surprising that recent theories continue to use them despite an overall increase in the sophistication of models (e.g., Palmer 1992, Abrams 1993). The mathematical convenience of LVMs evidently overrides their ecological relevance (see Hofbauer and Sigmund 1988: 66, Law and Blackford 1992: 575).

Versatility and interspecific competition

Perhaps the most influential of the early theoretical papers addressing the relationship between interspecific competition and resource utilization has been MacArthur and Levins (1967). This paper has provided the inspiration for many others since its publication. However, its unfortunate legacy has been an almost indiscriminate adoption of many of its restrictions, analytical short-cuts and assumptions. Some of these include: (1) the explicit or implicit assumption of constant resource availability; (2) gauging stability by using equilibrium analysis and invasibility; (3) the use of uniform niche breadths; (4) unidimensional arrays of resources ('coenoclines'); and (5) Gaussian utilization functions.

Roughgarden (1989) and Chesson (1990) pointed out that some of the assumptions of the MacArthur and Levins (1967) paper inadvertently have become identified as or mistaken for model predictions. The frequently cited expectation that local arrays of species should consist of populations with relatively similar niche breadths (i.e., largely nonoverlapping specialists) actually was a simplifying assumption rather than a prediction as such.

The empirical evidence for this form of coenocline is limited. Rosenzweig (1991) called such distributions *distinct preferences* because this assumption imposes specialities along gradients for each species (Fig. 7.1 a). Thus, in much of competition theory the emergence of resource partitioning is hardly surprising. Rosenzweig (1991) considered a differ-

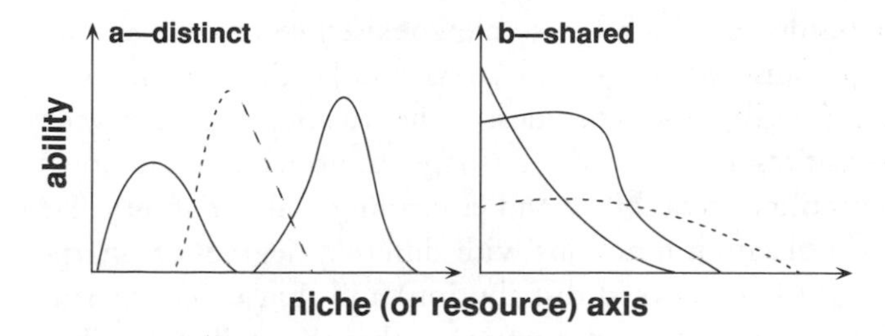

Fig. 7.1 The distinction between distinct (a) and shared (b) preferences of syntopic populations along a niche or resource axis (after Rosenzweig 1991).

ent form of response to gradients to be better supported by field measurements, namely, *shared preferences* (Fig. 7.1 b). Shared preferences encompass considerable resource overlap along a coenocline, although the shapes of the responses of populations may differ substantially. Such observations are commonplace along many gradients such as food size. The dietary range of the larger members of assemblages almost always subsumes that of the smaller members, although the right tails of the food-size distributions of larger animals are extended (e.g., Krzysik 1979; see Wilson 1975). Nevertheless, some theorists continue to reject uncritically the possibility that shared preference gradients might characterize many communities (see the discussion of the overlapping-niche model by Tokeshi 1993: 139).

May and MacArthur (1972) and May (1973) extended the MacArthur–Levins theory to incorporate stochasticity. Their analyses of uniformly sized niche breadths suggested that stable conditions might occur when the ratio of niche separation to niche breadth exceeded a certain amount. The effect of stochastic variation was to require an even greater separation of niche positions along the resource axis, with the increases reflecting the size of the fluctuations. The theory implies that arrays of comparatively specialized populations must use more-or-less distinct parts of a (unidimensional) resource axis, with separations being governed by the amount of environmental stochasticity.

However, by relaxing the constraints on the shapes of niches, especially of niche breadths, McMurtrie (1976) generated a different expectation altogether. His results suggested that the 'packing' of populations could be much tighter if niche breadths differed substantially between competitors. Thus, some sets of generalized and specialized populations in which the resources used by specialists were subsets of those used by generalists were found to be 'stable'. Others have derived similar results in which the stable coexistence of overlapping generalists and specialists occurs (e.g., Matessi and Jayakar 1981).

Roughgarden (1974b) also considered the relationships between different patterns of resource utilization (shapes of utilization functions) and packing, versatility, invasibility and environmental variability. He used a variety of utilization functions with different degrees of 'sharpness' – thick-tailed (sharp-peaked distribution) and thin-tailed (broad-peaked). The importance of this distinction is that thin-tailed distributions imply less reliance on resources from parts of the resource spectrum distant from the mean for a population, while thick-tailed distributions imply a greater dependence on those resources. Roughgarden (1974b) showed that thick-tailed distributions permit closer packing than thin-tailed ones and that the former would be penetrated more easily by immigrant populations. He identified thick-tailed functions with generalized or opportunistic strategies and thin-tailed functions with specialists. He also maintained that if environments were unpredictable any advantages in handling or processing efficiency that specialists might enjoy over generalists would be diminished. Thus, thick-tailed functions (generalists) should characterize unpredictable environments and, consequently, communities in these environments should both be more susceptible to invasion and exhibit less differentiation and higher overlaps.

The MacArthur–Levins approach, and all others deriving expected patterns of resource use by using dynamic models (e.g., limiting similarity theory), *depend* upon an explicit equivalence between resource overlap and competition coefficients (Roughgarden 1974b, J. M. Emlen 1981, Case 1984, Loman 1986; but see Brew 1984: 275). If this relationship is not tightly coupled, as seems likely in many cases, then the derived expectations are hard to justify. Although this has been recognized for many years, this critical assumption has been glossed over continuously (Abrams 1980b, 1980c, Mac Nally 1983a).

The foregoing ideas essentially are static expectations of the relationship between interspecific competition and patterns of versatility. Glasser and Price (1982) presented a more dynamic picture of exploitation competition – one avoiding asymptotic solutions and invasibility. Their method was based on the facultative use of resources and employed logistic models of several species using many resources. The theory has been discussed already in Chapter 5 although several issues are relevant to the relationship between versatility and interspecific competition. Limitations are important for circumscribing evolutionary and ecological theory. This prompted Glasser and Price (1982: 443) to impose the limitation that the 'total efficiency' of resource use be fixed. Thus, any increase in the efficiency of utilizing one resource must be offset by reduced efficiencies on one or more other resources. They also introduced the notion of the *malleability* of characteristics affecting the efficiencies of resource utilization. Thus, characteristics such as behaviour were deemed to be more malleable than others such as morphology. Their most important result was that exploitation strategies, resource requirements, and the efficiencies of utilization combine to affect the versatilities of populations (Glasser and Price 1982: 457). The patterns of resource use evident at any time and place then determine resource overlaps, which in turn determine the severity of interspecific competition. This reasoning means that *exploitation strategies determine the intensity of interspecific competition* rather than the reverse. I believe that this is a very useful picture and one that most probably represents the way in which most ecological communities function.

Much of the theory discussed to this point appears to be unsatisfactory to many plant ecologists, who believe that niche differentiation as an explanation of species coexistence is inadequate for plants (Agren and Fagerstrom 1984, Silvertown and Law 1987, Mahdi *et al.* 1989, Noble 1989, Taylor and Aarssen 1990, Zobel 1992; although see Grace and Tilman 1990, Gordon and Rice 1992).

There are several reasons why plant ecologists think that niche differentiation may be unlikely in plants. First, Aarssen (1983) proposed that competitiveness is a function of many attributes and genetic constraints. The interaction of these factors precludes any one genotype from becoming competitively dominant in a community, leading to complex systems of competitive intransitivity between species (i.e., situations in which A is superior to B and B is superior to C but C is superior to A; Aarssen 1989). In short, some genotypes of one species are competitively dominant to genotypes of another species but other genotypes of the latter are dominant to some genotypes of the first species. Thus, neither species can gain an absolute advantage. While the local exclusion of some genotypes might occur through competition, the exclusion of an entire population is unlikely. The theory also holds that the densities of genotypes change continuously, altering the immediate competitive environment for all genotypes at short time-scales. This effect is called *competitive combining ability* (Aarssen 1983) and is thought to become more pronounced when there are more species involved. There appears to be some experimental support for this idea

(Taylor and Aarssen 1990). This hypothesis would not appear to impose any pressure for niche restriction in plants. Indeed, the competitive combining ability hypothesis implies resource-like exploitation – generalized populations consisting of specialized components, in this case, sets of genotypes concentrating on different parts of the resource spectrum.

Other plant ecologists see niche differentiation of adults as an incomplete explanation for coexistence and view differences in competitive ability through time as a key element (e.g., differences in regeneration niches, Grubb 1977). This picture would lead to similar patterns of resource use among potential competitors (Agren and Fagerstrom 1984), but dissimilar competitive abilities at various stages during regeneration (Noble 1989). There seems to be some evidence for this viewpoint. For example, Lamont and Bergl (1991) investigated morphology and water-use patterns in three co-dominant species of *Banksia* in sand dune habitats in Western Australia. They found little evidence for any differentiation in either the above- or below-ground niches of the species (see Parrish and Bazzaz 1976). They attributed the apparently stable coexistence of the adults of the three species to interactions during regeneration. However, they saw these interactions in terms of the regulation of recruitment rather than as processes likely to lead to competitive exclusion. Such a process clearly would not lead to expectations of different degrees of ecological versatility arising from a competitive process, especially in the adults of terrestrial plants.

Overview

The literature on interspecific competition and its impact on patterns of resource use is enormous (Cherrett 1989, Law and Watkinson 1989, Schoener 1989a). The theory is dominated by the results of LVMs that, because they are largely phenomenological models, require many critical and often unsubstantiated assumptions to relate community dynamics to patterns of resource use (e.g., the direct correspondence of resource overlap and competition coefficients, Cushing 1992). Most results also are derived by using either asymptotic equilibrium analysis or invasibility criteria, neither of which is particularly robust or relevant to natural communities. The relaxation of constraints necessary for LVMs and mathematical tractability indicate that many of the better-known predictions of how interspecific competition determines versatility no longer should be viewed as compelling.

7.2 Modelling

Having briefly reviewed how competition theory has been used to predict the patterns of resource use, I now turn to the main subject of this chapter. This is an investigation of how competitive dynamics are affected by populations displaying different exploitation strategies and tactics. It appears that comparatively few studies have considered the interactions between populations exhibiting different exploitation strategies (e.g., Glasser and Price 1982, Roughgarden 1974b, Haefner and Edson 1984, Milligan 1986, Wilson and Turelli 1986). This paucity of results means that there remains much scope for investigating the relative 'competitiveness' of rival strategies and the extent to which various tactics determine the outcomes of interactions.

The modelling of interactions between populations with different exploitation strategies is extraordinarily complex. Some of the factors that might be considered at even a relatively basic level include:

- the form of exploitation strategy (e.g., coherent, resource-like);
- flexibility of utilization rates (i.e., soft versus hard);
- immigration lags;
- influx densities; and
- the number of resources and the nature of variation in resource availability.

These points suggest that hundreds or thousands of different scenarios may need to be explored for species *pairs* alone, each with many replicates or realizations. While simulations to cover this range are easy enough to generate automatically, the more difficult task is to distil and convey the pith of the results. Wilson and Turelli (1986) among others have lamented the difficulties involved in selecting parameter values for ecological models, particularly when few field data exist to guide that selection. They elected to restrict their attention to a set of apparently representative results rather than examine selected parameter values spanning the range of interest.

What other options are there? Optimal experimental design (e.g., Atkinson and Donev 1992) and sensitivity analysis (e.g., Kleijnen 1992) are two common methods. For example, Kleijnen chose to use various forms of least-squares and regression models to address strategic issues in simulation programs. Unfortunately, neither optimal design nor sensitivity analysis offers much when almost none of the parameters in models are fixed by observational or experimental information. Therefore, I look at some of the more interesting questions such as the effect of hard or soft strategies on interacting populations, the stability and/or coexistence of alternative strategies of resource exploitation and several ideas emanating from theoretical studies in the literature. Clearly, this is a piecemeal approach but one that nevertheless might shed a new light on existing ideas.

Parameter-space spans and the glyph representation

The results of simulations of interacting populations are known to be very sensitive to the choice of parameter values. To assess the impact of parameter choice, one normally generates many realizations that more or less 'span' the parameter space (e.g., Schoener 1976, Wilson and Turelli 1986). Parameter-space spans often show quite different results in different regions of the parameter space (Pacala and Roughgarden 1982).

Many interactions are studied in this chapter by using parameter space spans. Thus, for instance, the utilization rates of interacting populations may be varied over some range, and the results of many realizations then determined. This produces a matrix of results, with each result representing the distribution of outcomes of the set of realizations corresponding to a given selection of parameter values. Typically, 500 realizations of each selection were used to generate the spans presented in this chapter.

Displaying the results of parameter-space spans can be even more of a problem than producing them, especially if the number of combinations of parameters is comparatively large. But which results need to be shown and in what way? The crucial results seem to be the probability of coexistence of populations and the likelihoods that a given type of population persists while the other becomes locally extinct. Each result in the parameter-space span therefore is multivalued, and a simple representation of the results is not possible. This is where *glyphs* are useful. Glyphs are graphical devices for showing multidimensional data in fewer dimensions (Crawford and Fall 1990).

An explanation of the form of glyph used in this chapter is given in Fig. 7.2 a. The glyph consists of a solid circle and an associated vector. The area of the circle (stippled in Fig. 7.2 a for clarity) shows the relative proportion of a set of realizations in which coexistence occurred. Thus, larger circles indicate higher probabilities of coexistence than smaller circles. If there were no instances of coexistence then the area of the circle

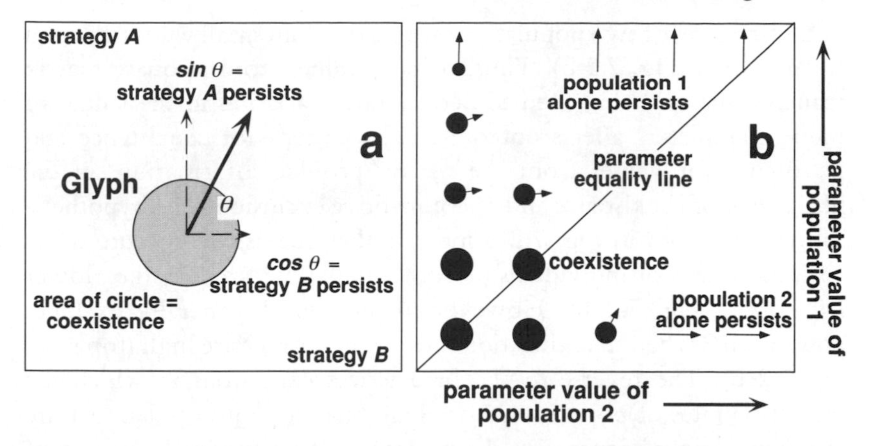

Fig. 7.2 (a) explanation of the glyph used throughout Chapter 7 to register the results of modelling interactions between heterospecific populations. The area of the stippled circle indicates the relative frequency of coexistence of the two populations, while the length and angle of the associated vector show the patterns of competitive exclusion. (b) an interpretation of different glyphs in an exemplary parameter-space span showing that coexistence is common when both populations have low values for the parameter being studied (e.g., maximum utilization rate; lower left corner) and the competitive dominance of the population with the substantially larger parameter value (lower right corner; upper row).

would be zero. The vector provides information on the frequency with which each strategy persisted when the alternative strategy did not (i.e., competitive exclusion). The length of the vector indicates the proportion of cases in which only one strategy persisted. If the strategies always coexist then the length of the vector would be zero. The angle of the vector, θ, indicates which of the strategies persist. Thus, a vector pointing vertically indicates that the strategy to which the arrow points (i.e., strategy *A* in Fig. 7.2 a) persisted alone; similarly, a horizontal vector means that strategy *B* in Fig. 7.2 persisted alone. Intermediate angles mean that one strategy persisted alone in some realizations and the other strategy persisted alone in other realizations. Thus, one might find a result in which, say, coexistence occurred in 40% of the realizations, strategy *A* persisted alone in 35%, and strategy *B* persisted alone in 25%. Such a result would yield a glyph resembling the one shown in Fig. 7.2 a.

Glyphs therefore allow the compact presentation of results of simulations performed with many combinations of parameter values. For example, one might want to explore the pattern of results when the utilization rates of two populations are varied from small values through to large ones (Fig. 7.2 b). Thus, pairs of values from a square matrix spanning the space are used to deduce how variation in the values of model parameters affects outcomes. The patterns of coexistence and persistence for values from the matrix provide information on the likelihoods of coexistence and of competitive exclusion. The hypothetical results shown in Fig. 7.2 b indicate that coexistence occurs when utilization rates of individuals of both populations are low (i.e., lower left corner, Fig. 7.2 b). However, population 1 generally excludes population 2 when the utilization rates of the former are high (top row, Fig. 7.2 b). The reverse result characterizes conditions in which the utilization rates of population 2 are high but those of population 1 are low (lower right corner, Fig. 7.2 b). Intermediate levels of coexistence and exclusion occur elsewhere in the parameter-space span.

It is worth emphasizing that glyphs in parameter-space spans can show only the relationships between two types of strategy and ranges of parameter values. The impact of different resource regimes or of a third (or more) strategy on the *target* pair can be shown by using multiple plots, one for each of the different regimes or parameter variations used for the *background* population(s).

Thus, the context for parameter-space spans is a pair of populations interacting in resource-based environments. The 'standard' conditions pit together evenly-matched populations in which the initial densities and utilization rates are equal. These standard conditions are referred to as standard values. For strategies consisting of differentiated populations (**Rh**, **Rs**) the initial densities were distributed evenly among phenotypes. Only three phenotypes were used for resource-like populations.

The effects of different combinations of maximum utilization rates (rate-spans), initial densities and lags are considered below. The parameter in question was varied over seven increments for each target population so that there are $7 \times 7 = 49$ glyphs represented in each parameter-space span.

For utilization-rate spans, the maximum utilization rate was varied between one-half and twice the standard rate (Table 7.1). This means that the maximum utilization rate of a target population may be as much as four times as great as that of the other target population in some realizations. The range of 0.5 to 2.0 was traversed in increments of 0.25 (Table 7.1).

Initial densities were varied by orders of two, from one to 64 times the standard initial density (Table 7.1). Note that although initial densities

Table 7.1. *Range of values for parameter-space spans. These values correspond to left-to-right and bottom-to-top progressions in the spans. Values are either multipliers (utilization rates, initial densities) or absolute values (immigration lags (in timesteps))*

Variable	Minimum	→	···	···	···	→	Maximum
Utilization rate	0.5	0.75	1.0	1.25	1.5	1.75	2.0
Initial density	1	2	4	8	16	32	64
Immigration lag	0	25	50	75	100	125	150

may have been specified at extremely high levels, the algorithms are structured to ensure that sustainable densities are achieved immediately (see §6.2).

Lags correspond to the influx of a target population at some time after the 'commencement' of a realization. Thus, a population might arrive at the beginning (no lag, which is the standard condition) or be delayed until timestep 150. The increment for lags was 25 timesteps (Table 7.1).

The majority of spans presented here are utilization-rate spans. Although several strategies are considered in one context or another, the interactions between the coherent strategists, **Cs** and **Ch**, are used as the standard example. This focus on **Cs** and **Ch** provides a thread of continuity while contrasting the patterns of hard and soft strategies.

Regrettably, this introduction to the modelling methodology has been rather long. Nevertheless, the scene now is set for investigating how some characteristics of resource use affect the competitive dynamics of populations. It is important to remember that the results presented in the rest of this chapter refer only to exploitation competition because interference competition has not been modelled (see Chapter 6).

7.3 Persistence and invasibility

Some factors influencing persistence and coexistence

Most of the ideas as to which combinations of species should coexist have been derived by using LVMs (e.g., May 1974b, Abrams 1975, Yoshiyama and Roughgarden 1977). Equilibrium analyses indicate the regions in parameter space in which the coexistence and exclusion of competitors occur (e.g., Roughgarden 1979: 415–416) or where invasion is possible or prohibited. The logistic nature of LVMs apparently enforces equilibrium solutions that are functions of carrying capacities (K_i) and competition coefficients (α_{ij}). These results have been used widely to predict the limiting similarity between competitors (e.g., MacArthur and Levins 1967, Roughgarden 1974b, Abrams 1983), which, of course, might affect patterns of versatility.

The degree to which the results of analyses based on LVMs are peculiar to these models should be questioned. I have already noted that carrying capacities exist in LVMs purely to enforce finite, logistic behaviour. The K_is almost always are assumed to be constant, but for any reasonable variation in resource availability such constancy seems unwarranted.

Competition coefficients, α_{ij}, generally are taken to represent the per capita depression of the densities of one species (*i* say) by individuals of another species (*j*) relative to the depression caused by individuals of species *i* on itself (Pomerantz and Gilpin 1979, Abrams 1980a). Symbolically

$$\alpha_{ij} = \left. \frac{\partial g_i / \partial N_j}{\partial g_i / \partial N_i} \right|_{\hat{N}_i, \hat{N}_j}, \qquad (7.1)$$

where g_i is the per capita growth rate of population *i*, N_i and N_j are the densities of populations *i* and *j* respectively, and the differential is to be evaluated at the equilibrium densities $\hat{N}_i$ and $\hat{N}_j$ (Abrams 1975: 363).

Competition therefore is viewed as a comparison of the relative intensities of interspecific and intraspecific competition and may be asymmetrical between competitor species. Note also that this form of competition coefficient addresses just one facet of fitness, population densities – other components are not included. This can lead to misleading inferences. For example, Addicott (1984: 439) commented that Hutchinson's (1978) interpretation of the results of experiments conducted by Wilbur (1972) led him to the view that several species of larval amphibians were effectively mutualists. This picture was based on analyses of densities alone, but when adverse effects on size and development rate were taken into account, the interactions probably were neutral or detrimental (see Cushing 1992 for a more complete approach to fitness-related quantities).

Several authors have proposed schemes by which the α_{ij} might be estimated by using matched densities of presumed competitors from a variety of locations (e.g., Hallett and Pimm 1979, J. M. Emlen 1981a, Chesson 1990). However, Taylor and Aarssen (1990) reported intense competition between populations of three perennial grasses (*Agropyron repens*, *Poa pratensis*, *Phleum pratense*) in which there were no significant correlations in density. They attributed this effect to intransitive competitive relationships between genotypes in each grass that prevented any one species from competitively excluding the others. Results like this one must erode confidence in expressions like (7.1) that accommodate neither all fitness components nor the potential for ecological differentiation within populations.

Competition coefficients comprise all possible competitive effects between populations, including interference as well as exploitation (Schoener 1983, Spiller 1986). For this reason they need not illuminate the mechanisms involved in competitive processes (Brew 1984, Vance 1985; see Ginzburg *et al.* 1988 for an illustration of how competition coefficients can become isolated from ecological mechanisms).

In principle, the α_{ij} should be continuously computable in the modelling used here by comparing the time-courses, timestep by timestep, of populations both alone and in combination with other populations in the same environment. The impact of the other population can be quantified by calculating the difference between the utilization pressures (i.e., density × utilization rate) exerted by each population when alone compared with the pressure exerted when in company with another population. The difference is the interspecific depression component (Mac Nally 1983a).

Some of these points can be demonstrated by looking at the depression caused by the interaction between populations of the soft coherent strategist, **Cs**, and of the hard coherent strategist, **Ch**, in the variable environment (Fig. 7.3). The reductions in the utilization pressures and densities can be calculated by comparing the values for each population when alone with corresponding values for interacting populations (Danielson 1992). If the efficiencies of conversion of resources into fitness are the same for both populations, then the depression of utilization pressure equates to the loss of fitness due to the effect of the other species.

As expected, both populations suffered substantial reductions in the utilization pressure that they exert when they were interacting compared with that exerted when they were isolated (Fig. 7.3 a). However, by timestep 350 this loss was largely overcome in the **Cs** population. The pressure exerted by **Ch** remained much reduced throughout the course

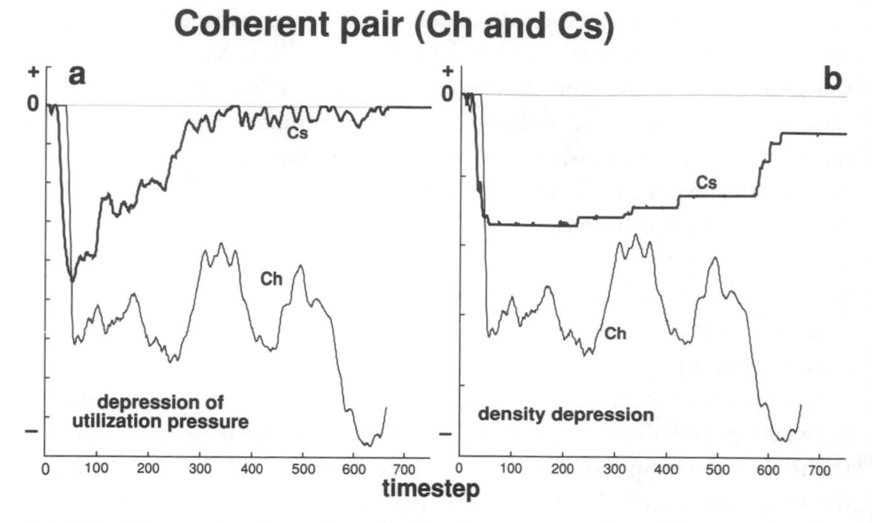

Fig. 7.3 The relative depression of utilization pressures (a) and densities (b) of interacting populations of **Cs** and **Ch** compared with isolated populations of each in the variable environment. **Ch** became locally extinct at about timestep 680.

of the interaction, ultimately falling to zero when that population became extinct (Fig. 7.3 a).

The importance of considering aspects of fitness in addition to density is evident by comparing the patterns for utilization pressure with those for density (Fig. 7.3 b). While the pattern for **Ch** is unaffected, given the inflexible utilization rates, the interaction resulted in a long-term depression of density (but not utilization pressure) in **Cs** compared with that for an isolated population (Fig. 7.3 b). Thus, in varying environments the impact of populations on one another must depend upon their respective exploitation strategies. The detail to which Abrams (1980a) went to show that competition coefficients in exploitation competition are unlikely to be constant seems unnecessary – they must vary substantially as both population densities and resource availabilities change (see Wilson 1992: 1986).

It is worth repeating that these remarks do not apply to interference competition, in which utilization pressure may not be maximized owing to some individuals excluding others from access to resources by agonistic or territorial behaviour (e.g., Schoener 1976, Nunney 1981, Maurer 1984). Vance (1985) noted that the effects of interference can depress population growth rates (reduced recruitment or overgrowth), diminish consumption rates (e.g., allelopathic effects), lessen encounter rates with prey, modify light regimes or by direct aggression reduce searching rates. These effects can lead to an underutilization of resources and hence to an inhibition that is not functionally related to population densities (see Schoener 1976, Vance 1985; see Chapter 6).

Note also that resource overlap, which always is close to unity between the obligatory generalists considered here (**Cs**, **Ch**), bears no functional relationship to the depression of fitness. These results mean that competition coefficients in natural communities cannot be estimated by using resource overlap but *must* rely on manipulation experiments (Mac Nally 1983a; see also Levins 1968, Schoener 1974b, Abrams 1975, Case 1984, Spiller 1986).

As we have seen in §7.1, many studies on limiting similarity and invasibility depend upon an exact correspondence between resource overlap and the intensity of interspecific competition (e.g., MacArthur and Levins 1967, May and MacArthur 1972, Roughgarden 1972, Abrams 1975, Turelli 1981). However, high levels of resource overlap may be associated with competition that is 'intense' *or* 'mild' – low overlap may reflect the results of intense competition or may be due to other factors entirely (Abrams 1980c). The results of the current modelling also show that there are unlikely to be any consistent relationships between resource overlap and the intensity of competition. It seems that theoretical expectations derived by using a direct correspondence between resource overlap and competitive intensity should be regarded cautiously.

These findings suggest that the predicted outcomes of interspecific interactions, most commonly derived by using LVMs and other logistic-based models, are likely to be specific to these models (e.g., MacArthur's 1969 minimization principle). The functions of carrying capacities and competition coefficients do not appear to yield meaningful expectations for the results of interactions. If they do not, which factors might determine the outcomes of exploitation competition?

The influence of utilization rates

The perception of coexistence and exclusion depends upon the length of time for which populations have interacted with each other and the stage of an interaction at which an observation is made (see Hubbell and Foster 1986). Because an ecological measurement is effectively a 'snap-shot' of a system at a particular time, the current coexistence of a pair of species may be illusory – one species ultimately may be destined for extinction given a sufficient length of time. This means that the term 'coexistence' in a simulation should be used in the sense that both model populations persist for the duration of the simulation. 'Coexistence' at 1000 timesteps may become strict exclusion by 2000 timesteps for some parameter values.

Populations of **Cs** and **Ch** are used to explore the dependence of outcomes on the duration for which an interaction is studied. For shorter durations, **Ch** could coexist with **Cs** provided that its maximum utilization rate were greater, or if the rate of **Cs** were not too high if the rates were equal (Fig. 7.4 a). There were low probabilities of coexistence even if the rate for **Cs** were greater than that for **Ch**, but not by too much. As the duration of realizations was extended to 2000 or more timesteps, coexistence was increasingly restricted to situations in which the maximum utilization rate of **Ch** was significantly higher than that of **Cs** (Fig. 7.4 b, c, d).

It is worth looking at one of these spans in a little detail so that the results of later analyses will be clearer. Consider Fig. 7.4 d, the span in which 10000 timesteps were run. For most combinations of utilization rates, **Cs** persisted and **Ch** became locally extinct. Such results characterized cases in which the maximum rates were equal (i.e., along the equality line). There was a low probability of coexistence when the maximum rates were equal, provided those rates were low (glyph in the bottom left corner of the span). The bottom row corresponds to the smallest maximum rates of **Cs**. Coexistence typifies this row – **Ch** could prevail if its rate were four times that of **Cs** (Fig. 7.4 d). As the maximum rate of **Cs** increased (i.e., rows higher in the span) the probability that **Ch** could persist decreased, even if its own rate were significantly higher than that of **Cs**. Note than none of the vectors is oblique, which indicates that no combination of maximum utilization rates (of those surveyed) for these strategies led to the persistence of only one strategy in some realizations and only the other strategy in other realizations.

There is an inverse relationship between utilization rate and density, which is a consequence of the definition of utilization pressure and the way in which excess pressure is reduced. Recall that the utilization pressure on resources depends upon both numerical and functional responses of populations, so that increases in utilization rate must be at the expense of densities and vice versa. This principle may not be particularly clear in soft strategists but is obvious in hard strategists such as **Ch**.

The density trajectories of two isolated **Ch** populations in the variable environment are shown in Fig. 7.5 a, b. Let the term 'alpha' designate a

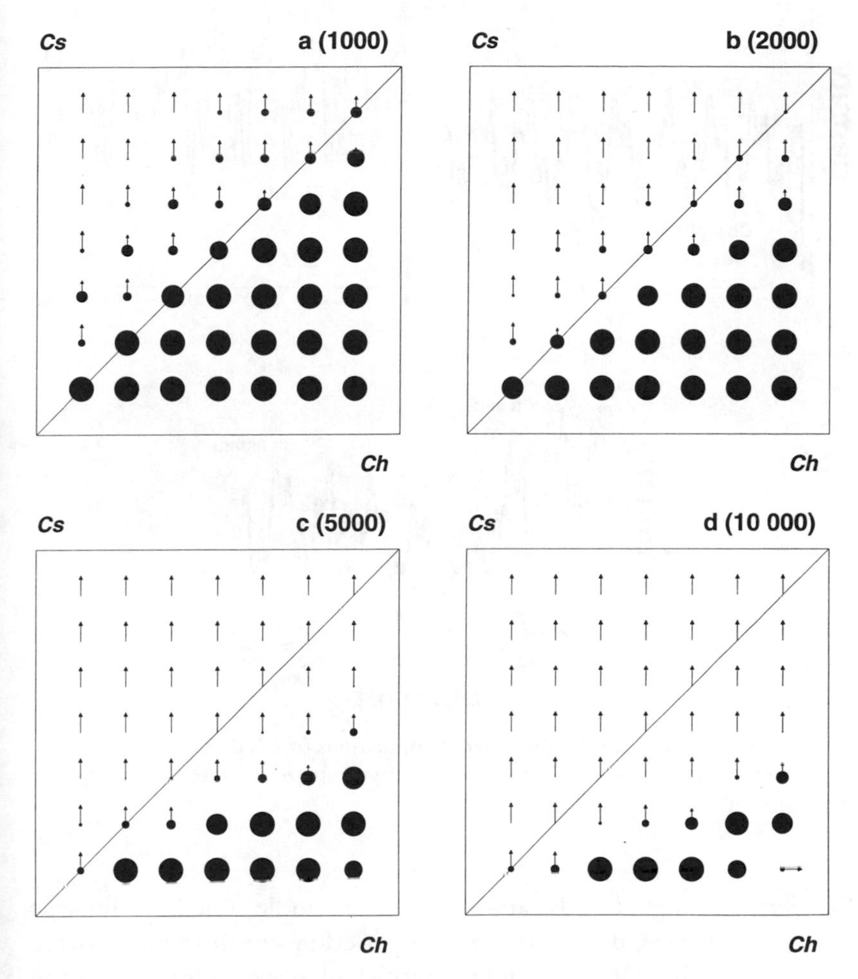

Fig. 7.4 Spans of utilization rates (rate spans) for interactions between populations of **Cs** and **Ch** in the variable environment: (a) 1000 timesteps; (b) 2000 timesteps; (c) 5000 timesteps; and (d) 10 000 timesteps.

population with the higher maximum utilization rate and 'beta' denote the population with the lower maximum rate. Then, although the amplitude of the density changes of the slower utilizer, **Ch beta**, was substantially higher, the mean density also was much greater than that of the faster user, **Ch alpha** (Fig. 7.5 a, b). The interaction between **Ch alpha** and **Ch beta** in the variable environment led to the exclusion of

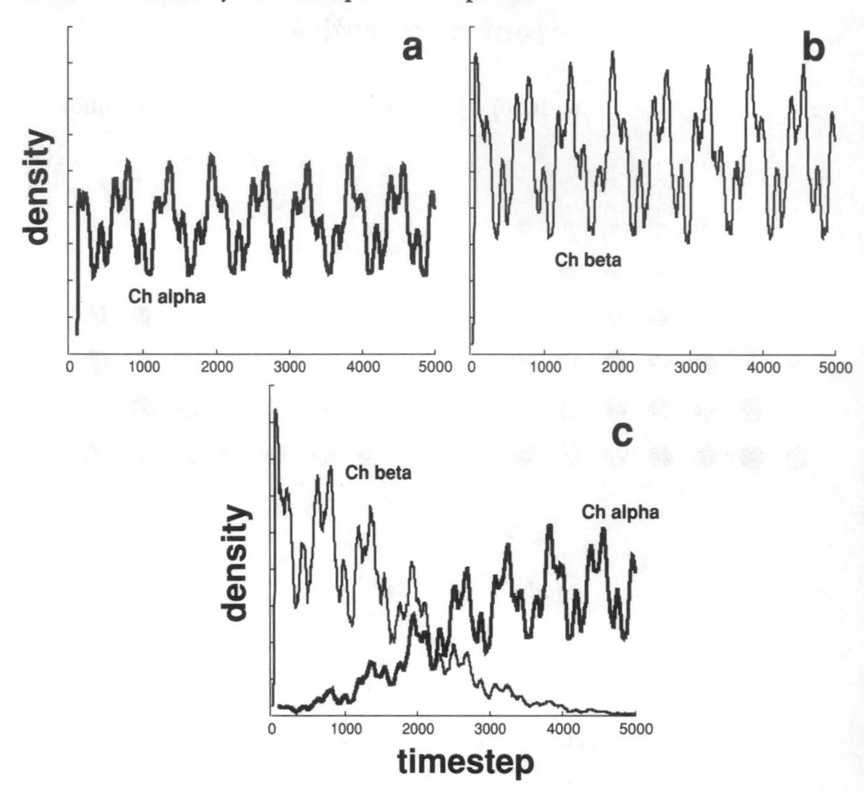

Fig. 7.5 Density trajectories for isolated populations of **Ch** differing in utilization rate (a, b) and for an interaction between them (c) in the variable environment.

Ch beta although the dynamics were not simple. **Ch beta** quickly established its peak density during the period in which resources were not overtaxed but **Ch alpha** did not reach high densities for a long time (Fig. 7.5 c). Nevertheless, the densities of both populations followed inexorable paths – **Ch beta** decreasing and **Ch alpha** increasing. The densities of the latter eventually settled into the same pattern as the isolated population, especially when **Ch beta** became very rare (after 4000 timesteps, Fig. 7.5 c).

This theme can be developed further by looking at the utilization rate-spans of pairs of populations of the same strategy. If utilization rates determine 'competitiveness', then populations with the greater rates should on average persist more often. Results for the hard coherent generalists (**Ch**) showed that the alpha population normally persisted,

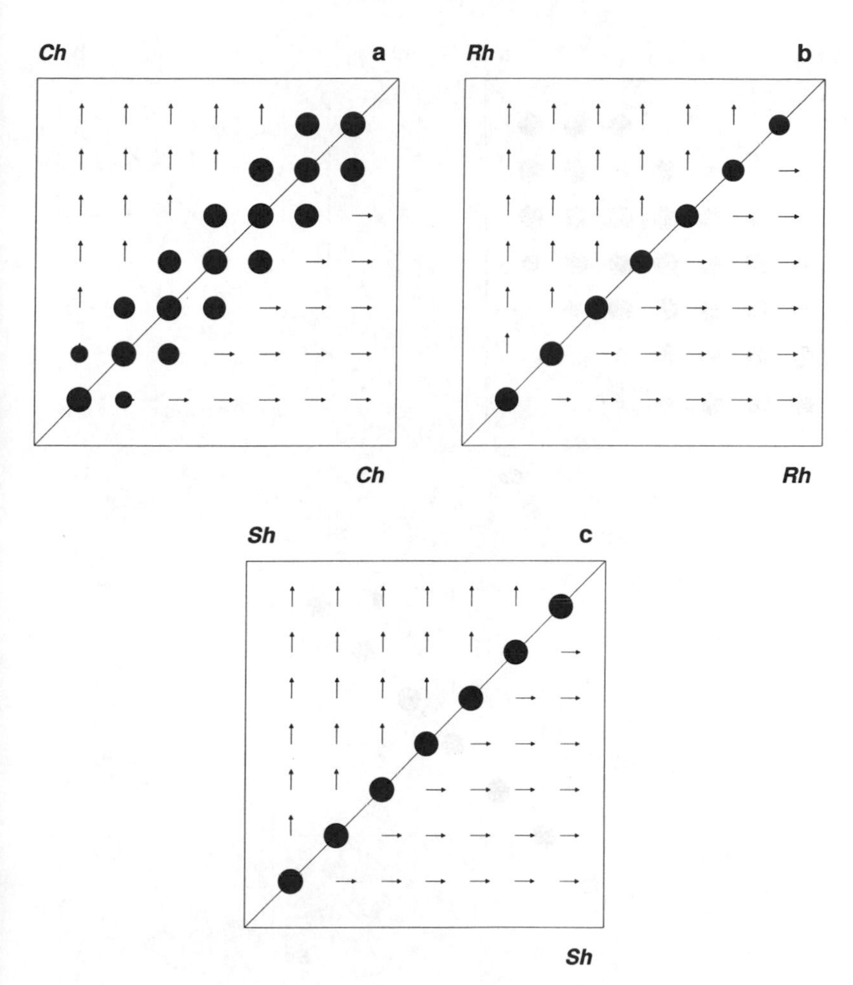

Fig. 7.6 Rate-spans for the interactions between pairs of hard strategists: (a) **Ch**; (b) **Rh**; and (c) **Sh**. (Variable environment, 5000 timesteps.)

out-competing the beta population. If the rates were equal, then the populations coexisted (Fig. 7.6 a). Coexistence along the equality line was 'spread', suggesting that pairs of **Ch** populations might coexist for long periods despite having somewhat different utilization rates. Pairs of the other hard strategists, the resource-like **Rh** and the specialized **Sh**, also showed the same pattern except that the spreading effect along the line of equality did not occur (Fig. 7.6 b, c). Thus, in pairs of populations

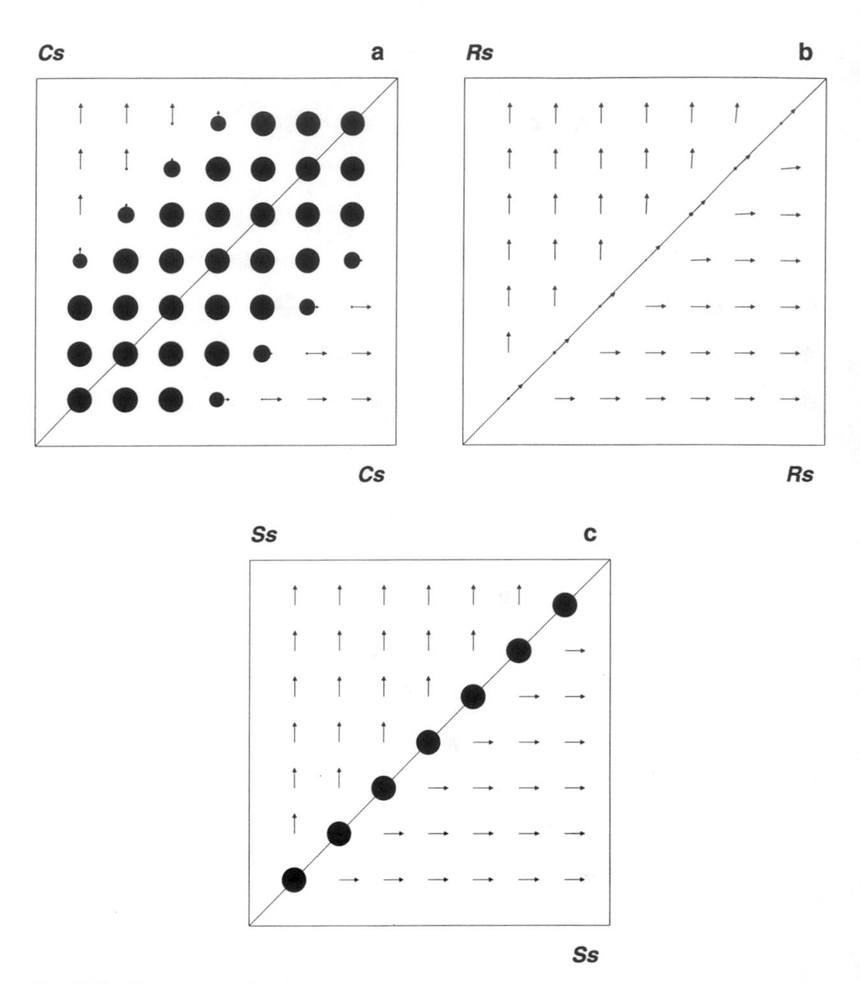

Fig. 7.7 Rate-spans for the interactions between pairs of soft strategists: (a) **Cs**; (b) **Rs**; and (c) **Ss**. (Variable environment, 5000 timesteps.)

having the same hard exploitation strategy the population with a higher utilization rate usually excluded that with a lower rate, although coexistence frequently occurred in **Ch** pairs (Fig. 7.6 a).

Given that soft strategists adjust their utilization rates, the relationships between pairs of soft strategists might be expected to be less clear-cut than for hard strategists. The spreading evident for **Ch** interactions around the equality line is even more pronounced for **Cs** (Fig. 7.7 a).

Thus, there are high likelihoods that pairs of **Cs** populations with extremely dissimilar maximum utilization rates may coexist for long periods. Although soft utilization enhanced the likelihood of coexistence in **Cs** pairs with dissimilar maximum utilization rates, a similar effect was not evident for the other soft strategies. In **Rs**, alpha populations always out-competed beta populations. Coexistence occurred only infrequently when the maximum utilization rates were equal – in general, only one or the other population persisted (Fig. 7.7 b; see also Schoener 1976). On the other hand, the picture for pairs of soft specialized populations (**Ss**) was identical to that for **Sh** (see Fig. 7.6 c). Alpha populations always excluded beta ones and coexistence characterized the situation when the rates were equal (Fig. 7.7 c).

Thus, coexistence (at least to 5000 timesteps) was common in pairs of the coherent generalists, **Cs** in particular. Populations with very dissimilar maximum utilization rates coexisted for long periods. This behaviour of **Cs** (and to a lesser extent **Ch**) is of theoretical interest because it runs counter to conventional thinking on patterns of resource utilization. Recall that both of these strategies are obligatorily generalized so that there is no resource partitioning. Resource overlap is complete and matched, timestep by timestep. Results from the standard LVMs assert that no pair of obligatory generalists can coexist for long because this would make the competition matrix singular and hence unstable (Case and Casten 1979, Chesson 1990). This difference is discussed in greater detail below (§7.5). The coexistence of pairs of **Rh**, **Sh**, and **Ss** strategists along the equality line (i.e., equal maximum utilization rates) is of less interest because an exact equality of rates is unlikely to occur or be maintained in natural communities.

To conclude, maximum utilization rates appear to be the most important determinants of the outcome of interactions mediated by exploitation competition. Gatto (1990), who developed an extended version of MacArthur's (1969) minimization principle, derived a similar result for exploitative competitors in a constant environment. He suggested that such systems should consist of species showing high voracities (which equate to utilization rates for food resources).

Hard and soft strategies

One of the main problems with classical LVMs is that utilization rates implicitly are fixed (i.e., hard strategies). This inflexibility is produced by the linear relationship between the intrinsic rates of increase and resource availability (see Armstrong and McGehee 1980). It reduces the sorts of situations that can be modelled and does not appear to conform well with natural systems in which consumption and utilization rates are known to vary appreciably (e.g., Lambers and Poorter 1992). Theorists have found that flexible resource utilization rates (i.e., soft strategies) produce qualitatively different results for systems of competitors. Matessi and Gatto (1984) have explored the influence of flexibility of utilization rates with what they call 'plastic' and 'nonplastic' strategies – flexible utilization rates played an important role in their theory. Abrams (1984) showed that fluctuating utilization rates may promote coexistence, while Cushing (1992) developed models in which utilization rates varied as functions of population densities, both of conspecific and heterospecific populations. This increasing focus on varying utilization rates clearly is a useful extension beyond the confines of the basic LVMs. However, more attention needs to be paid to the limits of utilization rates, which must have maxima enforced by morphology and processing capacity (Holling 1965, Belovsky 1986; see Chapter 6) and probably also minima due to the economics of resource exploitation, if search time and other factors make it unprofitable to seek out particular resources at low relative availabilities (Pyke 1984).

As was evident above, the results for the interaction between the hard and soft coherent strategies, **Ch** and **Cs**, suggest that the capacity to vary utilization rates may confer some advantages, at least in the variable environment (Fig. 7.8 a). Does this advantage of soft over hard exploitation extend to other strategies such as resource-like and specialized exploitation? It appears that the hard resource-like strategist, **Rh**, 'dominates' its homologue **Rs** (Fig. 7.8 b). Dominance in this sense means that at least some beta populations of one strategy (e.g., **Rh**) can out-compete alpha populations of the other (e.g., **Rs**). Thus, we see in some glyphs above the equality line in Fig 7.8 b that beta **Rh** populations excluded alpha **Rs** populations. Similarly, the hard specialist (**Sh**) is dominant to its homologous soft strategy (**Ss**) in the variable environment, although this dominance is expressed only when both populations have high maximum utilization rates (top-right corner of Fig. 7.8 c). Thus, one cannot state with much certainty whether soft or hard strategies will predominate in a given environment because other aspects of utilization strategies also influence outcomes (e.g., phenotypic differentiation, the degree of specialization).

While the capacity to vary utilization rates in environments with variable resource availabilities clearly confers an advantage on **Cs** over **Ch**, this effect is not evident in the other pairs of homologous strategies

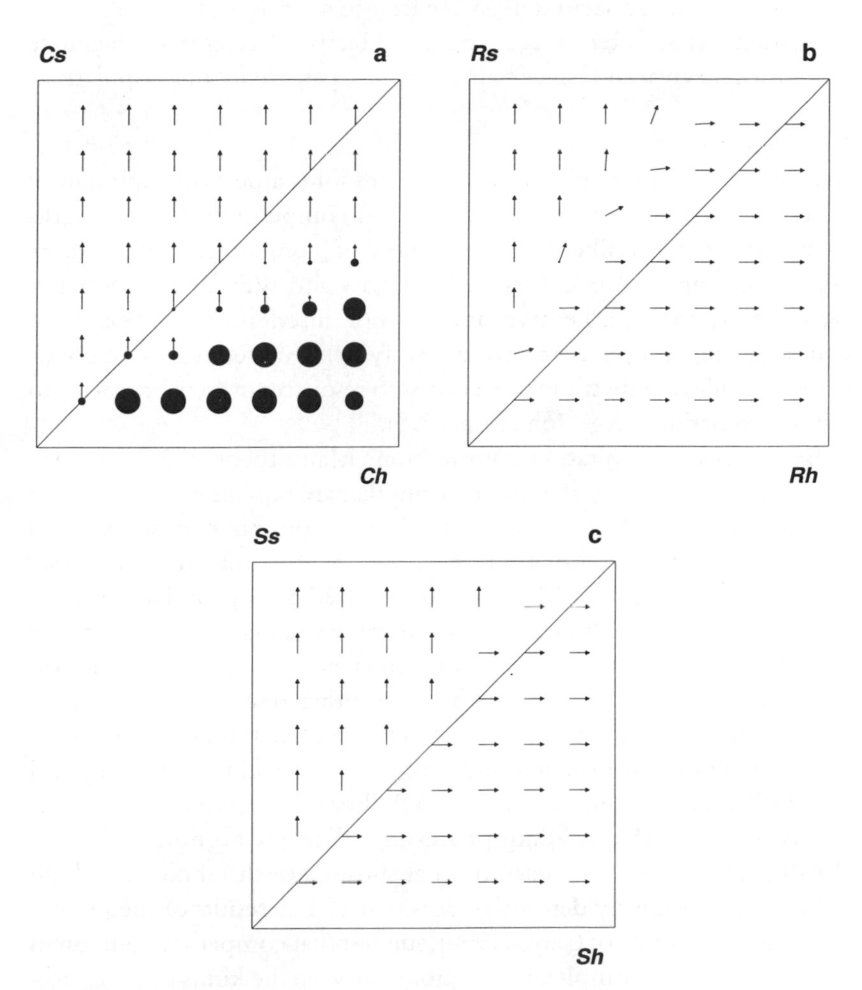

Fig. 7.8 Rate-spans for three homologous pairs of strategies in the variable environment (5000 timesteps). (a) coherent – **Cs** and **Ch**; (b) resource-like – **Rs** and **Rh**; and (c) specialized – **Ss** and **Sh**.

(**Rh** and **Rs**, **Sh** and **Ss**). However, soft strategists in the resource-like and specialized pairings (i.e., **Rs**, **Ss**) are not as subordinate to their homologous strategies as **Ch** is to **Cs**. Any of these strategies can prevail over a homologous strategy if their own maximum utilization rates are high enough whereas **Ch** at best forces coexistence with **Cs**. These results lead one to question whether individuals *must* always utilize resources at the maximum possible rate irrespective of resource availabilities. Some 'prudence', or more accurately, a capacity to alter dynamically resource utilization rates, often might be an effective strategy to facilitate individual survival and hence to promote the persistence of a population.

Competitiveness

In this section I have explored the impact of some aspects of exploitation strategies on interspecific interactions and competitiveness. Ecologists frequently use terms like 'competitiveness' or 'competitive ability' (e.g., Gill 1974, Hairston 1983, Aarssen 1989, Law and Watkinson 1989), but what constitutes competitive ability? For interference competition, competitiveness might correspond simply to how effectively one population excludes competitors from access to resources or to the capacity to injure competitors (e.g., Johanson 1993).

But what of exploitation competition? Many theorists have found, usually by using LVMs, that populations that are capable of surviving on the lowest levels of resource availability are the ones most likely to prevail in competitive situations (e.g., Armstrong and McGehee 1980, Persson 1985, Vance 1985). The relative efficiency of handling or processing resources also has been invoked, particularly in terms of the principle of allocation (i.e., increasing efficiency on one resource necessarily decreases the efficiency with which other resources can be used; Levins 1968, Cody 1974, Brew 1982). However, natural evidence often runs counter to that principle (e.g., Bryan and Larkin 1972, Huey and Hertz 1984, Bence 1986, Futuyma and Philippi 1987), which reduces the persuasiveness of the handling/processing efficiency argument. Gatto's (1990) analysis indicated that minimal subsistence thresholds and high utilization rates jointly determine persistence. The results of the present subsection, like those of Gatto (1990), suggest that competitive outcomes are determined by complex interactions between the kinds of characteristics considered here – utilization rates, population substructure, degree of specialization and variable utilization rates. Carrying capacities, such a dominant feature in solutions derived from conventional analyses, play no role (see Loeschcke 1985).

The two main conclusions are that: (1) utilization rates are likely to be a principal determinant of competitiveness in exploitation competition; and (2) general analytical solutions of the functional relationships between exploitation characteristics are unlikely to be found because of the intricate interrelationships between these characteristics and between the characteristics and resource fluctuations. That utilization rates should be an important consideration seems obvious – whichever population secures the most resources seems likely to prevail on average, although this does not necessarily mean that soft strategies are not effective ones in some circumstances. High carrying capacities, which imply low per capita utilization rates, intuitively do not seem to me to be consistent with the notion of competitiveness.

Invasibility

As outlined in §7.1, one of the main theoretical tools that has been used to investigate models of interspecific competition is invasibility analysis (e.g., MacArthur and Levins 1967, Hastings 1986, Case 1990). It is useful to consider spans in which invasions are lagged by varying amounts (see Table 7.1) and also to gauge the significance of invasions by large numbers of individuals compared with small numbers. The common, tacit assumption that densities of invaders will be low (e.g., MacArthur 1969, Roughgarden 1974b, Turelli 1981) probably is not true in general. In this subsection, I consider some of the effects of lagged arrival as well as the impact of different initial densities of invaders on populations that already fully utilize available resources. The latter has been implicated as a possible factor influencing invasibility (e.g., Ryti 1987).

In the variable environment and with the standard population parameters, resource utilization by **Ch** populations reaches total availability within about 45 timesteps (Fig. 7.9 a). Moderately large amounts of 'unutilized' resources are available until timestep 30 or so, but thereafter the population is resource limited (Fig. 7.9 b). These conditions suggest that the capacity of an immigrant population to invade a habitat occupied by a **Ch** population should be significantly higher if the lag period is less than 45 timesteps.

A span for the **Cs**–**Ch** interaction with immigration lags is presented in Fig. 7.10 a (equal maximum utilization rates). If neither population is lagged the interaction yields an almost complete dominance by **Cs** (i.e., point at the lower left-hand corner of the span, which is the interaction shown in Fig. 7.3). The effect of delayed immigration by **Cs** (rows higher in the span) is to increase the probability of coexistence between the two generalized strategies. Any lag that allows the **Ch** population to reach saturation densities yields a high probability of long-term coexistence. The asymmetry evident in the span indicates a sequence effect,

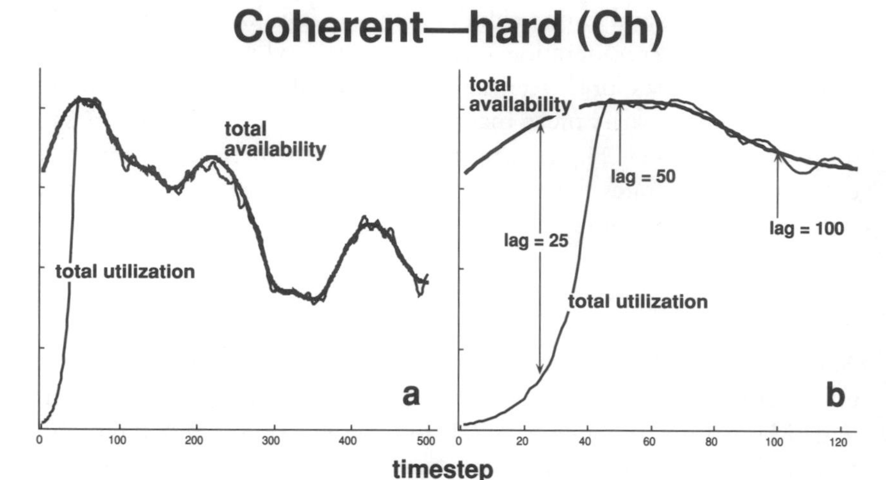

Fig. 7.9 (a) plots of utilization pressure and resource availability for a **Ch** population in the variable environment. (b) focus on the first 120 timesteps of (a) showing the levels of free resources at several stages (25, 50 and 100 timesteps).

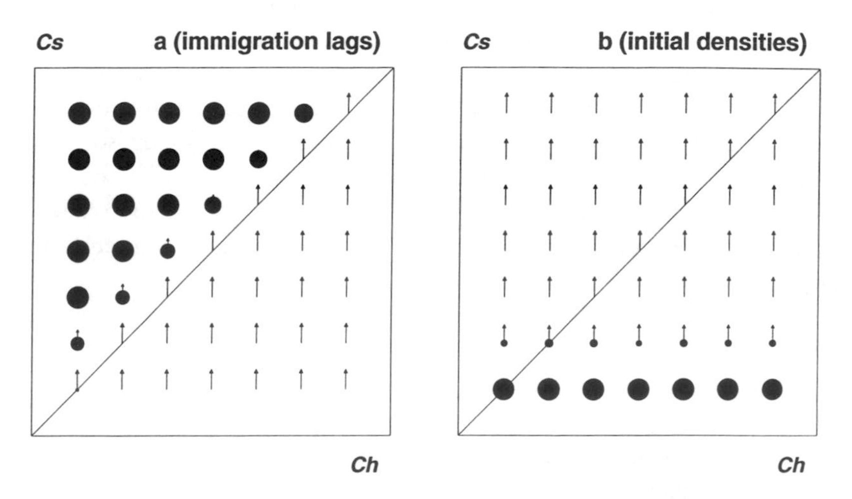

Fig. 7.10 Lag-span (a) and initial density-span (b) for the interaction between populations of **Cs** and **Ch** (5000 timesteps in the variable environment). **Cs** invaded after 50 timesteps in (b).

which means that the outcomes are sensitive to the order of colonization (see Wilson 1992). These results all are based on the standard low influx densities of immigrant populations.

How does the influx of high densities of immigrants affect invasibility in the **Cs–Ch** interaction? The results for different initial densities of populations in situations where the influx of **Cs** was delayed by 50 timesteps are shown in Fig. 7.10 b. The ability of **Ch** to resist invasion by **Cs** seemed to be independent of the initial density of **Ch**. That is, the results for each row were similar irrespective of the column (Fig. 7.10 b). If low densities of **Cs** invaded (bottom row) then long-term coexistence was likely. The ability of populations of **Ch** to persist was reduced sharply as the density of the invading **Cs** population increased (rows higher in span). Schoener (1976) reported similar sensitivities to initial conditions, which he called *priority* effects. The effects of priority have been documented in many natural systems (e.g., Hart 1992).

Overview

What do the results of this section show? Conventional results derived from LVMs (whether by stability or invasibility analysis) are functions of carrying capacities and competition coefficients (e.g., MacArthur and Levins 1967, Yoshiyama and Roughgarden 1977). Loeschcke (1985) demonstrated that the ultimate carrying capacities provide little indication of the likelihood of the success of an invading population because success is governed by initial rates of increase. These rates are determined largely by utilization rates. Loeschcke also noted that the persistence of invaders will be a function of the speed with which a certain population size is reached. The common-sense expectation that utilization rates should be a major determinant of successful invasion, given an equality of other factors, is supported by the results of this section. This does not necessarily mean that soft strategists are competitively inferior to hard ones. Invasibility is sensitive to influx densities, the degree to which resident populations fully use resources and how long the first population has had to establish itself before the arrival of the invader.

7.4 Environmental variability

Environmental variability has been regarded for a long time as an important factor in determining the patterns of resource use. For example, MacArthur (1970) used heuristic arguments to predict that 'constant environments' should permit very fine subdivision of resource axes, thereby permitting the close packing of extreme specialists. Some formal theoretical support for this idea was subsequently provided by Gatto (1990). Such close packing is not expected to occur if resource fluctuations are too great, so niche breadths must be larger, commensurate with the degree of environmental variability. Ebenhöh (1988) also showed the importance of variation in resource availability by formulating a model of several algal species using a single resource. Ebenhöh showed that the variation in resource availability potentially could allow the coexistence of any number of consuming populations, while a single species persisted if there were no variation.

Nonlinear terms in dynamic equations may function as 'resources' that might be exploited differentially by distinct populations (Levins 1979). The most likely sources of such nonlinearities are fluctuations in resources, or covariances between them, which arise in stochastic or variable environments. Levins (1979) noted several types of ecological system that were especially likely to express these effects because they involve the introduction of nonlinear terms in model dynamics. For example, if the rates of resource utilization increase as organisms gain greater experience then the utilization rate may be related nonlinearly to resource availability. Thus, greater experience can lead to improved hunting or handling skills so that individuals become more adept at utilizing a given availability of a resource (e.g., Bence 1986). Utilization rates are often facilitated by group action, for example, when group cooperation is required to hunt successfully a given prey (e.g., wolves on caribou, lions on zebra, and ants on larger insects; see also Charnov *et al.* 1976: 254). Too few hunting individuals make certain prey effectively unavailable to the predator, so that there is a threshold group size below which the use of that resource cannot occur. Ontogenetic shifts in resource utilization involve the dependencies between the resources used in each stage, which introduces a covariance between the resources. And last, if individuals have diverse diets or nutritional requirements then the relative changes in the availabilities of each resource also involve covariances. These may be complicated, especially if the resources are imperfectly substitutable or complementary. In each of these seemingly common situations, the variation in resource availabilities generates nonlinearities that might promote species coexistence at varying densities.

Variation in resource availability can arise from systematic changes that are 'predictable' as well as from random or stochastic variation.

Stochasticity often is represented by using comparatively simple statistical approaches, such as Wiener processes (Polansky 1979), to characterize the environmental fluctuations (e.g., May and MacArthur 1972, May 1973, Chesson 1988). Abrams (1984) believed that the forms of stochasticity used in theoretical studies have been vague and often unrelated to the relationships between consumers and resources. For example, May (1974a) and Turelli (1981) both implemented 'environmental' unpredictability by introducing stochasticity into the carrying capacity parameters of competitor populations. Species-specific stochasticities were uncorrelated, which means that at any time the environment is perceived differently by each competitor population.

May and MacArthur (1972), and May (1973) suggested that environmental variability should force resource overlap to lower levels than in deterministic environments. However, other analyses indicated that, given rather restrictive assumptions, environmental variability may have little impact on the limiting similarity of resource use between competitors (e.g., Turelli and Gillespie 1980). These studies show that unpredictable variation in resource availabilities has been considered as an important theoretical factor in resource use for many years, although there seems to be little consensus as to what the likely effects of environmental stochasticity will be. Perhaps the vagueness noted by Abrams (1984) accounts for the range of viewpoints (see also Chesson 1988).

In Chapter 6, I defined two 'environments' – the constant and variable environments. Each consisted of six resources (Fig. 6.5). Given the interest in stochastic environments that I have just related it will be useful to introduce a third form of environment in which each resource varies stochastically. Each of the six resources in this environment varies randomly about its long-term average, but none ever becomes unavailable at any time (Fig. 7.11).

Now, having three environments with which to work, the pair of strategies **Cs** and **Ch** again are used here to investigate the sensitivity of interactions to resource characteristics. **Ch** showed a weak dominance in the constant environment almost always excluding **Cs** whenever the two had equal maximum utilization rates (Fig. 7.12 a). Only rather special conditions led to coexistence over extended periods (Fig. 7.12 a). The results for the variable environment have already been discussed – in short, **Cs** was strongly dominant (Fig. 7.12 b). Neither **Cs** nor **Ch** was able to establish any dominance whatsoever in the stochastic environment – coexistence occurred over the entire rate span (Fig. 7.12 c).

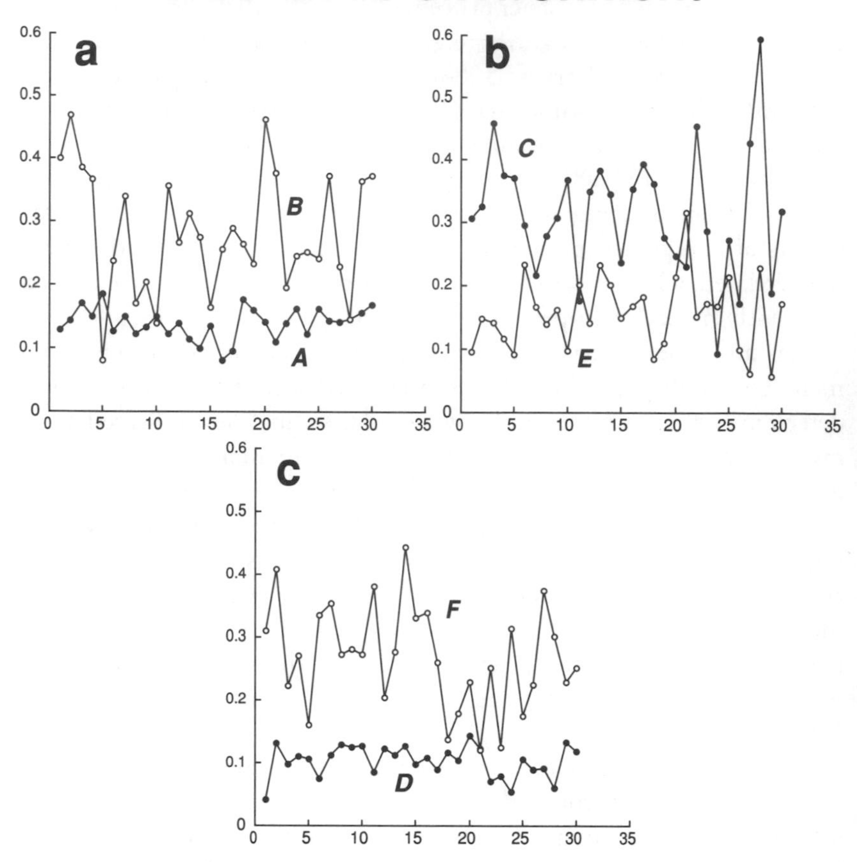

Fig. 7.11 Representative variation in resource availability in the stochastic environment over 30 timesteps: (a) resources *A* and *B*; (b) resources *C* and *E*; and (c) resources *D* and *F*.

Although the results are not presented here the enhanced level of coexistence in the **Ch–Cs** interaction in the stochastic environment compared with the variable environment occurred in neither the **Rh–Rs** nor the **Sh–Ss** pairs. The relationship of the resource-like pair was little affected by the increased stochasticity while the dominance of **Sh** over **Ss** was extended. Therefore, stochasticity *per se* does not appear to guarantee either greater or lesser likelihoods of coexistence – the nature of the strategies involved in each interaction has an important bearing on the outcome.

The way in which interactions turn out clearly is sensitive to the

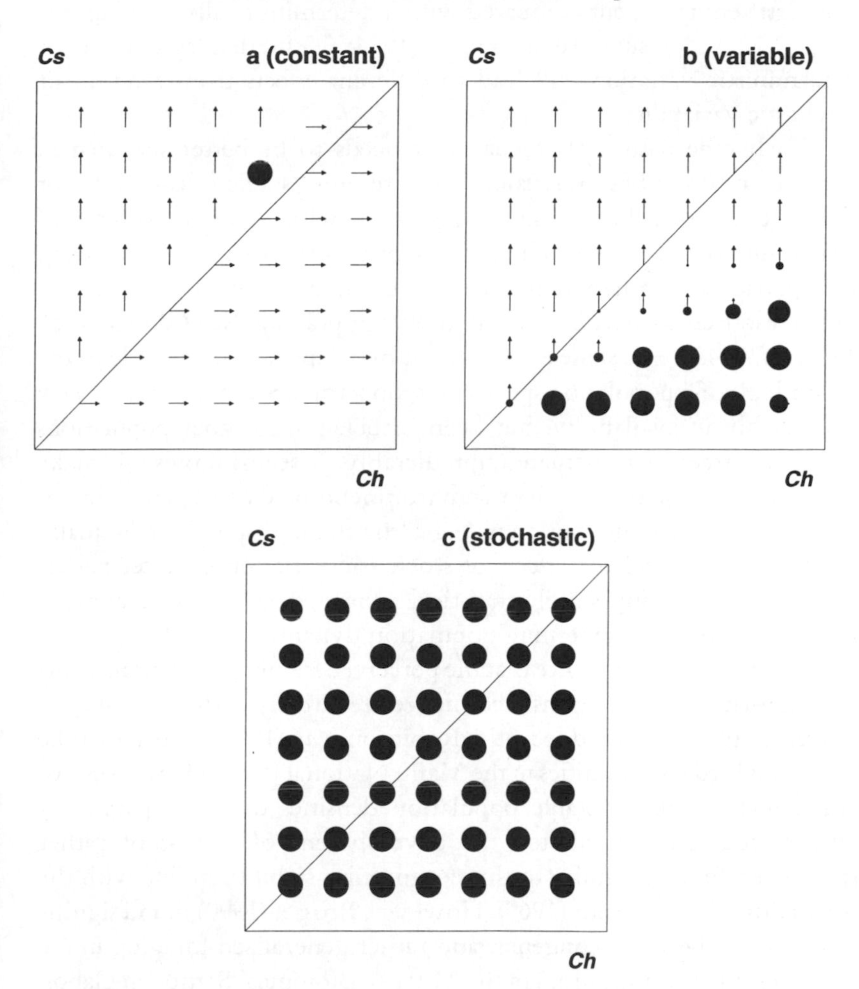

Fig. 7.12 Rate-spans for the interaction between populations of **Cs** and **Ch** in three different environments: (a) constant; (b) variable; and (c) stochastic. (5000 timesteps.)

nature of the fluctuations of resources (Levins 1979, Ebenhöh 1988). For example, severe, unpredictable variation in all resources prevents the establishment of the dominance of **Cs** over **Ch** found in the variable environment. The main message from these analyses is that one should be loath to generalize the results of interactions to other conditions, particularly if the patterns of variation in resource availability and the exploitation strategies are not known. The different results for a purely stochastic environment compared with a deterministically varying one (Fig. 7.12 b, c) supports Abrams' (1984) view that systematic or deterministic variation will lead to different effects from random or stochastic variability.

Clearly, the nature of stochasticity needs to be better specified in natural communities. Environments are not 'stable', 'constant', or 'variable' as such, but consist of resources with different patterns of availability through time. The stability or otherwise of an environment for a particular population must depend upon the nature of the variation of the resources used by individuals of that population (see Addicott *et al.* 1987: 342–343). The same environment may 'appear' to be stable to the individuals of a population specializing on a resource that does not vary appreciably in availability, but 'seem' stochastic to other populations using resources that fluctuate unpredictably. It seems unwise to make glib statements on the stability of environments based on a perception of their overall apparent constancy (e.g., 'the stable tropics'), and equally foolhardy to model the effects of stochastic variation on interspecific interactions by using simple statistical processes or stochastic components of parameters governing population dynamics.

Some important evidence that the perceived stability of environments (and constituent populations) need not relate directly to the versatility of resource use has emerged recently. Robinson *et al.* (1990) argued that the rainforest bird communities in the Manu National Park in Peru (western Amazonia) exhibited stable population densities over comparatively long times. They attributed the development of a host of rather specialized foraging guilds to this community stability in line with the prediction of MacArthur (1969). However, Brosset (1990) found significant overlap between congeners and rather generalized foraging in the rainforest bird communities in the M'Passa Biological Station in Gabon (central-west Africa), despite community stability of a similar order to that found in Peru. Such biogeographic differences in the relationship between community stability and patterns of ecological versatility mean that: (1) interspecific competition does not necessarily produce ecological specialization *even under the most favourable circumstances* for such developments; (2) (unrecorded) historical processes are at least as important as contemporary ones in shaping the patterns of resource use; and/or (3) contemporary processes other than competition (e.g., nest predation in birds, Brosset 1990) operating at different intensities in various places may sufficiently affect population densities to abolish any consistent patterns in large-scale comparisons.

Overview

The results presented in this section suggest that the outcomes of interactions between populations employing particular exploitation strategies depend upon the precise nature of the variation in resource availability. Of course, this comes as no surprise. Nevertheless, these results highlight the importance of distinguishing between stochasticity and deterministic variability of resource availability (Abrams 1984).

7.5 Many populations

Up to this point, only the interactions between pairs of populations have been considered. This is a common way of studying the relationship between interspecific competition and resource use, but evidence is accumulating that interactions between species-pairs are sensitive to the impact of other species (e.g., Bock *et al.* 1992, Werner 1992, Wootton 1993). How do such relationships depend upon context, that is, the set of background populations? And what happens when many species are studied together? This section provides a limited examination of these questions.

The impact of background populations – context-specificity

Despite the diversity of effects that might affect outcomes of competitive interactions, the impact of background populations is potentially crucial (Thompson 1988b). An example is presented to demonstrate how the variation in a single parameter of just one population can completely alter community structure.

The example consists of three populations exploiting resources in the variable environment. The three populations are **Ch**, **Rh** and **Sh** strategists. Recall that both **Ch** and **Rh** are obligatory generalists, with **Rh** consisting of three ecologically differentiated phenotypes. The **Sh** population specializes on resource *F* alone (see Fig. 6.5). Otherwise, the standard values were used for each population. The dependence of community structure on the variation in a single parameter, namely the utilization rate of **Sh**, formed the basis of this analysis.

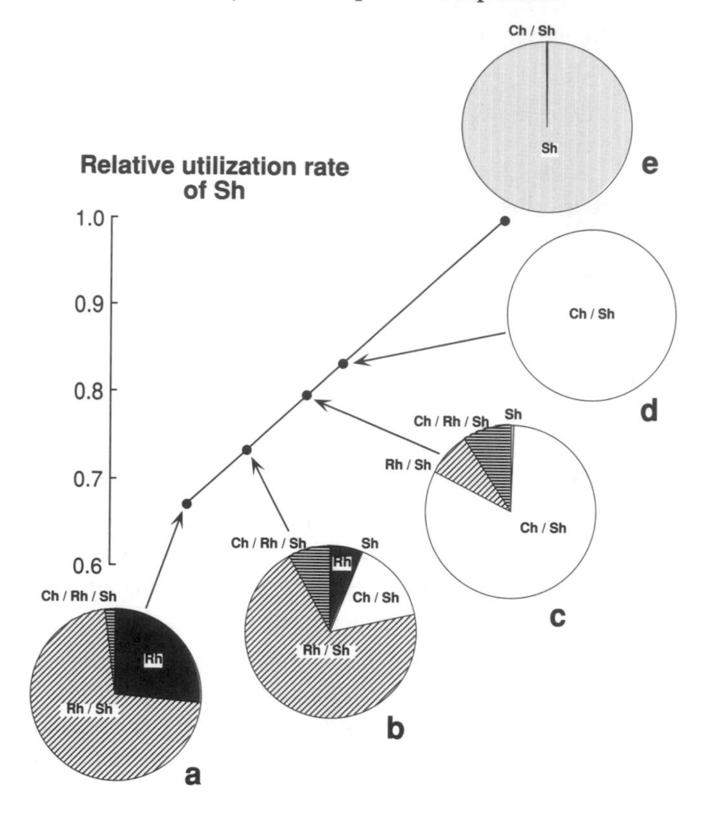

Fig. 7.13 The influence of the variation in the utilization rate of **Sh** on community composition. The rate of **Sh** was lowest in (a) and highest in (e). Data are the proportions of realizations for each rate leading to particular outcomes. Coexistence is indicated by joint combinations, like **Ch/Rh/Sh**, or **Ch/Sh**. There were 500 realizations, each of 5000 timesteps, in the variable environment.

At the lowest utilization rate of **Sh** the most common result was the coexistence of **Rh** and **Sh** (Fig. 7.13 a). **Rh** excluded both of the other populations in about 25% of the realizations while three-way persistence was rare. As the utilization rate of **Sh** increased (Fig. 7.13 b, c, d) the competitive dominance of **Rh** over **Ch** declined, ultimately leading to the exclusion of **Rh** by the other two strategists and their coexistence (Fig. 7.13 d). A further increase in the utilization rate of **Sh** tipped this balance in the favour of **Sh** to such an extent that both generalists were almost always excluded (Fig. 7.13 e). The distribution of outcomes evident in each case clearly shows the true stochastic nature of the modelling structure.

These results are presented mainly to demonstrate the futility of trying to deduce patterns of resource use by competitors on the basis of stability analysis, even in isolated communities in environments that do not change in their long-term patterns of resource availability. This is particularly true for the competition between pairs of species (e.g., Armstrong and McGehee 1980). Outcomes are just too sensitive to parameter values to make this a worthwhile approach (León and Tumpson 1975, Nold 1979). In fact, I suspect that the general expectations of competition theory, particularly the specialization on distinct resources or on distinct parts of resource axes are *trivial* results of stability and invasibility analyses in mathematical rather than ecological terms. It is a trivial result that six specialized populations may stably coexist on six distinct resources (provided none of the resources becomes unavailable), but this is not necessarily the outcome of coevolution for stability *per se* (Hall 1988). There are likely to be limitless combinations other than sets of extreme specialists that allow long-term coexistence or permanence, given the range of strategic and tactical options that are likely to be available to species. This interpretation is reinforced by the modelling results of the next section.

The *N* populations – *N* resources problem

The problem of how many populations can *stably* coexist (without recolonization or supplementation) on a given number of resources has been of perennial interest to theorists (e.g., Rescigno and Richardson 1965, Strobeck 1973, Schoener 1976, Abrams 1988a). Armstrong and McGehee (1980) provided a historical review of this idea, which can be summarized as: *N* species cannot stably coexist on fewer than *N* resources (or 'niches' or 'limiting factors'). The 'stably' qualifier here is crucial, with the meaning 'at fixed densities' (Armstrong and McGehee 1980).

Whether even *N* populations can stably coexist on *N* resources itself has been questioned. For example, Nold (1979) believed that such stability depends upon a fortuitous combination of parameter values and is likely to be relatively unstable in the face of small changes in these values. Different classes of resources (e.g., substitutable, complementary, etc.) may require certain combinations of consumption rates and returns from utilization to permit the stable coexistence of *N* populations on *N* resources (León and Tumpson 1975: 199–200).

It turns out that there are many conditions under which more populations might coexist than there are resources. These conditions include the relaxation of the requirements that persistence be at fixed densities and/or that population growth rates be linear functions of resource densities (Zicarelli 1975 cited in Armstrong and McGehee 1980). Cyclical population densities and saturating (i.e., non-LVM) responses also permit the coexistence of competitors, although the functional responses of populations in cycling systems may have to differ from one another (Abrams 1984). Syntopic populations also might persist on fewer resources than there are populations if the consumption rates vary greatly or if there are negative correlations between the consumption rates of populations depending upon the levels of resource availability (Abrams 1984). Predators might induce the coexistence of prey having identical patterns of resource utilization provided that there are refuges and frequency-dependent predatory responses (Vance 1978, Gleeson and Wilson 1986, Holt and Kotler 1987). As we have seen already, Levins (1979) suggested that populations might exploit nonlinearities in resource availabilities (e.g., the variances of single resources or the covariances between resources) so that more than *N* populations might persist at nonconstant densities on fewer than *N* nominal resources. Abrams (1983) and Vance (1985) summarized many of the conditions under which the number of consumers can exceed the number of resources (i.e., $N_{pops} > N_{res}$). And last, if individuals 'compete' more strongly with conspecific individuals than with similar heterospecific individuals (e.g., stronger intraspecific rather than heterospecific territoriality or interference), then each population will be largely self-regulating. This should not impose any strong pressure for resource differentiation (Chesson 1991). Thus, there are many theoretical conditions under which the coexistence of more populations than there are resources can occur but these rely on variable densities, non-logistic dynamics of some of the populations, variable utilization rates, and so on.

To further investigate the *N* population–*N* resource problem, the dynamics of many (10, 20 or 30) populations of either **Ch** or **Cs** were simulated in the stochastic environment. The stochastic environment was used because of the conflicting implications of existing thought on the impact of stochasticity on resource use. As we have seen, May and MacArthur (1972) and May (1973) suggested that environmental stochasticity should reduce overlap to an extent that reflects the degree of

Table 7.2. *The range of utilization rates for the many-species* ***Ch*** *and* ***Cs*** *simulations in the stochastic environments. Combinations consisted of either* ***Ch*** *or* ***Cs*** *(10, 20, or 30) populations, or of 10* ***Ch*** *and 10* ***Cs*** *populations in a mixed-strategy analysis. In the latter, the maximum utilization rates were alternated so that the lowest maximum rate was a* ***Ch*** *population, the second-lowest* ***Cs****, the third-lowest* ***Ch****, etc. (see text)*

No. of populations	Lowest maximum rate	Highest maximum rate	Increment
10 **Ch** *or* 10 **Cs**	0.050	0.140	0.010
20 **Ch** *or* 20 **Cs**	0.050	0.145	0.005
30 **Ch** *or* 30 **Cs**	0.0500	0.1466	0.0033
10 **Ch**	0.050	0.140	0.010
and 10 **Cs**	0.055	0.145	0.010

stochasticity (hence greater differentiation than in predictably varying environments). Roughgarden (1974b) on the other hand, believed that stochastic variation should lead to greater overlap and more generalized resource use. Thus, the stochastic environment is an appropriate setting to consider further the many-population question.

All of the populations shared the same parameters except for maximum utilization rates, which were different for all populations and spanned an almost three-fold range (i.e., the fastest user potentially could acquire resources three times faster than the slowest, see Table 7.2). The effect of the number of resources was explored by considering not only the usual set of six resources but also environments consisting of three, four, five and twelve resources as well. The base levels and random variabilities were adjusted in each case to maintain the overall total availability and variation of resources. The simulations were run for 10000 timesteps – coexistence over so many timesteps surely can be regarded as being effectively asymptotic coexistence for any natural system.

The results for these two generalized strategies, **Ch** and **Cs**, differed markedly but much the same pattern emerged irrespective of the initial number of interacting populations (Fig. 7.14 a, b, c). In almost every realization, all **Cs** populations coexisted if there were five or more resources. However, the number declined rapidly as the number of

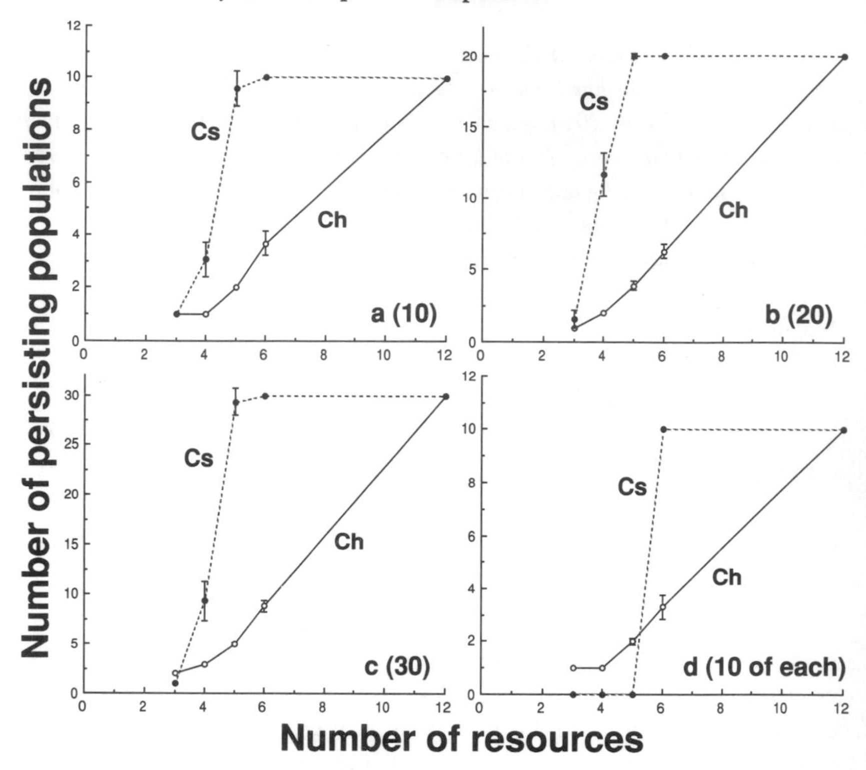

Fig. 7.14 The numbers of populations of **Cs** (solid circles, dashed lines) and **Ch** (open circles, solid lines) persisting on between three and 12 resources in stochastic environments, with 100 realizations of 10 000 timesteps each. (a) 10 initial populations of either **Ch** or **Cs**; (b) 20 initial populations of either **Ch** or **Cs**; (c) 30 initial populations of either **Ch** or **Cs**; and (d) initially 10 populations of **Ch** interacting with 10 populations of **Cs** (see Table 7.2). Data are mean ±1 standard deviation.

resources decreased below five so that just one population persisted on three resources irrespective of the initial number of populations. On the other hand, relatively few **Ch** populations coexisted on six or fewer resources. With six resources, about four on average of the 10 or 20 initial populations coexisted (Fig. 7.14 a, b), while almost nine of the 30 initial populations did so (Fig. 7.14 c). When there were twelve resources, all of the original populations coexisted.

These results suggest that there may be a nonlinear dependence of the number of species that can coexist on a given number of stochastically varying resources. Soft exploitation (**Cs**) potentially allows extremely high loadings with at least 30 populations coexisting on just five resources. However, the nonlinearity is pronounced when there are fewer than five resources with rapid declines in the diversity of supportable populations. It is important to remember that both **Ch** and **Cs** populations are obligatory generalists with total resource overlap. Even so, providing there is a diverse array of resources (≥ 5 for **Cs** and >6 for **Ch**), many more populations can coexist than there are distinct resources.

The reader may be concerned that these results reflect properties that are peculiar to highly stochastic environments, which they no doubt do. For example, of 20 initial **Cs** populations with the same characteristics as in the stochastic environment, an average of almost ten (9.9 ± 1.0) persisted in the variable environment (six resources) compared with all 20 in the stochastic environment. Only one, the population with the highest maximum utilization rate, persisted in the constant environment (also six resources). Thus, the nature of environmental variation also is crucial to the pattern of coexistence.

A comparison of the results for the stochastic, variable and constant environments shows that unpredictable environments, if not actually favouring generalized strategies as suggested by Roughgarden (1974b), may not prevent the coexistence of many generalized populations. This is especially true for **Cs** strategists and in environments consisting of moderate to large numbers of resources.

The analyses to this point concern the outcomes of interactions between strategists of the same form (e.g., 10 **Ch**, or 30 **Cs**), with the strategists differing only in relation to their respective maximum utilization rates. An important question however, is whether there are similar results when sets of homologous strategists interact with each another. There may be the suspicion that in an environment in which so many populations are interacting, any strategist failing to utilize resources at the fastest rate of which they are capable may be eliminated. In other words, in many-species interactions can soft strategists survive? Is prudence as viable a strategy as it was for **Cs** populations interacting one-on-one with **Ch** populations in the variable and stochastic environments (see §7.3)?

To investigate this question, a similar procedure was used as before – many realizations of the interactions of 20 populations in the stochastic environment. The difference however, was that now 10 populations were **Ch** strategists and 10 were **Cs** strategists. Populations of the two strategists were assigned maximum utilization rates alternately in ascending order. Thus, the lowest maximum utilization rate was a **Ch** population, the second-lowest a **Cs** population, the third-lowest a **Ch**, and so on. Thus, the average maxima were about the same for **Ch** and **Cs** populations. Otherwise, all other characteristics were the same.

When many populations of homologous species interact with one another, the results differ to some extent from the corresponding outcomes for the single-strategy simulations. The numbers of **Ch** populations persisting on 3, 4, 5, 6 and 12 resources differed very little from the single-strategy case (Fig. 7.14 a, d). That is, only one **Ch** population (the fastest exploiter) persisted on three or four resources, two on five, about three on six, and all 10 populations on 12 resources (Fig. 7.14 d). No **Cs** populations persisted in environments consisting of fewer than six resources when competing with **Ch** populations while all 10 did on six or more resources despite the presence of **Ch** populations. These results indicate that at least for stochastic environments, **Cs** strategists may be driven to extinction by **Ch** populations in simple environments ($\leq$5 resources) but not in more complex ones. This change is very sudden, showing a phase transition between environments comprised of five and of six resources.

Overview

Two main points emerge from these analyses. First, the classical argument that N resources can support N populations, each of which is specialized on a single resource, need not be a necessary outcome of exploitation competition. The results for **Ch** in a stochastic environment show that six obligatorily generalized, completely overlapping populations might persist for very long periods on six resources. However, there was a nonlinear decline in the number of coexisting populations as the number of resources decreased. This was particularly marked for environments consisting of three or four resources. Second, if populations show soft utilization rates, like **Cs**, then the number of populations that can be supported may far exceed the number of resources for long periods. However, if the number of resources is small (three or four, say) then even soft utilization may not sustain this imbalance between the number of populations and the number of resources. In environments consisting of few resources ($\leq$6) the number of coexisting populations of **Cs** greatly exceeds the number of **Ch** that can persist together under the same conditions. These nonlinear relationships between the number of persistent populations and the number of resources appear to be novel theoretical findings.

These many-species studies also show that interspecific competition, at least as mediated through exploitation competition, is unlikely to impose any strong constraints on the patterns of resource use. That is, interspecific exploitation competition probably has little systematic effect on ecological versatility. Completely generalized, completely overlapping populations can persist for very long periods without necessarily excluding one another, even though their maximum utilization rates may be quite different.

7.6 Additional remarks

Several important issues warrant more comment on the basis of the results of the previous sections. The first is the importance of absolute values of variables such as utilization rates and densities. The patterns evident in the spans often show quite different behaviour in different parts of parameter space. Second, although the methods used here allow considerable flexibility in specifying exploitation strategies and tactics, they nevertheless do not begin to address the range of strategies that can be conceived and that have been reported in natural populations. Thus, it is premature to conclude much from the results except that some of the common, limited expectations from theories developed in the past few decades are unlikely to be relevant to many natural communities. And third, coexistence does not seem so unlikely in models in which diverse strategies of resource use are permitted compared with LVMs. Several factors that make coexistence even more probable have been deliberately excluded from the models developed here, and some discussion of this seems to be desirable.

Absolute rates

The spans presented in this chapter demonstrate the dependence of outcomes of interactions on the absolute values of variables rather than on their relative values. The interactions between syntopic species must be modelled with values reflecting rates, densities and resource availabilities that are pertinent to specific natural systems. Biologically meaningful models can be framed leading to quite different expectations depending upon natural values. *For, even if there were just one general process governing competitive interactions between syntopic populations and that process could be modelled appropriately, the outcomes of interactions must be sensitive to where in parameter space particular natural assemblages lie.* The common practice of 'rescaling' the components of dynamic models for simplification (e.g., Case and Casten 1979, Brew 1984, Law and Blackford 1992) obscures the dependence on absolute values and makes it difficult to gauge the generality of results. This is *the* central problem in the predictions of competition theory and those of LVMs in particular – these models are based largely on dimensionless variables.

This recognition of the critical importance of absolutism certainly is not new. Several authors commenting on competition theory also arrived at the conclusion that absolute rates and availabilities are crucial for deriving meaningful expectations in particular systems. For example, Schoener (1976: 310–311) noted that 'Mechanistic models . . . because their parameters directly correspond to biological quantities, they have two enormous advantages. First, predictions can be directly made as to how variations in these quantities affect the form and outcome of competition. Second, the models can be fitted to growth or equilibrium data, and the resulting computer-calculated parameter values can be checked against independent measurements of what the parameters purport to represent.' Vance (1985: 83) wrote: 'The main value of realistic models . . . lies in their potential use in empirical studies. Component functions of such a model can be estimated from field data.' Matessi and Gatto (1984: 359) provided perhaps the best general guidance: 'A much wiser alternative for evolutionary ecology . . . building and using models that, besides the phenomenology of population dynamics, explicitly incorporate a description of the basic physiology and ecology of individual organisms.' Thus, there is a critical need to view resource use in community ecology in these terms, which seems more likely to lead to operational definitions and testable predictions.

The richness of possibilities produced by mechanistic or component models that use absolute values measured in natural communities is a definite boon from this approach to understanding the impact of interspecific competition. It seems hardly surprising that LVMs appear to be such poor representations of nature (Vance 1985, Abrams 1987b) given that values for parameters rarely if ever are related to situations in natural communities (see however, Law and Watkinson 1989).

The modelling results reported here clearly are indicative only and do not represent natural systems in a definite way because of the absence of empirical information concerning rates and availabilities. The models have not been 'unpacked', as Rosenzweig (1991) has put it, into forms in which field measurements can be conducted. However, the models do incorporate several more realistic characteristics. Utilization rates in the current models have both upper (maximum) and lower (marginal) values. Stochasticity and deterministic variability in resource availability also are handled more explicitly than in most other models. I believe that the main value of the models lies in the demonstration of the diversity of outcomes that can arise from relatively simple alternative strategies of resource use.

Strategic and tactical options

The range of options studied here – six main strategies (**Ch**, **Rs**, etc.), variation in utilization rates and a variety of resource regimes – hardly taps the potential for investigating resource utilization strategies. For example, only obligatory generalists have been considered. Opportunists and facultative generalists have not been modelled. Facultative strategies can be based on a diverse range of criteria for selecting and dropping resources from repertoires. The 'environments' generally consisted of six resources and, given the nonlinear relationships between the number of coexisting populations and the number of resources (§7.5), there may be opportunities for great species diversity in environments with large numbers of resources (i.e., $N_{pops} \gg N_{res}$ for many resources). The limit on the diversity of model consumers is likely to be set by the densities of populations that on average decrease as diversity increases for a given level of total resource availability. Ultimately, populations will routinely be at such low population densities that extinction probabilities will be relatively high, creating a limit to species diversity (see Cornell and Lawton 1992: 5). Simplistic characterizations of specialists and generalists in most models cannot provide general expectations for feasible natural communities.

On coexistence and ecological versatility

Several decades ago, the prevailing mathematical models of interspecific competition seemed to offer serious problems for understanding why natural communities are so rich in species. Much of the difficulty can be attributed to the implications of LVMs, which are models in which competitive exclusion is implicit (Armstrong and McGehee 1980). When some of the assumptions of LVMs and other formulations are relaxed there appear to be many circumstances in which (local) coexistence appears to be more likely (§7.5). Notwithstanding these advances, equilibrium and invasibility methods remain standard modelling tools despite the limitations of each with respect to natural communities (e.g., Case 1990). If these methods were to be phased out then one suspects that models might be developed that could produce probabilities of coexistence over different durations (e.g., Fig. 7.4), which then could be related to meaningful time scales for particular communities. Given the natural variability in climate and resource availabilities, even essentially isolated communities (e.g., remote islands, mountain-tops, oceanic thermal springs) cannot be expected necessarily to reach the sorts of equilibria that asymptotic methods require. Thus, coexistence over specified periods is more biologically relevant than idealized, asymptotic results that probably will not be reached.

If coexistence is viewed in terms of the probability distributions of outcomes for different durations then one can say that effective coexistence is a common result of much of the modelling presented in this chapter. This is a very conservative conclusion. The models do not include interference competition, which facilitates coexistence in many models (Vance 1985). Resources are distributed randomly and there are no refugia. Both resource clumping and access to refuges have been thought to enhance local coexistence (e.g., Ives and May 1985, Shorrocks and Rosewell 1986, Czárán 1989, Danielson 1992) or delay competitive exclusion (e.g., Palmer 1992). The 'environment' used most often here, the variable environment, has a pronounced periodic element. Natural resources often have periodicity but also marked differences in the respective levels of availability from year to year (e.g. Dunham 1983, Pulliam 1986). Periodicities may not be evident at all, although resources might rise and fall in abundance. Both circumstances could alter competitive relationships in such a way as to promote coexistence (Levins 1979, Abrams 1984).

Although these exogenous factors are likely to increase the probability of coexistence between competitors, the characteristics of individuals and populations may have even greater potential to do so. Two possibilities that spring to mind have been deliberately suppressed in Chapters 6 and 7 to make the implications of the modelling clearer. The first is hybrid strategies in which individual organisms display alternative exploitation strategies depending upon circumstances (e.g., different resource regimes or milieu of competitors or natural enemies, e.g., Schoener 1971). Hybrid strategists may be difficult to exclude by competition because of their potential short-term adaptability and responsiveness. The other feature is the transfer of fitness between ecological phenotypes in which fitness increments associated with the resources accrued by one ecological phenotype appear to the benefit of others. This typifies many sexual populations with stage-specific use of resources (Werner and Gilliam 1984). Juveniles mature into ecologically distinct adults (effectively increasing the representation of the adult phenotype), while resources gathered by adults are converted into juveniles through reproduction. Maturation–reproduction is not the only mechanism by which transfers of fitness might be achieved – adaptive polyphenism and genotypic recombination are others.

Recolonization and dispersal have not been considered at all. Even at the local scale the movement of individuals between patches enhances coexistence (e.g., Tilman 1994). Schoener (1976: 329) suggested that immigration functions in a similar way to the existence of an exclusive resource for a population, and according to him this always facilitates coexistence. Comins and Noble (1985: 719) showed that coexistence might be enhanced in species with limited dispersal capacity if the dispersal distances are similar to the range of environmental correlation of factors affecting competition for establishment, particularly in terrestrial plants and other sessile organisms.

All of these factors suggest that the existence of relatively diverse biota need not be as surprising as it once was. The use of models with greater strategic and tactical variety coupled with the relinquishment of equilibrium methods have helped to produce this change in perception.

7.7 Summary

For many years, interspecific competition has been held to be a (or the) significant determinant of the patterns of versatility in natural communities. Competition theory has relied heavily on coevolutionary arguments between syntopic populations to make predictions, usually underpinned by Lotka–Volterra and related logistic models. Expectations have been found by using two main criteria: (1) asymptotic equilibrial stability (i.e., resource differentiation is such that densities remain constant); and (2) invasibility (i.e., the integrated patterns of resource use in coevolved communities should be effective at resisting the establishment of immigrant populations). Both criteria suffer from weaknesses and have restricted applicability to natural situations. Rather than modelling by using these standard techniques, the interactions between the exploitation strategies described in Chapter 6 (i.e., **Cs**, **Ch**, **Rh**, etc.) are used to document the sensitivities of interactions to resource regimes, exploitation strategies and tactics. (Note that these interactions are mediated by exploitation competition, and not by interference competition.) 'Coexistence' is reframed in terms of the probabilities of joint persistence over a given duration rather than as absolute stability, which is unlikely to occur given the changeability of environments and community composition. Under this definition coexistence seems easier to explain than by using conventional competition theory. Indeed, for example, many model populations of **Cs**, all completely overlapping in resource use, can persist together for very long periods. Thus, interspecific competition *per se* need not be expected to produce coherent patterns of resource differentiation ('partitioning', a putative manifestation of competition), although this does not deny that interspecific competition is an influential process in natural communities (i.e., the actuality of interspecific competition is not disputed). In the light of these results there seem to be no theoretical reasons to expect that interspecific competition consistently and predictably controls ecological versatility or necessarily is responsible for producing systematic ecological differentiation in resource use among syntopic species.

8 · *Ubiquity or habitat versatility*

One of the reasons I first became interested in specialization and generalization was due to the results of some of my work on communities of cicadas (e.g., Mac Nally and Doolan 1986, Mac Nally 1988b) and birds (e.g., Mac Nally 1989, 1990a, 1995), each involving distributions in relation to a variety of habitat types. The most remarkable feature was the range of 'habitat versatility' – ubiquity – displayed by species. For example, the cicada *Abricta curvicosta* occupied a wide range of forest, woodland and shrubland habitats, while several other species were 'tall forest specialists' (*Psaltoda moerens*, *Henicopsaltria eydouxi*, *Thopha saccata*) or 'shrubland specialists' (*Cystosoma saundersii*) (Mac Nally and Doolan 1986). Other species exhibited intermediate levels of habitat versatility. Similarly, birds such as the grey shrike-thrush (*Colluricincla harmonica*) and laughing kookaburra (*Dacelo novaeguineae*) occupied all five major forest and woodland types that were studied, whereas other species were restricted to only a single kind of forest or woodland (e.g., red-capped robin, *Petroica goodenovii*; yellow rosella, *Platycercus flaveolus*) (Mac Nally 1989). Species occupying practically any combination of the five habitat types were recorded. Such disparities in ubiquity among species, which typify many studies (e.g., Collins *et al.* 1982, Williams 1988, Woinarski *et al.* 1988), demand an explanation and certainly represent an important component of the degree of ecological specialization or generalization of species in its broadest sense.

Although recent work on biogeographic patterns is beginning to clarify the picture for the reasons for differences in the extent of geographic ranges among species (e.g., Brown and Gibson 1983, Hengeveld 1990), the determinants of variation in ubiquity remain less clear. This then is the purpose of this chapter – to consider some of the effects and theories on the origin of different degrees of ubiquity among species.

At the outset of this analysis, it is worth noting that no single explanation will be likely to account for differences in habitat versatility, for many factors involved in controlling ubiquity have been documented. For instance, natural enemies may exert a powerful effect on the habitat use of prey species (e.g., Werner *et al.* 1983, Smith *et al.* 1989, Pagel *et al.* 1991). From the predators' viewpoint, habitats differ in suitability for finding prey in terms of habitat structure and other attributes, which can limit ubiquity (Aldridge and Rautenbach 1987, Neuweiter 1989, Preston 1990). Avian nectarivores can be constrained in habitat use by biophysical limitations (e.g., the costs of flying) and the ways in which nectar sources are distributed in space (e.g., Feinsinger *et al.* 1979, Ford and Paton 1985). Body size has been invoked to explain different degrees of ubiquity. In some African ungulates, for example, larger body size appears to allow or even enforce a reduced dietary selectivity, which means that large animals spread more evenly across habitat types (du Toit and Owen-Smith 1989). On the other hand, Pulliam (1986) suggested that small body size is associated with greater ubiquity in western chipping sparrows (*Spizella passerina arizonae*) compared with other sympatric granivorous birds in Arizona. Only the chipping sparrow can economically harvest the small-sized, low-return seeds that are the only ones available in some habitats. Therefore, large body size may correspond to either high or low ubiquity depending upon the circumstances. Ubiquities can be increased by subspecific or clonal differentiation such that subspecies or clones are better adapted to particular physical regimes, which allows a greater penetration of habitat types by the species as a whole (see Hesse *et al.* 1951, Jaenike *et al.* 1980, Weider and Hebert 1987, Stevens 1989). Gender-specific ecophysiologies in many terrestrial plants (e.g., Dawson and Bliss 1989, Dawson and Ehleringer 1993) largely seem to account for the distinct differences in habitat use by males and females (Freeman *et al.* 1976, Bierzychudek and Eckhardt 1988, Cameron and Wyatt 1990). Thus, gender-related differences in habitat use may expand ubiquities in plants. And last, species may not be finely adapted to particular habitats in the sense of having had insufficient time to evolve restricted habitat use. For example, both Cody (1976) and Keast (1990) attributed a lack of habitat specificity in many species of holarctic forest birds to the recency of the establishment of these forests following the last main glaciation. Keast (1990) also suggested that a high potential for using different habitats would have been of great adaptive value during the vicissitudes of the Pleistocene. These ideas place ubiquity in a historical, biogeographic setting.

Perhaps the two largest sets of explanations for control of habitat use are based on ecophysiology and interspecific competition. Ecophysiological theories allow the direct experimental corroboration of ideas on how species successfully occupy particular habitats (e.g., McLean and Ivimey-Cook 1973, Good 1974, Dorgelo 1976, Wainwright 1980, Gilles and Pequeux 1983, Alscher and Cumming 1990, Storey 1990, Crawford 1992). Unfortunately, the more interesting question: 'why do species show such different physiological tolerances?' is less easy to answer. That is, we often know why species are restricted or ubiquitous (in the sense of Hesse *et al.* 1951) by virtue of their physiologies but not why species come to have such different physiologies. For example, Jackson (1973, 1974) showed that some species of infaunal marine bivalves in the Caribbean basin have relatively wide physiological tolerances (temperature, salinity, stagnation) and several other features consistent with broad habitat use such as respiratory pigments and the ability to occupy sediments of different grain sizes. These characteristics clearly are consistent with high ubiquity but did ubiquity foster physiological versatility or vice versa? Physiological constraints on the use of resources or habitats have been reviewed many times (e.g., Cloudsley-Thompson 1991) and, as I am considering the ecological basis for versatility and ubiquity, I will not consider ecophysiological effects very much in this chapter.

Interspecific competition is the other major source of ideas for explaining ubiquity, or rather, the restriction of habitat use (e.g., MacArthur 1972, Diamond 1973, 1975, 1978, Cody 1974, Jackson 1981, Pimm and Rosenzweig 1981, Bull 1991, Rosenzweig 1991). Although this may seem to be just an extension of the ideas of the Chapter 7, I have stressed throughout this book that 'habitats' ought not be regarded as 'resources' in themselves but rather as integrals of resources (and constraints). Thus, there is a qualitative difference between the issues of 'competition between species for resources [within a habitat]' and 'how interspecific competition affects habitat use'. The impact of interspecific competition on ubiquity is considered in this chapter, particularly in relation to ecological compression and release (see MacArthur and Wilson 1967) and in habitat selection.

Distributions – range and ubiquity

I think that it is worthwhile to review briefly some of the terminological distinctions made in Chapter 2 concerning habitat use and the geographic ranges of species. Ubiquity is not synonymous with the extent of geographic range, which is *cosmopolitanism* (see Fig. 2.2). Thus, wide-

Table 8.1. *The list of factors affecting geographic distributions provided by Grinnell (1917). I have divided the factors into 'abiotic' and 'biotic' groups*

Abiotic factors	*Biotic factors*
rainfall	vegetation
humidity	food supply
soil moisture	safety of breeding places
barometric pressure	refugia
atmospheric density	cover, shelter from enemies
ground surface (rocky, etc.)	interspecific competition
insolation	parasitism
cloudiness	habitat selection/preferences
temperature (means, extremes)	
water (for terrestrial species)	
land (for aquatic species)	

spread species are *cosmopolitan* while species limited to small geographic areas are *localized*. Species that occupy a narrow range of habitat types are *restricted*, while those occupying many or most of the available kinds of habitat are regarded as being *ubiquitous*. Therefore, populations of one species might be able to persist in only an extremely limited set of habitats, rocky outcrops say, yet because that habitat is distributed widely and patchily the species might have an extensive range. This would be a restricted (in habitat) yet cosmopolitan (in geographic range) species.

The relationship between ubiquity and range is an intricate one. The factors affecting geographic distribution have been important questions for ecologists and biogeographers for well over a century. The list provided by Grinnell (1917) is reproduced here in Table 8.1, yet, even despite its length, Grinnell himself considered it to be an incomplete set. The list undoubtedly reflects Grinnell's bias as a student of vertebrate ecology, for many other constraints on distribution can be imagined.

Ubiquities are often constrained by abiotic effects associated with altitude and weather (e.g., Terborgh 1985). The populations of a species might be able to occupy more types of habitat, or similar habitats of a particular type in other places, except for the impact of different patterns of rainfall, minimum winter or maximum summer temperatures, and so on, for which they are physiologically restricted. Of course in this view, abiotic factors such as weather are considered to be 'external' to the definition of a habitat. This clearly will not always be the case, for salinity defines the habitat type in aquatic systems (e.g., brackish water, Remane and Schlieper 1971) and plant communities are effective at modifying their physical environments (Tilman 1985, Wilson and Agnew 1992). So, while the controls on range or cosmopolitanism are not considered much here, it is worth keeping in mind that ubiquity is subject to these sorts of effects. A species may, in principle, be able to occur in a much more diverse set of habitats than it does, but is restricted by the impact of abiotic conditions. A species also may be prevented from occupying potentially useable habitats because those habitats are separated from the current range of the species by hostile, intervening habitats.

8.1 Niche pattern

I want to start the analysis by considering a unifying concept that links a variety of aspects concerning habitat use, namely, *niche pattern* (Shugart and Patten 1972). Niche pattern is a way of representing some of the main ecological attributes of species and can be used to characterize community structure (Dueser and Shugart 1979). In principle, niche pattern can also be used to infer the mechanisms responsible for distributions of ubiquities (Mac Nally 1990a).

In graphs of niche pattern, measures of niche 'breadth', which corresponds to ubiquity, niche 'position' and density are plotted together in a three-dimensional space (Fig. 8.1 a). But what are these quantities? Ubiquity has been referred to in some detail already (§2.5) – it is a measure of the variability of habitats occupied by populations of a species. However, the meanings of position and density require some explanation.

In the simplest case, a habitat gradient of some sort (e.g., altitude, nutrient concentration; i.e., 'coenoclines', Whittaker 1975), niche position simply is the absolute distance between the mean location of a species along a gradient and the mean of the gradient. Thus, if populations of a species occur in habitats characterized by extreme values (all very low soil phosphorus say) compared with the mean value of all habitats in the sample, then the niche position of the species is large. The species occurs in 'extreme' habitat types relative to those sampled. On the other hand, if all populations occur in habitats with values close to the mean then niche position is small (as is the position of a ubiquitous species occupying all sites). Large values of position generally are taken to indicate an overall maladaptation to the set of habitats in the sample (Dueser and Shugart 1979), although it is possible that large position

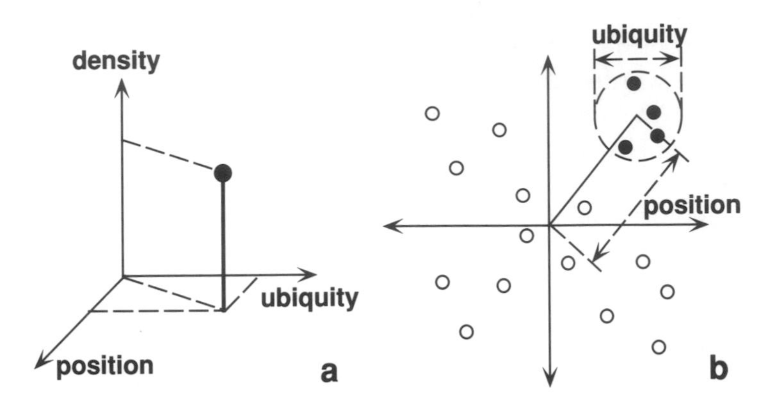

Fig. 8.1 (a) representation of the location of one species in the three-dimensional space defined by ubiquity, position and density. (b) the meaning of niche position and ubiquity in a multivariate setting. Position is the distance from the centroid of sites occupied by a species (*solid circles*) to the origin of the space. Ubiquity is a measure of the variability of sites occupied.

might indicate a high level of adaptation to the extreme habitat types (of those sampled). Position is not just an ecological curiosity because it plays an integral role in many community models (e.g., McNaughton and Wolf 1970, Roughgarden 1972, Rummel and Roughgarden 1985, Palmer 1992).

The concept of position can be extended to multivariate situations as well. If one had two gradients and representative habitats spanning these gradients, then niche position is the Euclidean distance of the average locations along each axis of habitats occupied by a species to the origin of the plane (see Fig. 8.1 b). Ubiquity can be calculated on the basis of the variability of occupied sites in the plane (Fig. 8.1 b).

In practice, habitats (especially terrestrial ones) are often represented in a multivariate space based on either the structural (*physiognomy*, Whittaker 1975) or the *floristic* properties of sites. Such habitat ordinations involve the measurement of many variables (Greig-Smith 1983). Variables may be physiognomic characteristics such as the spatial separation of trees or shrubs, vertical foliage profile or chemical constitution, or floristic measurements representing the taxonomic proportions of plant species (James 1971, Rotenberry 1985, Burgman 1989, Mac Nally 1990b). These measurements are then used in multivariate analysis and classification (e.g., principal component analysis (PCA), detrended correspondence analysis (DCA), or multidimensional scaling (MDS); see Dillon and Goldstein 1984 and Wiens 1989a, Chapter 9). This procedure yields an abstract 'ordination space' in which sites of similar composition or structure are located close to one another while dissimilar sites are widely separated. Sites located near the origin or centre of the ordination space are ones that are 'representative' of the whole collection of habitats – thus, they possess average attributes. Once this process has been performed, various measures of ubiquity and position can be computed, notwithstanding some technical problems in defining these measures (e.g., Carnes and Slade 1982, Van Horne and Ford 1982, Burgman 1989).

The third component of niche pattern, density, can also present difficulties because one wants to summarize much information (values at possibly many sites at different times) in a single statistic for a species. A mean value of some form clearly is needed (Van Horne and Ford 1982) but what is the correct form? Should information for sites from which the species is absent be included? A variety of arguments indicate that density estimates should be based only upon sites at which a species occurs (e.g., Carnes and Slade 1982, Hanski 1982, Van Horne and Ford 1982, Collins and Glenn 1990). Note that density refers to individuals per unit area in most species of animals, but often means percentage cover in plants (e.g., Collins and Glenn 1990) and encrusting animals (Tokeshi 1993).

Ubiquity, position and density are all *sample* estimates based on the collection of habitats used to derive them. Unless habitat use by every population of a species is known, which is most unlikely, these variables are sample rather than global values. On the other hand, geographic range, which is not a feature of niche pattern, is a global measure even though every population within the range need not have been located. Determining the range of a species involves just finding a polygon enclosing the most extreme geographic records of its populations (Rapoport 1982).

In the following sections, I describe some empirical findings of relationships between elements of niche pattern. These results are used to illustrate some of the inferences on habitat use and ubiquity that can be derived by using niche pattern.

Ubiquity and niche position

A key result of analyses of niche pattern is that ubiquity and position are usually found to be related inversely, at least if ordinations are based on physiognomic variables. There were statistical and methodological problems with some of the early analyses showing this relationship (see Carnes and Slade (1982) and Van Horne and Ford (1982) for criticisms of the method advocated by Dueser and Shugart (1979)). However, more refined field methods and computer modelling (including null models) have shown that this relationship is an ecologically meaningful effect for situations as diverse as nine species of cicadas in habitats within a 20 km radius of Port Macquarie, New South Wales, Australia (Mac Nally 1988b) and an avifauna of 92 species in a 250 km transect through central Victoria, Australia (henceforth referred to as the *Victorian transect*, Mac Nally 1989, 1990a). The relationship between ubiquity and position for the avifauna is presented in Fig. 8.2.

Factors contributing to this inverse correlation are: (I) no species exhibit both small ubiquity and small position; and (II) species with large positions and small ubiquities are comparatively common (Fig. 8.2). Thus, no species are specialists on habitats that are structurally 'average' with respect to the set of habitats as a whole (point (I)) but many species specialize on particular habitat types yielding point (II).

Point (I) most likely is a result of the fact that individual habitat plots (in this case woodlands and forests) that are 'average' in structural terms usually are extreme examples of their type (Fig. 8.3 a, Mac Nally 1989). This means that a set of structurally similar sites near the origin of the ordination space (exclusive occupation of which would yield small values of both position and ubiquity) consists of distinct habitats in terms of vegetation and possibly geographic location. Many species of animals show some degree of floristic or geographic fidelity, which means that they are *unlikely* to occupy *only* those plots near the origin of ordination space (Fig. 8.3 b).

What appears to happen is that species either occupy a limited number of habitat types including sites of those types near to the origin (Fig. 8.3 c, d) or are habitat generalists (Fig. 8.3 e). The only way in which low values of position and ubiquity could happen concurrently is if distributions are governed exclusively by structural criteria, which does not appear to happen. Habitat specialists (point (II) above) appear to require particular floristic associations or structural features, which generally leads to restricted ubiquities and high values of niche position.

The inverse relationship between position and breadth and the relatively even distribution of species along the major axis of the ellipse in Fig. 8.2 show that there is a continuum of ubiquities. Thus, in the Victorian transect, there are ubiquitous species occupying all woodland

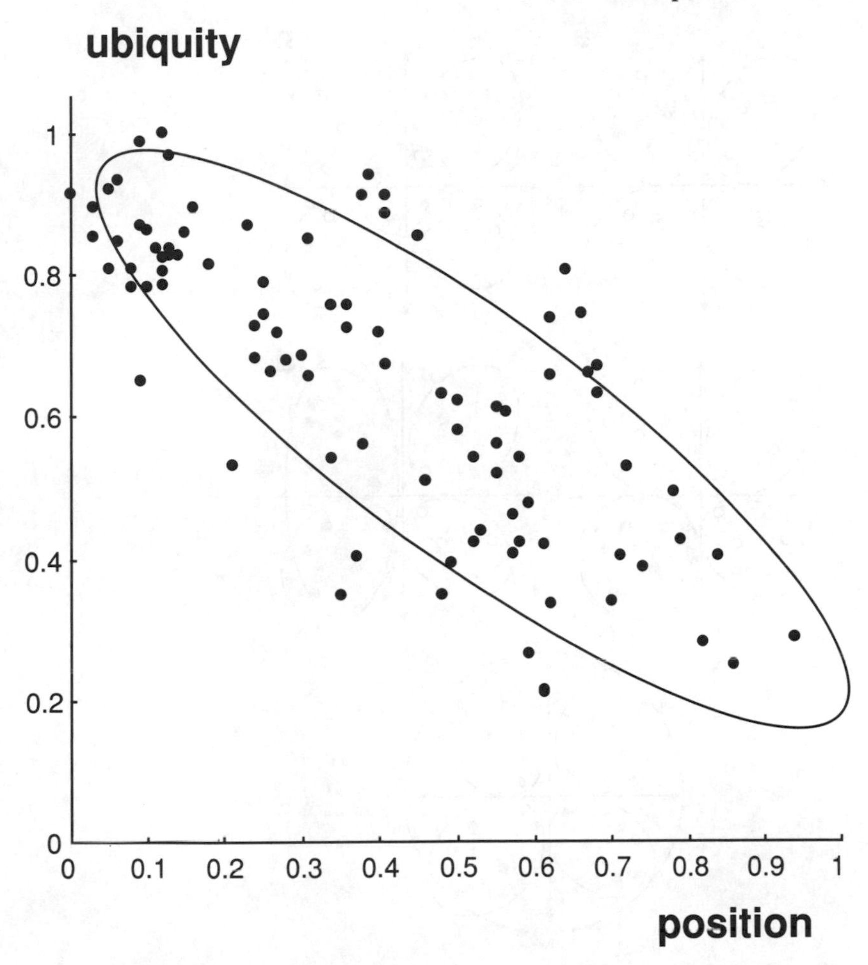

Fig. 8.2 Empirical relationship between niche positions and ubiquities for 92 species of birds along a 250 km transect in Victoria, Australia ($r = 0.70$, $P < 0.001$).

and forest types and virtually all sites (e.g., grey shrike thrush (*Colluricincla harmonica*), golden whistler (*Pachycephala pectoralis*), grey fantail, (*Rhipidura fuliginosa*)) and habitat specialists appearing in just one habitat type in the transect (e.g., red-capped robin in ironbark (*Eucalyptus tricarpa*) woodlands and the yellow rosella in river red-gum–grey box (*E. camaldulensis–E. microcarpa*) riparian woodlands; Mac Nally 1995). As I mentioned at the beginning of this chapter, there are species occurring in virtually every floristic combination of forest and woodland types in this transect. Thus, one finds a complex arrangement in which there is a

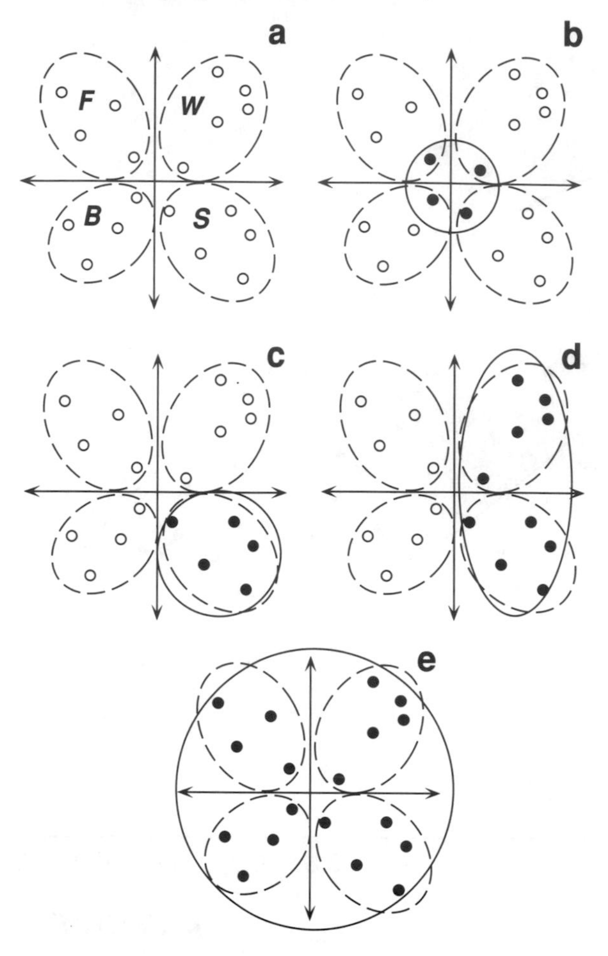

Fig. 8.3 (a) hypothetical distribution of representatives of four types of habitat (*F*, *W*, *B*, and *S*) in the first two dimensions of an abstract space based on habitat ordinations. (b) Habitats occupied by a hypothetical species (*solid circles*) specializing on structurally intermediate habitats. This pattern was not observed for any species of cicada or bird (see text). (c) and (d) Examples of the common observation for species occupying structurally intermediate habitats, either extreme habitat restriction (c), intermediate ubiquity (d) or high ubiquity (e). Ellipses with solid outlines in (b), (c), (d) and (e) are measures of ubiquity.

relatively even distribution of ubiquities, with a predominance of neither restricted nor ubiquitous species.

The process of quantifying ubiquity is complicated by scale issues, especially in relation to the geographic extent over which habitats are sampled (Wiens *et al.* 1987). As one increases the area of sampling (say, from 500 km^2 to 10000 km^2), a greater diversity of habitats can potentially be incorporated into the survey, and this will affect the estimates of ubiquity. Thus, a ubiquitous species in an area in which, say, three distinct habitats occur probably will be judged to be less ubiquitous as the sampling area is extended and more habitat types included. This effect will be exacerbated by factors such as the inclusion of increasingly different types of habitat (e.g., adding savanna and grassland habitats to what was a set of forest and woodland habitats) and when the extension of the survey area crosses into other biogeographic zones. In a sense, this scaling problem in estimating ubiquity corresponds to the difficulty one has in defining the resource basis for estimating ecological versatility (see §2.3). As with resources, the solution is not to seek an illusory ideal scale but rather to state clearly the scope and domain of the habitats surveyed; no absolutely appropriate scale exists – ubiquity is a *sample estimate*.

Density and distribution

Many ideas have been put forward to relate the extent of geographic range to commonness and rarity (e.g., McNaughton and Wolf 1970, Rabinowitz 1981, Brown 1984, Bock 1987, Williams 1988, Burgman 1989). The consensus seems to be that abundant species generally are more widespread or occur in more habitat types than rarer species (see cited studies in Hanski 1982, Brown 1984, Williams 1988, Gaston and Lawton 1989). Some authors suggest that this relationship may be scale-invariant (Bock and Ricklefs 1983, Collins and Glenn 1990), although others have found a variety of relationships (positive, negative, neutral) at different spatial scales (e.g., Wiens *et al.* 1987).

Several workers have reported no relationship between density and range or density and ubiquity. For example, Rabinowitz *et al.* (1986) and Burgman (1989) found little evidence to support the correspondence between density and distribution in plant communities in the British Isles and south-western Australia respectively. Seagle and McCracken (1986) reported that breadth and abundance were unrelated in several forms of animal communities in the north-eastern United States. Regional-scale (hundreds of kilometres) analysis of woodland bird communities in south-eastern Australia also failed to show a strong coupling between density and distribution (Mac Nally 1989). Gaston (1994) suggested that the absence of a correlation between density and the extent of geographic range is a widespread phenomenon. In animals, there may be complications because densities depend upon body size (Peters 1983,

Damuth 1991, Nee *et al.* 1991, Tokeshi 1993) and also upon trophic level (Schoener 1968, Juanes 1986). Whether these dependencies are sufficient to account for the absence or inconsistency of relationships between distribution and density is unclear.

In summary, niche pattern is a useful tool for viewing three of the main components of habitat use, namely, ubiquity, position and density. The existence of the inverse relationship between ubiquity and position appears to be a real phenomenon but high densities do not always correspond to ubiquity nor vice versa.

Although it is possible to use niche pattern for the inference of mechanisms (see §8.4), it is largely a descriptive tool for summarizing the variety of strategies evident in communities. Therefore, it seems appropriate at this point to move on to consider more mechanistic explanations for patterns of ubiquities, which are the bases of §§8.2–8.4. These explanations take a number of distinct forms with some of the factors involved being: (1) the temporal variation in food availability; (2) habitat selection; (3) interspecific competition; and (4) the degree of spatial complexity.

8.2 Temporal variation in food availability

One factor that immediately springs to mind when trying to account for patterns of ubiquity in animals is temporal variation in food availability (Wiens 1989a). Therefore, the effects on ubiquity of differences in availabilities between seasons (e.g., Fretwell 1972) and in the same season between years (e.g., Pulliam 1986) are considered in this section.

Seasonal variation

Fretwell (1972) presented several hypotheses about how seasonal variation in food availability might influence the ubiquity of mobile animals. Of course, in the most extreme cases, species abandon regions altogether so that their regional ubiquities fall to zero (i.e., seasonal emigration, see Herrera 1978a), but there are many places in which species remain as residents throughout the year, so Fretwell's hypotheses can be tested on them.

Fretwell's hypotheses state that for resident species, the main expectations are (1) that species should become more restricted in their habitat use when their food resources become scarce, and (2) that a rearrangement of densities among alternative habitats should occur in the season prior to the onset of the most demanding conditions. In the latter prediction, species are predicted to behave as though they 'anticipate' the changes in resource availability and react accordingly (Fretwell 1972).

Although Fretwell (1972) presented some evidence to support these ideas, a detailed study of bird communities of the lower Colorado River valley provided only equivocal support (Rice *et al.* 1980). The reasons for this prove to be cautionary. Several species showed idiosyncratic ecological behaviour that was inconsistent with the patterns evident for most other species, which highlights the range of ecological options that animals have had available to them to contend with natural environmental variability.

Fretwell (1972) considered winter to be the most demanding season, but Rice *et al.* (1980) generalized this argument by noting that there are both summer and winter residents, where the distinction hinges on the season in which the maximum densities of each species occur (see Alatalo 1981). So, for winter residents, summer is the season in which a contraction of ubiquity is hypothesized to occur, and vice versa for summer residents. Thirteen of 14 summer residents showed greater habitat restriction in winter than in summer. The exception was the house finch (*Carpodacus mexicanus*), which switched its diet from insects in summer to berries in winter with an associated change in the actual habitats used but without much effect on the diversity of habitats used. Thus, the magnitude of ubiquity in this species is largely unaffected from season to season but is based on a different set of habitat types. Similarly, one winter resident (of eight species), Say's phoebe (*Sayornis saya*), maintained a similar ubiquity year-round. These results indicated that the first of Fretwell's hypotheses appears to be sound for most species – habitat versatility generally declined during the seasons of greatest food stringency.

The case for anticipatory reductions in ubiquity before the season in which food levels are least was less clear, although six (of nine) winter residents and nine (of 14) summer residents did show this pattern. The exceptions often displayed major demographic rearrangements among habitats but without appreciable changes in ubiquity.

At larger scales (e.g., regions) the availabilities of a food resource may be asynchronous in different habitat types throughout the year or at even shorter time scales (Herrera 1978a). This means that 'resource-tracking' species following food blooms might appear to exhibit restricted levels of ubiquity at any one time (one type of habitat say) yet have wide tolerance over an annual cycle. Therefore, very mobile species might not show the Fretwell responses, merely because they manage to negate the sorts of stringencies that Fretwell (1972) envisaged by exploiting the asynchronous availability of food resources in regional-scale habitat mosaics.

Seasonal change in food availability probably strongly influences ubiquity in many animals, but this is not the whole story. Dispersal, demographic changes, large-scale rearrangements of densities and switches in feeding niches are alternative reactions to the stringent circumstances that can occlude the Fretwell patterns (Rice *et al.* 1980).

Between-year variation

Most workers who have monitored the availability of food resources have reported substantial differences in the corresponding seasons between years (e.g., Dunham 1983, Pulliam 1986, Holmes and Schultz 1988). For example, the availability of seeds in summer may vary by an order of magnitude between successive years. Thus, regional patterns of ubiquity that are evident in one year probably will not necessarily reflect the characteristic extent of ubiquity or restriction that each species undergoes within a region over many years.

Studies of wintering sparrows in Arizona illustrate the impact on habitat use of different amounts of food availability between years (see Pulliam 1986). If seed production were too low in a particular habitat (two forms of grassland and riparian woodland) due to the amounts and the timing of summer rainfall, no species of sparrow could overwinter there. Thus, between the winters of 1972–73 and 1973–74 the number of species occupying riparian woodland went from none to four as seed availability increased by a factor of more than 40. Pulliam (1986) noted that the ubiquity of each sparrow species could be related directly to seed production – low availabilites meant restricted occurrence (or absence), while in years of high production the ubiquities of all species were at their respective maxima.

Flexibility of habitat use arose because the sparrows could not exhaust food supplies in very productive years so that, according to Pulliam (1986), several species could co-occur without competition for food arising. Pulliam believed that each species nevertheless is better adapted to one particular habitat in which it enjoys a competitive superiority to other species, perhaps because of a more appropriate body size or better crypsis. This superiority is expressed mainly in lean years. Thus, ubiquity is tightly coupled to massive differences in the levels of food availability between years, which will be a powerful influence on the realized ubiquity of animal species in variable environments.

In summary, there is some evidence to support the idea that the diversity of habitats used both during and between years is governed by changes in food availability, at least in terrestrial birds. Some species may wax and wane in ubiquity throughout the year in response to these changes while others substitute one set of habitats for another yet maintain similar levels of ubiquity. The patterns may be complicated further by between-year differences in the levels of food availability, which can also affect the diversity of the habitats used by species.

8.3 Habitat selection and competition

The ubiquity of species must often depend upon the reasons for and the mechanisms by which organisms select alternative types of habitat (Hutto 1985). Habitat selection lies at the interface between ethology, population and community ecology and evolutionary biology (Rosenzweig 1985). Ethologists have attempted to unravel the neurological–behavioural mechanisms by which animals select places in which to live (e.g., Lack 1937, Klopfer 1963, Wecker 1963, Klopfer and Ganzhorn 1985). The relationship of territoriality to habitat selection has had a long history (e.g., Fretwell and Lucas 1970), as have the impacts of intraspecific competition (e.g., Hildén 1965), interspecific competition (e.g. Svärdson 1949, Slagsvold 1975, Pimm and Rosenzweig 1981) and predation (Rosenzweig and MacArthur 1963, McLaughlin and Roughgarden 1992). Evolutionary issues have also been addressed in detail (e.g., Templeton and Rothman 1981, Rausher 1984, Pease *et al.* 1989, Brown and Pavlovic 1992). This pivotal position accounts for the voluminous literature on habitat selection.

I now consider some of the recent work and highlight its relationship to the main theme of this chapter. Note that most of the following discussion refers to habitat selection in animals but some comments on the process in plants are provided later in the section. Note also that I generally use the term 'habitat selection' in its realized sense (which habitats a species ends up occupying) rather than in the mechanistic or behavioural sense (the sensory and motor processes by which individuals identify or distinguish habitats and achieve occupation, Sweatman 1985, Sale 1990).

Before considering habitat selection itself it is worth referring to some of the ideas on interspecific competition in relation to habitat use. Many reviews and models regard interspecific competition as an important component in habitat selection (e.g., Svärdson 1949, MacArthur and Levins 1964, Hildén 1965, Pimm and Rosenzweig 1981, Rosenzweig 1991).

Compression and release

One of the main predictions of competition theory is that there will be a constriction of habitat use rather than of food use when a resident species encounters an invading species, at least on proximal time scales (MacArthur and Wilson 1967). This is called the *compression hypothesis*. Thus, ubiquity is deemed to be more malleable than the use of food resources over short time scales (MacArthur and Pianka 1966). Over much longer contact periods (i.e., evolutionary timescales), however, specialization of both habitat and food use is expected to occur (Schoener *et al.* 1979). The corollary of the compression hypothesis is that species increase their densities and expand the range of resources used when competitors are absent (see Terborgh and Faaborg 1973, Yeaton and Cody 1974, Cox and Ricklefs 1977). This is referred to as *ecological release*.

Some empirical support for the compression hypothesis and ecological release in animals was documented by MacArthur and Wilson (1967). Terborgh *et al.* (1978) cited cases in which the absence of certain bird species in the montane rainforests of some Lesser Antillean islands apparently allows the expansion of habitat use by ecologically analogous species from coastal scrub. Thus, the Antillean crested hummingbird (*Orthorhynchus cristatus*) expands in the absence of the blue-headed hummingbird (*Cyanophaia bicolor*) and similarly, the Caribbean elaenia (*Elaenia martinica*) occupies habitats left vacant by the Lesser Antillean pewee (*Contopus latirostris*). Schoener *et al.* (1979) and Pulliam (1986) also reported results that were largely consistent with the compression hypothesis. Pulliam stated that during times of food scarcity, species of sparrows generalized their use of food but contracted their habitat use, in line with the predictions of compression theory.

McNaughton and Wolf (1970) studied the patterns of relative abundance in several types of community (e.g., forests, grasslands, shrubs and birds). They concluded that increased species diversity leads to reduced ubiquity (p. 135). That is, in comparing communities consisting of few species with richer ones, species seem to be accommodated into the latter by reductions in habitat versatility. Although McNaughton and Wolf (1970) considered mainly plant communities, their results resembled the compression hypothesis – the diversity of habitats occupied is compressed rather than the resources within habitats being subdivided more finely. The compression hypothesis may be especially applicable to plants because of the limited set of (local) resources that almost all plants require or use. This means that the subdivision of resources locally is not possible in the same way as in animal communities (Aarssen 1989).

Notwithstanding these observations, there are several reservations concerning the validity of the compression hypothesis. First, the best tests of theories unquestionably are ones couched in terms of comparisons of the predictions of rival explanations for the phenomenon under study (Wiens 1989b, Niemelä 1993). One study in which the compression hypothesis was tested against the predictions of possible alternative explanations for habitat and food use was that by Dunham (1983). He found that other hypotheses such as optimal foraging behaviour were as good or better explanations than the compression hypothesis for several species of lizards. And second, Terborgh *et al.* (1978), while discussing the cases of ecological release at length and favourably, mentioned without much comment several cases in which ecological analogues did not increase their ubiquities in the absence of rainforest species on some islands, thus contradicting the compression–release picture. For example, the yellow warbler (*Dendroica petechia*) failed to extend into habitats left unoccupied by its congener, the plumbeous warbler (*D. plumbea*), on St Kitts and Montserrat, and the grey kingbird (*Tyrannus domenicensis*) did not occur in typical habitats of the brown-crested flycatcher (*Myiarchus tyrannulus* [*oberi*]) on Montserrat despite the absence of the latter. Thus, despite some confirmatory evidence for the compression hypothesis, there are doubts about its prevalence in natural communities because of the lack of consideration of alternatives to account for the phenomenon in most cases (see also Wiens 1989a: 393–394) and the occurrence of counter-examples in which the anticipated release or compression does not occur.

The dominance model

One of the most influential competition-based models to account for the differences in ubiquity was proposed by McNaughton and Wolf (1970). They used the concept of *dominance*, which is the idea that some species exert an unduly strong influence on the occurrence of other species. Thus, ecologically dominant taxa are held to determine community composition and dynamics. Dominance was equated with *relative abundance* (or density).

McNaughton and Wolf (1970: 131) specifically considered ubiquity ('the ability of a species to maintain populations in differing types of environment'). Their measure of ubiquity, *W*, is the sum over all populations of a species of the weighted relative abundances (McNaughton and Wolf 1970: eqn. 4), where the weighting is derived from the position of the habitat of each population along an environmental gradient. Thus, '*W* measures constancy of relative abundance over a range of environments' (McNaughton and Wolf 1970: 133). In this usage, a species might occupy all habitat types yet is effectively restricted because of an uneven distribution of relative abundances between habitat types. A truly ubiquitous species therefore maintains its relative abundance irrespective of the set of habitat types considered.

Empirical results from a variety of plant communities (e.g., forests, grasslands) showed that 'dominant' species have the largest ubiquities while subordinates are more restricted. There are two alternative explanations for this observation: (1) dominants are generalists and are less likely to encounter constraints from limiting factors, while subordinates are specialists; or (2) all species are specialists but dominance reflects the abundance of resources upon which each species specializes. McNaughton and Wolf (1970) were inclined to favour the second explanation. This is their key finding – ubiquities differ because the niche dimensions underlying species-specific specializations have different abundances and carrying capacities. Thus, similar levels of resource specialization may be expressed as either restricted or ubiquitous distributions depending upon the availabilities of resources underlying each exploitation speciality. If this view is correct, then the ubiquity of a species must be a dynamic feature that responds to continuous changes in environments.

Habitat selection in animals

Although a steady stream of work has continued since the work of Fretwell and Lucas (1970) on how individuals should distribute themselves to maximize their fitness (the *ideal free distribution*), there has been a surge of interest in habitat selection recently. Much of this activity has centred on two forms of theory concerned with the control of habitat selection. Although not exclusive of one another, these theories can be regarded as: (1) models of *density-dependent habitat selection* (see Morris

1989, 1992); and (2) *habitat source–sink* models (see Pulliam 1988, Pulliam and Danielson 1991).

In models of density-dependent habitat selection, the densities of individuals in alternative habitat types are distributed so that the expected reproductive success of any one individual is the same as of any other (Morris 1992). Thus, there will be an equilibration of densities among habitats that reflects their relative suitabilities (although see Lidicker 1975, Van Horne 1983). At extremely low densities, all individuals occupy the best habitat but the suitability of that habitat declines as numbers grow. At some point it becomes advantageous for individuals to move into what were previously less advantageous habitats (Rosenzweig 1981). Thus, selectivity should erode as densities increase (Svärdson 1949, O'Connor 1987, Rosenzweig 1991).

An important limitation in the theory of ideal free distributions is that consumer density is perfectly correlated with resource availability (or habitat quality in a general sense) in alternative habitats, which implies that only hard strategists (in the sense of Chapters 6 and 7) are representable in these models (Rosenzweig 1991: S10). Some of the factors that may decouple the correlation between density and habitat quality are the seasonality and patchiness of habitats, the temporal unpredictability of habitats, social dominance and interactions and high reproductive capacities (Van Horne 1983).

Partly in response to these problems, the more recent versions of the model incorporate important limitations on habitat selection that are enforced by the scales and geometries of habitats and the idiosyncrasies of organisms (e.g., foraging and dispersal ranges, Morris 1992). The comparative profitability of selecting alternative habitat types at the scale at which an animal normally forages depends upon the spatial arrangement of habitats and whether selectivity is worthwhile given the lost opportunities in extended searching or greater exposure to natural enemies. Similarly, the 'decision' to emigrate to another habitat type altogether carries penalties that must be offset to make dispersal worthwhile. Thus, moving and establishing elsewhere means that reproduction in the current habitat is forsaken. There are risks to moving too, usually greater than in staying put (Morris 1987b). When these costs are considered, the comparative profitabilities of alternative habitats are altered, which leads to departures from the simplest expectations of the ideal free distribution.

Having considered the effects of the spatial configuration of alternative habitats and the dispersal capabilities of individuals, Morris (1992) concluded that different processes of habitat selection may be inferred depending upon location. Populations in ecotonal areas may display quite different patterns of habitat selection to populations occupying patchwork landscapes consisting of smallish mosaics, and the dynamic redistribution among habitats may not be evident at all if alternative habitats are separated by distances in excess of the dispersal capacities of individuals. If the dimensions of habitat patches exceed dispersal distances, then population dynamics are more likely to be controlled by factors operating within habitats than those at work between habitats.

Although density dependence is generally assumed to lead to reduced selectivity as populations grow (i.e., increased ubiquity), interspecific competitors may force the restoration of selectivity (Svärdson 1949). Rosenzweig (1985) described a graphical approach to habitat selection under the influence of interspecific competition, which he called *isoleg analysis*. The dynamic nature of this representation means that the patterns of habitat selection can be understood as the densities of each competitor species change, as they do almost continuously in natural communities. The more recent versions of isoleg theory incorporate greater powers of habitat discrimination by individuals, which often leads to profound changes in the expected patterns of habitat selection (Rosenzweig 1991). For example, depending upon circumstances, species may display partial preferences for habitat types – some patches of habitat of the 'same kind' may be occupied while others are not because of subtle (at least to humans) differences between patches. These ideas may contribute at least in part to explaining many seemingly mysterious cases in which apparently suitable sites for a species are not occupied, although stochastic recruitment undoubtedly plays a part in some situations (Sale 1990). In general, the models of density-dependent habitat selection assume or suggest that interspecific competition strongly influences patterns of ubiquity, normally but not always restoring selectivity lost during population growth.

There are several models of optimal habitat selection under the impact of predation with both single and several prey populations (e.g., Sih 1987). Many of the results seem to be consistent with viewing predators as cost factors that differ between habitats, and so can be incorporated neatly into the density-dependent framework (Rosenzweig 1991).

Models of the second kind, habitat source–sink models, are based on the idea that some habitats are especially suitable for a species while others are not so good (Pulliam 1988). In *source* habitats, reproductive success exceeds the numbers required for stable population sizes so that these habitats act as centres of emigration. *Sink* habitats are poorer to an extent that local recruitment is insufficient to maintain population densities and immigrants from source habitats can be accommodated.

The interplay between source and sink habitats has many consequences, one of which is to depress population densities over the entire ensemble of source and sink habitats if the ratio of sink to source habitats is high and if dispersal abilities are limited. The 'dilution' of source habitats by intervening sink habitats has important repercussions for the viabilities of species at the landscape scale (Pulliam and Danielson 1991).

When extended to many species, the interactions between nominal competitors in source–sink models can take on a variety of forms in not necessarily intuitive ways. Pairs of competitors that are poor habitat selectors may even show an apparent facilitation (Danielson 1991).

There is evidence for some species of plants that the source–sink idea is realized in natural systems. For example, Kadmon and Shmida (1990) and Kadmon (1993) have shown that recruitment from favourable habitats (specifically wadis and depressions) supplements population densities in less favourable ones (xeric slopes) in the desert annual *Stipa capensis* in Israel. Kadmon (1993) suggested that more than 90% of the individuals found in slope habitats emanated from wadis and depressions. Depending upon rainfall, between 75% and 99.9% of seeds were set from the plants growing in the wadis and depressions although these habitats comprised just 10% of the area occupied by the population.

Danielson (1992) extended the simple source–sink models to include 'unusable' habitats so that source and sink habitats are viewed as being embedded in a matrix of habitats in which reproduction cannot occur. Thus, unusable habitats impose burdens on dispersing individuals in the form of increased travel and search times, and as places where individuals generally are maladapted leading to greater exposure to predation and physiological stress. As habitat selection implicitly depends upon animals 'sampling' patches to determine their relative suitabilities, the amount and distribution of unusable habitat must greatly affect optimality arguments because of the effects on the ease with which appropriate habitats can be found (Taylor and Taylor 1979).

The dilution of usable (source and sink) habitat by unusable habitat may cause an effective contraction of ubiquity because dispersers may often fail to find sink habitats. This means that the densities in source habitats increase relative to sink habitats, which is equivalent to reducing ubiquity (McNaughton and Wolf 1970, Rice *et al.* 1980).

Again, the extension of the model by including unusable habitats to interacting species showed that unusable habitats may increase the chances of facilitation and reduce or eliminate the mutual depression of densities (Danielson 1992: Fig. 4). These models assume that the sources for one species are the sinks of the other and *vice versa*. Danielson (1992) concluded that dispersal through unusable habitats has significant effects on the interactions between potential competitors even if the costs of dispersal through such habitats are negligible.

Source habitats have been regarded as fundamental niches and some think that such habitats drive the adaptive evolution of species (e.g., Holt and Gaines 1992). Species characteristics are fashioned in source habitats simply because more individuals occur – and are produced – in them. Surplus individuals dispersing from source habitats carry the marks of adaptation to source habitats into sink habitats (Holt and Gaines 1992). However, sink habitats do contribute some individuals to recruitment, so that their effects are qualitatively different from unusable habitats.

In scrutinizing recent work, Rosenzweig (1991) felt that habitat selection theory had developed into a comparatively successful explanation for the distributions of animals among habitat types. Some applications of the theory seem to support this conclusion (e.g., Morris 1989). The implications of models with source, sink and unusable habitats are particularly interesting. As workers consider more grades of suitability of habitat classes and greater habitat fragmentation (e.g., Palmer 1992), these models must increasingly rely on computer simulation rather than on mathematical modelling. The problems then will rest with empirically estimating the appropriate parameters for natural systems (e.g., fitnesses, diffusion rates, actual differences in habitat suitability, how individuals gauge these suitabilities, etc.; see §6.5: Metapopulations).

Habitat selection in animals – physiognomy and floristics

An ecologist's perception of the ubiquity of species depends upon the criteria used to characterize the similarity of habitats. It also depends upon the basis by which organisms themselves choose habitats in which to live. In ornithology, there has been much debate as to whether birds use the structural or floristic features of vegetation to select habitats (Rotenberry 1985, Wiens *et al.* 1987). Some workers have highlighted physiognomic responses, while other have demonstrated the fidelity of some birds to particular species of plants. For example, I recorded red-capped robins only in box–ironbark (*Eucalyptus microcarpa – E. tricarpa*) woodlands within the Victorian transect, but found white-plumed honeyeaters (*Lichenostomus penicillatus*) in both Gippsland manna-gum (*E. pryoriana*) and river red gum–grey box woodlands, possibly because both are open with almost savanna-like structure.

Rotenberry (1985) listed several examples of studies showing either floristic or physiognomic correlates with habitat use in bird communities in many parts of the Americas and similar examples can be found on other continents (see Mac Nally 1990b). Rotenberry believed that physiognomic influence is couched mainly at the biome scale and that floristic effects are manifested at smaller spatial scales. However, both can be shown to be influential at the same scale (Mac Nally 1990b). It seems likely that many species select habitats for structural reasons, others for floristic elements and yet others for both reasons. There seems little to gain in contesting the respective merits of physiognomy and floristics in habitat selection (Mac Nally 1990b).

The dichotomy between physiognomic and floristic effects in habitat selection has important repercussions for ubiquity. Our picture of the diversity of habitats occupied by a species depends upon whether similarities of habitats are gauged on floristic or physiognomic grounds. Floristically distinct combinations of tree species may be structurally rather similar. For example, the physiognomically similar foothill woodlands of the Great Dividing Range in Victoria consist of a variety of mixtures of boxes, stringybarks, peppermints (all *Eucalyptus*) and *Acacia* species. On the other hand, particular examples of the one type of floristic association (e.g., river red gum–grey box woodland) may be structurally more dissimilar from one another than are the floristically diverse foothill woodlands (Mac Nally 1989: Fig. 1). Quantifying ubiquity becomes problematic under these conditions because the breadth of habitat use may be deemed to be moderate to large if one were using physiognomic criteria but much smaller if floristics were used, and vice versa. The way in which ubiquity is measured and determined, as described in §8.1, masks this dilemma. Clearly, some unified approach to the floristics–physiognomy problem is needed.

Floristics not only directly determine the occurrence of some species of animals but can also shape community structure. B. J. Traill of Monash University (pers. commun.) has studied the relationship between distribution and floristics in several species of Australian honeyeaters, the regent (*Xanthomyza phrygia*), fuscous (*Lichenostomus fuscus*) and yellow-tufted (*L. melanops*) species. Traill was particularly interested in finding the reasons for the puzzling distributions of these species in relation to woodlands dominated by red ironbark (*Eucalyptus sideroxylon*) in Victoria. The three species of honeyeaters are common in ironbark-dominated woodlands in north-eastern Victoria but are rare, sporadic, or absent in the large remnant tracts of ironbark-dominated woodlands spread across central Victoria.

It turns out that the ironbarks in the two regions are different species although they have been regarded as only subspecifically differentiated until very recently (Hill and Johnson 1991). *E. sideroxylon sideroxylon* ranges from north-eastern Victoria to southern Queensland and *E. s. tricarpa* occurs in central Victoria and south-coastal New South Wales (Bramwells and Whiffin 1984). The latter taxon has been reclassified as *E. tricarpa* (Hill and Johnson 1991).

The upshot of Traill's work is that the three species of honeyeaters all either breed or occur in high densities in woodlands consisting of *E. sideroxylon* (now mugga ironbark) but are rare or sporadic in *E. tricarpa* woodlands. There are no apparent differences in physiognomy between the two forms of woodland and other characteristics such as understorey and edaphic features do not seem to account for the differences in bird distributions. Traill attributed the difference in attractiveness between the two species of ironbark to nectar production, with *E. sideroxylon* being the more consistent or generous producer of nectar.

Thus, floristics may control the distributions of these three species of honeyeaters rather than physiognomy, at least in ironbark-dominated forests. This dependence appears to be translated into community-level effects, for fuscous and yellow-tufted honeyeaters pugnaciously exclude smaller species of carnivorous birds (e.g., thornbills, *Acanthiza* spp.; robins, *Petroica* spp.; fantails, *Rhipidura* spp.; gerygones, *Gerygone* spp.) from *E. sideroxylon* woodlands, but obviously not from *E. tricarpa* woodlands (Traill, pers. commun.). Hence, the influence of habitat choice based on the floristic composition of woodlands by some species of honeyeaters ultimately reduces the ubiquities of several other species of birds.

What can we conclude about habitat selection and ubiquity in animals? Some models involving density-dependent behaviour appear to be useful descriptions of the ubiquity of particular species (e.g., *Peromyscus maniculatus*, Morris 1992). However, the work of Rice *et al.* (1980) sounded a warning because of the idiosyncratic behaviour of species occurring within the same region. They doubted whether models of habitat selection based only on densities are applicable to most or even many species. The diversity of seasonal responses they reported indicates that many other factors are involved and that the reasons for particular species showing the patterns that they do remain unclear. Rice *et al.* (1980) also found that the optimality of habitats varied seasonally and inconsistently between species, which means that models of habitat selection appealing to optimality criteria also must incorporate a seasonal dependence.

Habitat selection in plants

The (unconscious) choices by organisms involved in theories of habitat selection mean that most ideas have little or no direct application to plants. As Bazzaz (1991) put it: 'characteristics of the habitat determine which species . . . [that] disperse into the habitat become established . . . choice is made proximally by the habitat but ultimately by the plants through natural selection.' It seems difficult to envision a dynamic equilibration of densities in the sense of the ideal free distribution in plants.

Most of the exposure of plants, particularly terrestrial ones, to alternative habitats is mediated through the juvenile (= seed) or germ (= pollen) phases (similar arguments also apply to obligatorily sessile marine animals). To this end, plants have evolved a host of characteristics to overcome the limitations associated with the sedentary nature of adults. These include the wide dispersal of many propagules, directed dispersal to suitable habitats (perhaps by using animal vectors), large nutrient reserves for propagules, morphological differentiation, clonal spreading and the synchronization of life-history events with suitable habitats through time. Thus, the lack of adult mobility has been counteracted by a high acclimation potential, phenotypic plasticity and ontogenetic changes that reduce the need for habitat selection in adults (Bazzaz 1991). So, ubiquity in plants amounts to the degree to which populations can maintain their presence in different habitat types and choice, as such, is restricted to the capacity to disperse to favourable habitats.

Terrestrial plants appear to have comparatively wide ubiquities for several reasons. These include a restricted set of basic resources (light, water, a small set of nutrients), high genetic diversity and phenotypic plasticity, gender-based habitat specificities and an ongoing record of previous generations as manifested in the seed bank in the soil (Bazzaz and Sultan 1987, Dawson and Ehleringer 1993). The seed bank may negate the impact of selection events in the intervening years since seeds were deposited and also reduce tendencies for specialization to occur. It should be pointed out that this picture of wide ubiquities in plants is by no means a universal view. For example, Tilman (1985) believed that plant succession could be attributed to the specialization of different species on certain combinations of nutrient and light concentrations. Thus, competitive exclusion in Tilman's view occurs serially as specialists at each stage alter the physical and biotic environment to make conditions unsuitable for themselves and more suitable for specialists of later successional stages.

8.4 Other models of ubiquity

Although there has been extensive modelling of habitat use in terms of habitat selection, this is by no means the only theoretical framework in which modelling has been developed. In this section, I review some of the alternatives to the density dependence and source–sink models of habitat selection. Some of the constraints or functional relationships in most of the models considered in §8.4 are derived from empirical observations. The models then describe the consequences on the patterns of ubiquity by using these empirically derived constraints. For convenience such models are referred to here as *contingency* models because they are 'contingent' on certain empirical observations. The capacity of some of these models to account for many observations is promising, even though some facets of the theory remain unexplained and arise from or depend upon field measurements.

Modelling the ubiquity-position relationship

Recall from §8.1 that ubiquity and niche position are usually found to be inversely related (Fig. 8.2). Ubiquitous species show small positions while restricted species show large values of position. I have developed site-selection algorithms to try to account for these relationships between position and ubiquity in studies of both cicadas (Mac Nally 1988b) and birds (Mac Nally 1990a). If any one of the algorithms reproduces the details of the patterns, then the basis by which sites are chosen in that algorithm may reflect the ecological processes leading to those patterns. It turns out that most of the details of the position–ubiquity relationship can be generated for both the cicada and bird systems by using rather simple algorithms.

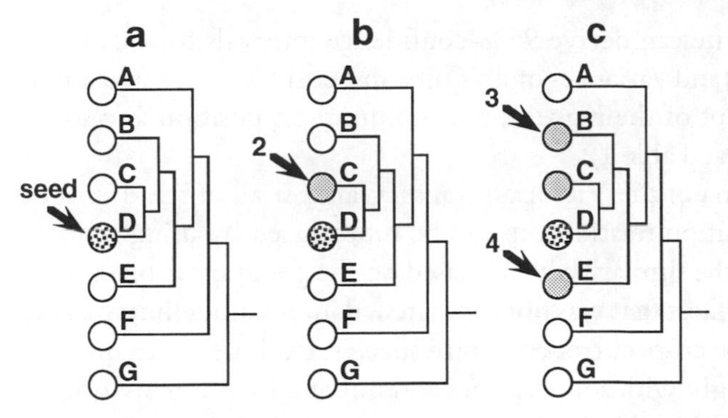

Fig. 8.4 Conceptual representation of the way in which sites are selected in algorithms for modelling the ubiquity–position relationship. The sites are linked by using a dendrogram such that sites that are most similar to one another are clustered together first, in the usual way. The 'similarity' of sites may refer to physiognomic or floristic similarity, or geographic proximity depending upon the algorithm involved. For simplicity, let there be seven sites (A–G). Also, say that four sites have to be selected to represent the site occupancy of a model species. The first site, the 'seed', is randomly chosen, in this case the seed is site D (a, coarse stippling). The next site to be selected is the site that is most similar to D, namely, C (b, fine stippling). The remaining two sites, here B and E, are selected in order of similarity to the seed site (c).

In the site-selection algorithms, each 'real' species has a counterpart 'model' species, which is used to represent it. The algorithm is based on a three-step formula (Mac Nally 1990a). First, the number of sites (S say) to be occupied by each model species is randomly drawn from a distribution based on the observed numbers of sites per species. Second, the initial or 'seed' site for each model species is randomly selected (Fig. 8.4 a). And third, the remaining ($S-1$) sites are selected in terms of their 'similarity' to the seed site. Thus, the second site to be selected is the one most similar to the seed site, the third site chosen is the one that is next most similar to the seed and so forth up to the number of sites the model species is to occupy (S, from the first step) (Fig. 8.4 b, c). The 'similarity' of sites may refer to physiognomic or floristic similarity, geographic proximity or the similarity of climate or elevation.

Once all S sites are selected, estimates of the ubiquity and position for the model species can be calculated as though the model species were a real species. This procedure is repeated for all of the model species in one simulation run and once the ubiquity and position values for each are calculated, a relationship such as the one depicted in Fig. 8.2 can be computed. Clearly, the entire process can be repeated as many times as desired, so that one can derive 95% confidence intervals for quantities such as the means and variances of ubiquity and position, the correlation, slope and intercept of the regression of ubiquity on position and so on (Mac Nally 1990a: Table 1).

For the avifauna of the Victorian transect, almost all of the details of the ubiquity–position relationship can be reproduced by using such an algorithm, with the similarity being based on the geographic proximity of sites. The only facet that was poorly matched in this modelling was the distribution of the co-occurrence of bird species. Real species frequently overlapped strongly with some species and not at all with many others. The most successful (and indeed all) algorithms produced very few such strongly overlapping species-pairs and rather fewer nonoverlapping pairs of species than observed. Thus, there appear to be sets of bird species that share significantly stronger preferences for particular habitat types than would be expected by chance associations. Similar results have been reported for many years (e.g., Grinnell 1917, Gilpin and Diamond 1982, Cody 1983, Wiens 1989a, Schoener and Adler 1991).

Two important points must be made here. First, the algorithm, although reasonably successful at reproducing many of the details of the position–ubiquity relationship for the avifauna as a whole, nevertheless is intrinsically unsatisfying because it requires the observed distribution of sites per species to be supplied (step 1 of the algorithm). One would want to be able to do away with this dependence in a more complete understanding of the distributions of the birds involved. In a sense, at least some of the variation in ubiquity arises from this dependence on an empirically-derived distribution of sites per species.

The second point is that although the most successful algorithm is one based on geographic proximity of sites, this probably will not be a general result. For example, the best 'fits' in the models for the cicadas depended upon the structural similarity of sites rather than on their geographic relationships (Mac Nally 1988b). Although differences may be expected for organisms as different as subtropical, short-lived (as adults) insects such as the cicadas and temperate zone birds, there are several reasons why different results may emerge apart from these taxonomic and climatic ones.

The spatial scales over which the studies were conducted were very dissimilar. The cicada work was based on locations separated by distances of the order of tens of kilometres at the most. In some cases, two or three structurally dissimilar plots were in relatively close proximity to one another (hundreds of metres apart). Given the fidelity of some of the cicada species to habitats of a certain physiognomy (e.g., *Cystosoma saundersii* in shrubland but never in forest), the 'best' algorithms were those that used the structural features of habitats rather than those that employed geographic criteria because structurally different sites often were quite close to one another.

On the other hand the habitats of the bird study were distributed over a transect hundreds of kilometres long, with rather less 'mixing' of structurally different habitats. There was some geographic assortment of sites, with red gum–grey box woodlands and box–ironbark forests concentrated in the northern part of the transect and Gippsland manna gum woodlands in the south. The significance of this segregation of sites consisting of different plant associations is that if many species of birds show geographic localization in the northern, southern *or* central parts of the transect, then geographic algorithms will capture this zonation within the transect and, therefore, perform better than structurally based algorithms. This is what seems to happen.

These comments show that one has to be wary about the conclusions one draws given the potential impact of the spatial scale and sampling geometry of sites included in a study. In some respects, the geometric arrangement of study sites drew out the physiognomic dependencies in the cicada study and the geographic influence in the bird work. Thus, although the ubiquity–position relationship could be modelled well by using certain simple algorithms in both cases, the sampling design had a major influence on the form of algorithm that performed best in each case.

Contingency models of ubiquity

Three other contingency models are considered here to provide an overview of the variety of approaches and the sorts of success that these models have had in accounting for patterns in natural communities. The models are: (1) the *marginal mosaic* model of Williams (1988); (2) the *hierarchical* picture of Kolasa (1989); and (3) the *fractal landscape* idea of Palmer (1992).

As these are regarded as contingency models, what are the empirically derived elements in each case? The marginal mosaic model requires response surfaces for densities such that densities decline from range centres to range margins, in line with some empirical observations (see Hengeveld and Haeck 1982, Brown 1984). The hierarchical model assumes that differences in ubiquity exist *ab initio* and then attempts to account for the abundance and diversity of ubiquitous and restricted species. The fractal geometry model imposes a coenocline along which species have different physiological specializations. Thus, none of these approaches is entirely self-sufficient and each appeals to empiricism in one form or other.

Of course, the three models are by no means an exhaustive list of contingency models (see for example, Hall *et al.* 1992) but they should give an indication of the variety of ideas that have arisen in this area. The subsection also includes a brief discussion of the *core and satellite model* of Hanski (1982). This model applies only to relatively uniform habitats and therefore, has little direct relevance to ubiquity *per se*. However, the Hanski model is pertinent to niche patterns within habitats so is worth reviewing briefly here for completeness.

Note that almost all contingency models, including those of this subsection, are pitched at the *landscape* scale of spatial resolution. The landscape scale encompasses many habitat types. It may be defined as the scale over which the dynamics of the set of populations of a species influence one another; in practice, this means that these populations communicate with each other by dispersal (Danielson 1992). This intercommunication means that the local dynamics of a population can be influenced by the influx of individuals from other populations, and also that local dynamics can be altered because individuals are able to disperse. The landscape scale is integral to the models discussed here because they depend upon interchanges between the populations of a species in different habitat patches.

The marginal mosaic model (Williams 1988)

Williams linked geographic range and local density together to account for the ubiquity–density patterns evident in bumble bee communities in Kent, UK. The significance of this work lies in its refinement of previous ideas on the range–ubiquity–density relationship (Andrewartha and Birch 1954, Hengeveld and Haeck 1982). According to Williams (1988), habitats provide different levels of resources for organisms, which in turn affect their carrying capacities. Assuming that extinction probabilities increase as densities decrease then the relative viabilities of populations also must depend upon habitat type. Populations maintain higher densities in some habitats than others, which means they may be less prone to local extinction through stochastic events or the Allee effect (see Hopf and Hopf 1985). At range centres, however, species are assumed to be best suited or matched to the climatic conditions (Brown 1984) and

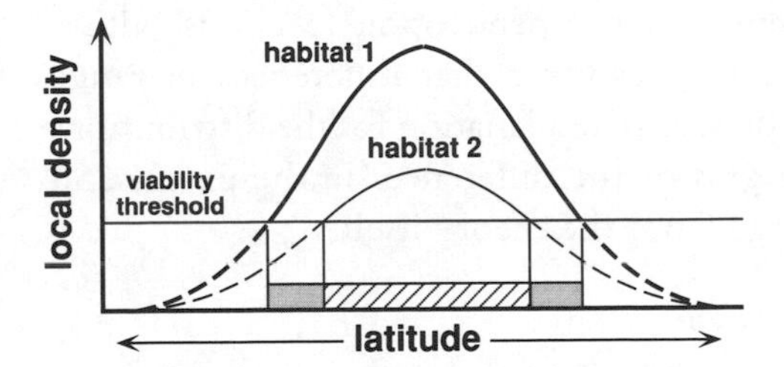

Fig. 8.5 Rationale behind the marginal mosaic model. The density of a species in any habitat is assumed to increase from the margins to the centre of its geographic range. Population viability is held to be the same throughout the range and in all habitats so that populations cannot be supported when the carrying capacity is less than the viability threshold (dashed lines). Different habitats support different carrying capacities so that the species is differentially lost from habitats at locations away from the range centre in a habitat-specific fashion. The species occupies habitat 1 over a larger spatial domain (hatching and stippling) than habitat 2 (hatching only). Thus, ubiquity declines from range centres to range margins. After Williams (1988).

the densities are at the highest levels (Fig. 8.5). Ubiquities should be maximized at range centres irrespective of the relative 'carrying capacities' of different habitats (given that other factors are equal) because the carrying capacities of all habitats are assumed to be at their highest levels there (Brown 1984). However, away from range centres the carrying capacities of habitats are differentially reduced so that some habitats no longer can support viable densities. This should lead to patchy distributions at range margins where populations persist in only the 'best' habitats for the species. Of course, this argument requires that the ecological attributes of species are similar throughout their ranges (i.e., *ecological invariance*, see Brown 1984, Hengeveld 1990: 131).

The marginal mosaic model predicts a span of ubiquities within a 'region' depending upon the mix of species and the overlap of their geographic ranges. Thus, regionally ubiquitous species occur because the region being studied happens to coincide with their range centres while the degree of restriction in habitat use is more-or-less inversely related to the distance from range centres. Species at their margins should appear to be habitat specialists at the regional scale although they may well be regarded as ubiquitous nearer to their range centres (Williams 1988).

This approach clearly has great relevance to ubiquity because it generates distributions of ubiquities within regions. However, the marginal mosaic model requires the existence of species-specific numerical and environmental responses that depend upon locations within ranges. In other words the model assumes that differences in range distributions exist and then produces an explanation for the distributions of regional ubiquity. The origins of the differences in the extent and centres of ranges do not emerge from the theory itself.

The hierarchical model (Kolasa 1989)

In this model, Kolasa bypassed a detailed knowledge of the effects of ecological processes within communities (e.g., competition, predation, habitat selection, physical constraints) to account for the patterns of ubiquity and density. The model intrinsically links together the degree of habitat versatility and density and predicts the occurrence of distinct clusters of species, with the clusters having significantly different ubiquities. These clusters result from abrupt changes in the environmental granularity to which species react.

The empirical component in this model is that species have been shown to display a wide range of ubiquities – how this differentiation arises is not explained. Different levels of ubiquity may be due to species-specific physiology, behaviour, and regimes of natural predators and competitors, but the precise factors in each case are not needed in or generated by the model.

However, by using this assumption Kolasa (1989) asserted that hierarchically structured systems should consist of a pyramid of species (Fig. 8.6). There should be few abundant, ubiquitous species (habitat generalists), somewhat more species showing intermediate habitat versatility and abundance and many rare, restricted species.

The link between habitat versatility and density depends upon the amount of resources to which species have access. Thus, generalized species occupy more habitats giving them access to greater resources and, consequently, producing higher abundances. The differences in density between ubiquitous and restricted species should also be amplified and reinforced by various mechanisms that limit the efficiency of habitat specialists. Some of these effects are increased travel times and exposure to natural enemies in moving between favourable habitats, coping with the barriers between habitats, higher likelihoods of stochastic extinction in fragmented landscapes due to small numbers and so on.

The hierarchical model appeared to be able to account for the ubiquity distributions in a variety of assemblages including turbellarians, chironomids, foraminiferans, birds, rodents, water bugs and mayflies (Kolasa 1989). The hypothesized hierarchical divisions also appeared to be well

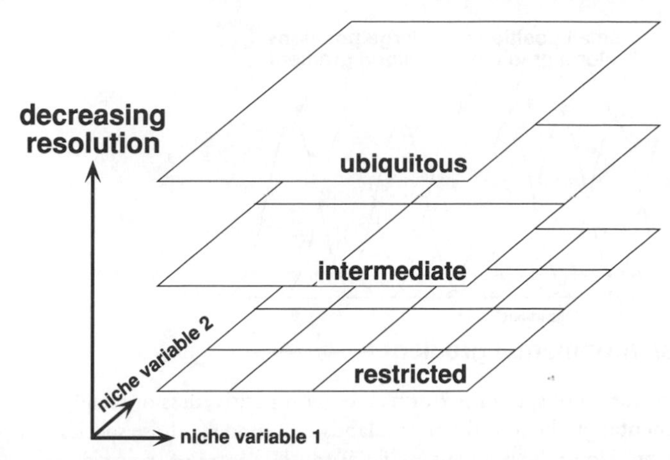

Fig. 8.6 The basis for the hierarchical model. Species use habitats according to the degree to which they can or do resolve habitat heterogeneity. Species showing poor responsiveness to heterogeneity are ubiquitous, while intermediate species distinguish some degree of heterogeneity. At the lowest level are restricted species that display the strongest resolving power. After Kolasa (1989).

supported. By using the simple heuristic that the densities of any two species within communities be proportional to the square of their respective ranges, Kolasa also was able to account for patterns of densities. However, the model requires rather than generates differences in ubiquity among species, making it more contingent than one would like in order to account for niche patterns.

The fractal landscape model (Palmer 1992)

Palmer used fractal landscapes to explore the effects of spatial variability on ubiquity, density, and species richness at both the local and landscape scales. The fractal aspect in this case does not refer to the self-similarity of microhabitat, macrohabitat and habitat structure (see Collins and Glenn 1990) but rather, the spatial dependence ('autocorrelation') in the variation in the landscape. For example, consider a landscape defined by a single constituent, soil phosphorus concentration ([P]) say. If the spatial autocorrelation of [P] is high then the fractal dimension, D, is close to 2 (i.e., density dependence). If adjacent points are uncorrelated then D is 3 (i.e., density independence). Different levels of spatial autocorrelation can be represented by using various values of D between 2 and 3. Thus, D represents or characterizes the spatial structure of the landscape, including patch size and the separations of similar patches.

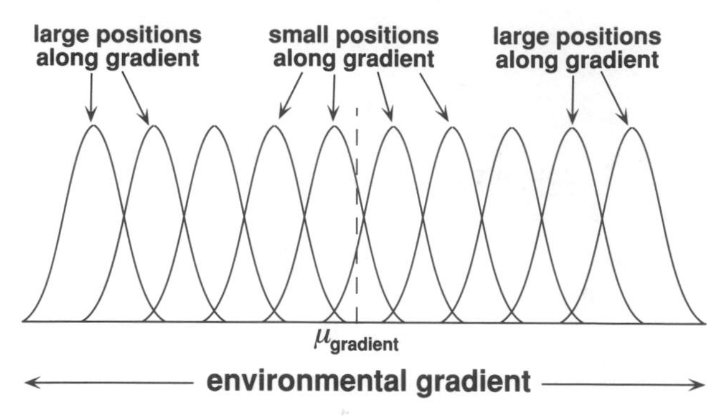

Fig. 8.7 The distribution of species performances (competitiveness or fecundity) along an environmental gradient in the fractal landscape model. Those species whose mean locations along the gradient are far from the centre of the gradient ($\mu_{gradient}$) have large niche positions while those species whose means are near to the centre of the gradient have small positions. After Palmer (1992).

However, the fractal dimension is only one of two key aspects of spatial distribution, the other being the variance of the environmental factor. If the variance is zero then the same, completely homogeneous spatial structure occurs regardless of the fractal dimension. On the other hand, the most heterogeneous landscapes have both a large variance and high D (see Palmer 1992: Fig. 1).

Palmer (1992) implemented the performance of species with respect to the environmental gradient (e.g., of [P]) in the model as a set of offset, partially overlapping Gaussian curves (Fig. 8.7). Thus, the survivorship and reproduction of each species at any point in the landscape depended upon the species' relative performance ('competitiveness' or fecundity in different models) given the value of the environmental variable at that point. This coenocline also explicitly defines niche position for each species as a function of the position of the curve along the gradient (Fig. 8.7). Recall however, that this form of coenocline may not be particularly common in natural communities (see §7.1; Rosenzweig 1991).

In this setting Palmer (1992) found that niche pattern was significantly affected by fractal dimension. As the spatial autocorrelation declined (high D), densities decreased, with species having small niche position increasing in relative abundance compared with species with extreme niche positions. Although overall densities declined, species with smaller positions decreased less than species with larger positions, so that species with smaller positions became *relatively* more abundant. Ubiquity also increased with larger values of *D*. Thus, the effect of high spatial independence in landscapes is to spread species more evenly among different habitat types, thereby increasing ubiquity. A higher variance in the environmental factor also increased ubiquity.

Palmer (1992) attributed the increase in ubiquity in landscapes with high fractal dimensionality to the fact that any local habitat plot is small and is surrounded by dissimilar habitat types, and is also distant from other similar habitats. This leads to frequent extinction (small populations in small habitats) and a recurrent cross-colonization between habitat types (small distances to other habitats, Malanson 1985, Shmida and Wilson 1985). Thus, although many of these surrounding habitats are suboptimal and unsuitable for successful reproduction, species can maintain a presence in them through dispersal, which increases their ubiquity (see §8.3: Habitat selection in animals).

One of the most interesting outcomes of the study was that although each species had equal physiological breadths as gauged by the range of tolerances along the coenocline, the physiological equality was not expressed as an equality of realized ubiquity in these highly fractured environments. Species with small niche positions (i.e., close to the middle of the coenocline) acted like habitat generalists, while species at extreme positions performed more like habitat specialists. Thus, this model produced results that are consistent with the inverse coupling between ubiquity and position reported in field studies even though the potential ubiquities were equal (see §8.1: Ubiquity and niche position). If this effect is ecologically meaningful, then the realized ubiquity in different landscapes will depend upon the absolute ranges of environmental factors in each landscape even if species maintain the same intrinsic physiological and ecological characteristics from place to place. This reflects the fact that niche position corresponds to the overall level of adaptation of a species to the set of habitat types encountered. A species may fall in the mid-range of the span of an environmental factor in one landscape and at an extreme part elsewhere. Perhaps this partially explains why ecological performance (density, productivity, etc.) at range peripheries is usually significantly poorer than at range centres – positions with respect to one or more resource dimensions are likely to be more extreme in most of the habitats at the range margin of a species than at its range centre (Hengeveld 1990, Hall *et al.* 1992).

The simulations of Palmer (1992) provided a context in which many aspects of niche position and range could be interpreted together in the light of spatial heterogeneity. The emergence of a numerical dominance of species with small niche positions and increased ubiquities of all species, as spatial autocorrelation decreased, was a key result.

Overview of contingency models

The contingency models discussed here (including the models of the ubiquity–position relationship) apply very different methods to deal with the problem of ubiquity. In one form or another, each model except for Palmer's (1992) requires the prior existence of differences in ubiquity, which means that they are at best partial explanations for observed patterns. Palmer's fractal landscape approach is promising because differences in ubiquity arise from similar breadths of physiological performance along a coenocline. Thus, the diversity on one characteristic (habitat use) emerges from a uniformity with respect to another characteristic (physiology). McNaughton and Wolf (1970: 137) came to a similar conclusion: 'What determines niche width ... frequency and carrying capacity of the exploitation speciality [of species].' Whether similar results hold for other configurations of physiological distributions is unclear at present but the model seems amenable to field testing. In a similar vein to the source–sink theories of habitat selection the fractal landscape model implies that populations of species may occur in habitats in which they are not particularly suited, thereby increasing the versatility of habitat use. Ubiquity in such a picture must be dynamic, reflecting the spatial and temporal frequency of habitat changes in landscapes and stochastic rather than deterministic interactions between species (Wilson 1992).

The core and satellite species hypothesis (Hanski 1982)

I mention in passing an important model that bears some relationship to the contingency models of ubiquity. This is the *core and satellite species hypothesis*, which was suggested by Hanski (1982). A clear statement of the model assumptions, which are rather restrictive, is provided by Gaston and Lawton (1989: 762). The model seeks to account for the proportions of sites occupied within more-or-less homogeneous habitats by using the rates of local extinction and colonization. A key relationship is that the probability of (regional) extinction is inversely related to the numbers of sites currently occupied by a species.

Two outcomes are predicted: (1) a bimodal distribution in which there are widespread, abundant species (the *core* species) and localized, rare species (*satellites*); or (2) all species are widespread (Gotelli and Simberloff 1987). Which of these alternatives is to be expected in any particular case depends upon the values of model parameters but most work has concentrated on the bimodal prediction. Although the core–satellite hypothesis generates predictions similar to earlier explanations of distribution and abundance it provides a unique combination of predictions (see Gotelli and Simberloff 1987, Williams 1988, Gaston and Lawton 1989, Collins and Glenn 1990). Several plant communities seem to show patterns that are consistent with the predictions of the model (Gotelli and Simberloff 1987, Collins and Glenn 1990), but the most searching test on animal communities failed to support it (Gaston and Lawton 1989). The reason why this model does not assume a higher profile in this chapter is that one of its main restrictions is that sites are more or less the same type of habitat (e.g., prairie (Gotelli and Simberloff 1987) or bracken (Gaston and Lawton 1989)), which means that it cannot account for ubiquity *per se* (Williams 1988). Nevertheless, Hanski's model provides a way of exploring the overall distribution and abundance of some species. Patchy distributions within habitat types also contribute to our perception of the versatilities of species and need to be explained by this and rival models (see Rabinowitz 1981).

8.5 Versatility and ubiquity

Having considered many issues influencing ubiquity I now move on to the issue of whether ubiquity and local versatility are functionally related. One might naïvely expect that ecologically versatile species also would be relatively ubiquitous by capitalizing on their generalized capacities of resource use to occupy a diverse set of habitat types (e.g., Jacnike *et al.* 1980: 202, Bowers 1988). Organisms that are obligatorily dependent on specific resources, such as some parasites and insect herbivores, must be constrained to just those habitats in which such resources (hosts) occur.

Of course, this simple argument has flaws, some of which were discussed by Fox and Morrow (1981). For example, although individuals of each population might be specialized in a local sense, alternative resources may be substituted elsewhere allowing the species to occupy many habitat types (Fox and Morrow 1981, Nitao *et al.* 1991). The mistletoebird (*Dicaeum hirundinaceum*) of Australia is a case in point being a specialized consumer of mistletoe berries in whichever habitat types that mistletoes occur. Thus, similar sorts of resources can occur in different habitat types and from different sources, which means that the same specialty can be supported widely by using a variety of vehicles.

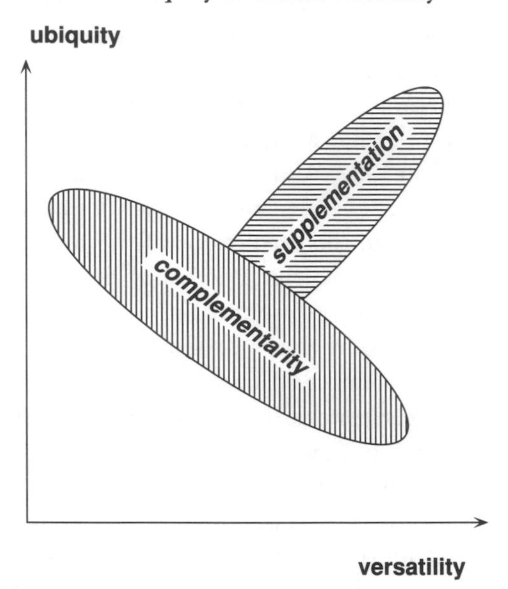

Fig. 8.8 Representation of complementarity and supplementation between ubiquity and versatility in the model of Cody (1974).

Although the link between local versatility and ubiquity would appear to be an obvious target for both theorists and field workers, there seems to have been little systematic assessment of the degree to which local versatility assists or hinders species occupying many kinds of habitat. However, Cody (1974) considered the ubiquity–versatility relationship in some detail. He noted a trend for ubiquitous species to be food specialists (e.g., parulid warblers) and for restricted species to be food generalists (e.g., emberizid finches). In this scheme, local versatility would be inversely related to ubiquity. Cody called this *complementarity*. He envisaged that complementarity most likely occurs under stringent competition in ecologically saturated conditions. Cody (1974) also recognized the possibility that some species might show a positive relationship between ubiquity and versatility, in line with the simple or naïve expectation mentioned at the start of this section. This positive relationship, *supplementation*, might arise in situations in which interspecific competition is less severe or not influential or when extremely competitive situations occur for some part of the year (e.g., in wintering grounds).

Complementarity and supplementation are depicted in Fig. 8.8 – note the Y-shaped distribution between ubiquity and versatility arising from the union of the two relationships. A serious problem with this picture is that it is a difficult one to falsify – in other words, how can a Y-shaped distribution be distinguished statistically? Unless one teases apart *ab initio* faunal (or floral) groups that are likely to show either complementarity or supplementation, a random selection of species probably will yield a non-significant relationship between ecological versatility and ubiquity.

In any event, there is a dearth of matched estimates of ubiquity and ecological versatility in the literature so that definitive tests of the complementarity and supplementation ideas remain to be conducted. However, surrogates for ecological versatility have been used by some workers to provide a preliminary indication of whether ubiquity is influenced by versatility. The most common surrogate is foraging versatility, especially in birds. It is important to emphasize that versatility in foraging does not necessarily correspond directly to versatility of resource exploitation (Wiens 1989a: 333, Chapman and Rosenberg 1991, Mac Nally 1994a) – the use of foraging versatility to estimate true versatility requires a good deal of faith, particularly when testing the predictions of community theory (e.g., Fretwell 1978, Rusterholtz 1981a,b). As Sherry (1990: 348) commented: 'an organism can eat a broad array of foods in a stereotyped way', which is indicative of a non-monotonic relationship between dietary and foraging versatility.

Notwithstanding these reservations, H. A. Ford of the University of New England, New South Wales, Australia, analysed whether 'widespread' species display greater foraging versatility than 'localized' species (Ford 1990). He addressed this issue in terms of two contrasting hypotheses:

(1) foraging generalists can exploit a wide variety of habitats and foraging sites and foods; and
(2) habitat generalists are foraging specialists.

Essentially, hypothesis (1) corresponds to supplementation and hypothesis (2) to complementarity, if we assume that 'widespread' and 'localized' in a geographic sense also indicate something about the variety of habitats used.

Ford (1990) related three aspects of geographic range (which are assumed to reflect ubiquity): total range, breeding range, and local range (occurrence in 10 minute squares in the New England area) to four components of foraging versatility (foraging method, substrate use, plant species use and foraging height) for 40 species of birds occupying eucalypt woodland in New England. He found that neither hypothesis (1) nor (2) was supported – of the 12 range–foraging relationships only one was statistically significant. Even this relationship, between breeding range and the variety of plant species used accounted for just 8% of the variation (Ford 1990: Table 1.3). Given the large number of pairwise comparisons on related data, it is doubtful whether this relationship should be regarded as ecologically significant.

In a similar vein to Ford (1990), I have also explored whether ubiquity and foraging versatility are related in birds of eucalypt forests of eastern Australia. Rather than use range as a surrogate for ubiquity, I have calculated estimates of the versatility of habitat use more directly, based on habitat use itself (see Mac Nally 1989). Foraging versatility was based on the categories defined in Mac Nally (1994b: Table 1) and both foraging versatility and ubiquity were estimated by using the technique of Mac Nally (1994a).

I found that practically any combination of ubiquity and foraging versatility can be found among the 63 species (Fig. 8.9). However, the majority displayed moderate to high ubiquity (0.6–1.0) and low to moderate foraging versatility (0.2–0.4). There were examples of species showing both high ubiquity and high foraging versatility (e.g., crested shrike-tit, *Falcunculus frontatus*), low ubiquity and versatility (e.g., yellow rosella, *Platycercus flaveolus*), low foraging versatility and high ubiquity (e.g., striated pardalote, *Pardalotus striatus*) and restricted habitat use and moderate foraging versatility (e.g., yellow-tufted honeyeater, *Lichenostomus melanops*) (Fig. 8.9). Therefore I found little evidence for consistent complementarity or supplementation, as did Ford (1990), but I had better measures of ubiquity.

Overall, this analysis of the ubiquity–foraging versatility relationship indicates that very little in the way of consistent patterns could be discerned. This result almost certainly reflects the influence of idiosyncratic or species-specific patterns in a similar way to those reported by Rice *et al.* (1980). Other factors that might obscure patterns include the use of foraging versatility instead of true versatility and the practical impossibility of sampling all types of habitat found within a region, which probably would alter sample estimates of both forms of versatility for many species. Further work undoubtedly is needed to appraise whether ubiquity and ecological versatility are functionally related in this and in other systems, although at this stage it seems unlikely that ecological versatility has a consistent influence on the diversity of habitats used by species.

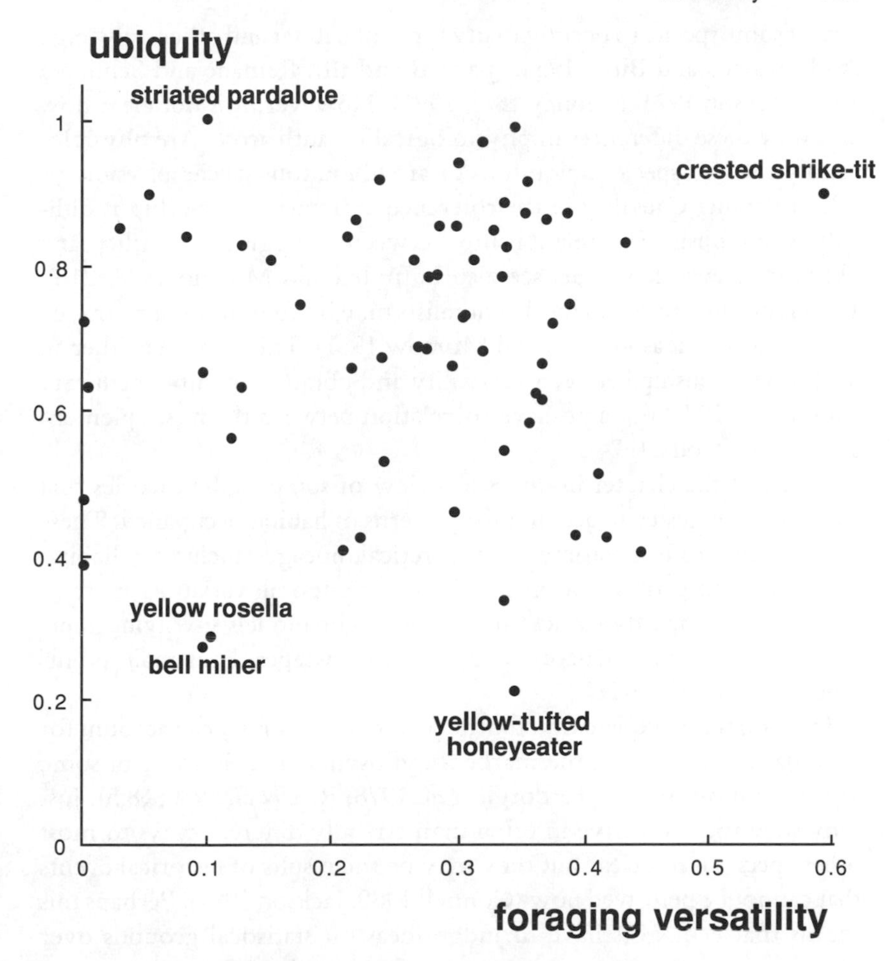

Fig. 8.9 Patterns of ubiquity and foraging versatility in birds from the Victorian transect. Positions of several extreme strategists are indicated (e.g., striated pardalote).

8.6 Summary

In this chapter, I consider some of the ideas on the mechanisms responsible for the differences in the versatility of habitat use – ubiquity – between species. Habitat use is influenced by many factors. Physiological tolerances (climatic, physical, and chemical) and interspecific competition have been two of the most common explanations for the constraints on distributions, but there are many others, including natural enemies and differentiation within and among populations of species. Work in ecophysiology provides proximate explanations for the capacity of some species to occupy many types of habitat and others few (e.g., Andrewartha and Birch 1954: (parts II and III), Remane and Schlieper 1971, Jackson 1973, Mooney *et al.* 1991). However, it is not clear how and why these differences in physiological breadth arose. Are physiologically tolerant species ubiquitous or are ubiquitous species physiologically tolerant? Causality in the tolerance–ubiquity relationship is difficult to establish. The relationship between ecological versatility and ubiquity is even less clear (see also Futuyma and Moreno 1988: 218). Both local specialists and local generalists may be ubiquitous or restricted for a variety of reasons (Fox and Morrow 1981). This leads to neither an inverse relationship between versatility and ubiquity (complementarity *sensu* Cody 1974) nor a positive correlation between them (supplementation *sensu* Cody 1974).

Much of the chapter involves a review of some of the theories that have been suggested to account for patterns of habitat occupancy. These ideas come from a variety of theoretical lineages including habitat selection, competition or niche theory, temporal variation in food availability, range dynamics and more recent models involving hierarchical community structure and fractal landscapes. Four main points emerge from this review.

First, no single explanation for the patterns of ubiquity can account for all situations because of the markedly idiosyncratic behaviour of some species in most studies (Terborgh *et al.* 1978, Rice *et al.* 1980; §8.5). Just why such species behave in often diametrically different ways to most other species is not clear but they may be the results of historical events that cannot be perceived now (Connell 1980, Jackson 1981). Perhaps this means that ecologists have to judge ideas on statistical grounds over many species rather on intensive study of few species. Is a theory a success or failure if nine out of ten species conform to its predictions, or two out of three? Certainly, the species showing exceptional behaviour merit greater attention in their own right.

Second, almost all models place a high emphasis on density (or abundance) as a key factor in determining relative ubiquities. Some workers assume that the differences in density exist and then trace the impact of such differences on ubiquity and even on geographic range (e.g., McNaughton and Wolf 1970, Jackson 1974). Others do not necessarily assume inherent differences in density between species yet numerical disparities emerge from the model dynamics (e.g., Palmer 1992). However, there are complicating effects in the ubiquity–density relationship that may interfere with a simple correspondence (high densities mean high ubiquity and vice versa) such as body size and trophic level.

Third, the ubiquity of a species is a dynamic property that is subject to change at quite short time scales in some species, particularly in mobile ones (Fretwell 1972, Rice *et al.* 1980). Species may show pronounced changes in the diversity of habitats used over durations much shorter than generations. Measured ubiquities also may vary regionally – a restricted species in one region may be ubiquitous elsewhere in its range (Pulliam 1986, Williams 1988).

Fourth, many recent models no longer consider species to be as finely adapted to habitats in which they occur as conventional competition theory implies (e.g., Seagle and Shugart 1985, Danielson 1992, Palmer 1992). Species may occupy or even breed in some habitats at rates that are insufficient to maintain local densities, but emigration from more suitable (source) habitats can sustain such marginal or sink populations (e.g., Kadmon 1993). These observations suggest that if ubiquities were to be gauged in terms of the production of recruits associated with particular habitats rather than by the densities of individuals residing in them, then a better understanding of patterns of ubiquity must ensue (i.e., 'performance' rather than current density, Lawton 1993). This also would make the definition of the versatility of habitat use more consistent with that of ecological versatility generally by associating 'fitnesses' (here recruitment) with alternative 'resources' (here habitat types).

9 · *Recapitulation and commentary*

The ecological versatility (and ubiquity) of organisms clearly varies greatly both in terms of the overall magnitude of the differences between organisms (e.g., versatility of species $A >$ versatility of species B), but also secularly (e.g., versatility of species $A_{summer} >$ versatility of species A_{winter}) and spatially (e.g., versatility of population $A_{heath} >$ versatility of species $A_{woodland}$). Although some organisms display rather consistently limited exploitation of resources even in the face of the availability of a wide assortment of alternatives, it seems that the overwhelming majority of organisms either alter their resource use dynamically or are opportunistic to some extent. By analysing ecological versatility in a general way as I have done in this book, it seems that the claims of widespread, extreme specificity characterizing some sorts of organisms and interactions probably have been overstated. For example, as we have seen, many parasites exhibit strikingly different morphologies when they infect alternative hosts and, in the absence of independent genetic information, this phenomenon can misleadingly increase the perceived degree of host specificity (Downes 1990). Thus, the range and dynamism of ecological versatility have struck me most during the preparation of this volume.

Of course, these comments are just a crude summary of the *patterns* of versatility evident in natural communities – they do not refer to the *processes* leading to the patterns. Many of the propositions for explaining the variation in versatility and ubiquity have formed the subject matter of the preceding five chapters. What needs to be done now I feel is to provide a synthetic overview of the findings both in terms of general explanations for different degrees of ecological versatility among organisms and also in relation to interspecific interactions. I also think that it is necessary to offer some comments on the strengths and weaknesses of the knowledge base and the experimental and observational protocols used to collect information on versatility. Thus, the main objectives in this chapter are to:

(1) recount the factors that have been proposed to account for different levels of ecological versatility, some of which have been put forward as nonspecific solutions while others are particular to certain processes (§9.1);
(2) address in a general way the questions about our knowledge and understanding of ecological versatility raised in Chapter 1 (§9.2); and
(3) provide a commentary on several of the most important general observations arising during the course of this review, including some thoughts on the problems of improving our understanding of how ecological versatility is determined in natural communities (§9.3).

9.1 Overview of factors influencing ecological versatility

A list of factors gleaned from previous chapters is provided in Table 9.1. These factors are presented as 'contrasts' between the conditions likely to lead to, or be associated with, specialized resource use on the one hand, and generalization on the other. Thus, the factor 'intense competition' is often thought to result in greater specialization if the competition is between species but greater generalization if intense competition arises from within a population. The order of factors in the table generally reflects the sequence of topics in this book, except that general effects such as environmental variation are listed first. Some factors have been thought to favour or be associated with either specialization *or* generalization but have not been formulated as contrasts *per se*. These factors are also listed where appropriate in Table 9.1. Much of the remainder of this section involves fleshing out the contrasts and factors listed in Table 9.1.

General ideas

The most prevalent assumption in versatility theory is that there is a cost to generalization – somatic organization or behaviour contributing to the efficient exploitation of one resource necessarily diminishes an organism's capacity to exploit other resources (e.g., Emlen 1968, Glasser and Price 1982, Lynch and Gabriel 1987). For convenience, this is referred to as *antagonism*. Specialization has been seen as being advantageous in terms of the efficient use of resources (e.g., Emlen 1973) but more recent views see specialized organisms as being at a disadvantage, particularly in terms of reduced ecological opportunities. For example, Kolasa (1989) believed that specialization may be a consequence of

Table 9.1. *A selection of contrasts (double-headed arrows) and factors (single-headed arrows) that have been proposed to favour or be associated with ecological specialization or generalization*

Specialization favoured	Contrast/factor (sources)	Generalization favoured
	General ideas	
Large differences	← Efficiency trade-off between resources → (Levins 1968)	Small differences
Low	← Genetic diversity → (McNaughton and Wolf 1970, Steiner 1977)	High
Coevolution/coevolutionary arms-races (mutualism, parasitism, herbivory)	← Obligatory dependence (Ehrlich and Raven 1964, Feeny 1976, Rhoades and Cates 1976, Price 1980, Schemske 1983)	
Low	← Environmental saturation (population density) → (Glasser 1982, Gladfelter and Johnson 1983, Bennett and Branch 1990)	High
Poikilotherms	← Thermal physiology → (Schoener 1971)	Homeotherms
Low	← Trophic level → (Glasser and Price 1982, Glasser 1983, Pennings *et al.* 1993)	High
	Environment	
High	← Environmental predictability → (Roughgarden 1974b, Lynch and Gabriel 1987)	Low

Low end	Factor (references)	High end
Low (monomorphic specialist)	← Habitat heterogeneity in space and time (Levins 1968) →	High (monomorphic generalist or polymorphism)
Low trophic levels	← Fluctuating environments (facultative strategists) (Glasser 1982) →	High trophic levels
Ephemeral (carnivores)	← Habitat stability (Pfennig 1992) →	More permanent (omnivores)
Mild	← Seasonality of environments (Fretwell 1972, Emlen 1973) →	Pronounced
	Resources	
Not crucial (e.g., many carnivores)	← Nutrient balance in diet (McNaughton and Wolf 1970, Emlen 1973, Westoby 1978, Owen-Smith and Novellie 1982, Bernays and Graham 1988) →	Crucial (e.g., many herbivores)
Substitutable	← Resources (Futuyma and Moreno 1988, Pennings *et al.* 1993) →	Complementary
Clumped	← Resource distributions (Schoener 1971, Futuyma and Moreno 1988) →	Dispersed
Nearby	← Resource position (Schoener 1968, 1971) →	Distant
High and durable	← Resource predictability/durability (Jaenike 1978, Lacy 1984) →	Low and transient

Table 9.1. (*cont.*)

Specialization favoured	Contrast/factor (sources)	Generalization favoured
High	← General food availability → (Emlen 1966, MacArthur and Pianka 1966, Levins and MacArthur 1969, Schoener 1971)	Low
Low	← Relative food availability – residents → (Herrera 1978a)	High
High	← Relative food availability – nonresidents/migrants → (Herrera 1978a)	Low
	Behaviour	
Poor	← Sensory perception → (Futuyma 1983, Jermy 1988, Barclay and Brigham 1991, Fox and Lalonde 1993)	Good
	Mutualism	
Symbioses via host-parasite relationship	← Symbioses (Howe 1984)	
	Enemy deterrence via ants → (Atsatt 1981, Pierce and Elgar 1985)	Ant protection of insect herbivores

Parasitism

Dense/continuous	← Density/continuity of parasitoid host populations (Janzen 1981) →	Sparse/fragmented
Koinobiotic (hosts not killed immediately upon infection)	← Parasitoids: attack mode (Hawkins and Gross 1992, Hawkins *et al*. 1992) →	Idiobiotic (hosts killed on infection)
Direct	← Parasites: transmission mode (Garnick 1992) →	Indirect, free-living, vectors, high rates of cross-transmission between hosts
	Parasitism: polyhospitaly → (Garnick 1992)	Less vulnerable to fluctuations in host densities/resistances, greater genetic diversity and geographic range
Sedentary, widely dispersed, rare	← Parasites: vagility and abundance (Rohde 1979) →	Highly vagile and abundant
Specialized	← Parasites: degree of host specialization (Cameron 1964)	
	Parasitism? enslavement → (Bernstein 1978, 1979)	Ant–ant slavery
Living host cells (biotrophic)	← Plant parasites: infection mode (Callow 1977) →	Dead host cells (necrotrophic)
Limited	← Plant parasites (mistletoes): capacity to penetrate host bark (Kuijt 1969) →	Extensive

Table 9.1. (*cont.*)

Specialization favoured	Contrast/factor (sources)	Generalization favoured
	Predation	
Pursuit	← Foraging method → (MacArthur and Pianka 1966, Rosenzweig 1966, Schoener 1971)	Search
Large and/or unambiguous	← Differences between prey alternatives → (Westoby 1978, Hughes 1979)	Small and/or ambiguous
Some high, leading to better recognition and handling efficiency	← Encounter rates with prey alternatives → (Hughes 1979)	All low
High (carnivores)	← Specific prey availability → (Pfennig 1992)	Low (omnivores)
Abundant	← Availability of preferred foods → (Emlen 1968, Schoener 1971, Glasser 1982, Glasser and Price 1982, Pyke 1984, Sherry 1984)	Scarce
Mandibular morphology	← Morphology (Freeman 1979)	
	Foraging → (Greenberg 1983)	Maladaptation
High	← Resource neophobia → (Greenberg 1983, 1990a, 1990b)	Low

High	⟵ Predation pressure ⟶ (Greenberg 1983)	Low or ameliorated (e.g., social behaviour)
	Herbivory	
Abundant/predictable/easily located	⟵ Availability of hosts ⟶ (Fox and Morrow 1981, Jaenike 1990)	Rare/unpredictable/difficult to locate
• High chance of finding known/suitable • Low fraction unknown/suitable • High fraction of unknown/unsuitable	⟵ Knowledge and suitability of plants ⟶ (Levins and MacArthur 1969)	• Low chance of finding known/suitable • High fraction unknown/suitable • Low fraction of unknown/unsuitable
Rare or never	⟵ Shifts in optimal foods and sampling ⟶ (Westoby 1978, Owen-Smith and Novellie 1982)	Frequent
Parasitic	⟵ Mode of insect feeding ⟶ (Thompson 1988a)	Grazing
Small ⇒ small guts/selective	⟵ Body size in noninsect terrestrial herbivores ⟶ (Caughley and Lawton 1981, Belovsky 1986, du Toit and Owen-Smith 1989)	Large ⇒ large guts/nonselective
Low	⟵ Handling : search time ratio ⟶ (Verlinden and Wiley 1989)	High
Stenophagy	⟵ Impact of generalized predators and parasites (Bernays and Graham 1988, Bernays 1989)	
Yes – survival much better on plants producing sequesterable allelochemicals	⟵ Ability to sequester plant chemicals for self-defence? ⟶ (Hay *et al.* 1989, Denno *et al.* 1990)	No – no differential survival on various hosts

Table 9.1. (*cont.*)

Specialization favoured	Contrast/factor (sources)	Generalization favoured
Host-specificity associated with vector-specificity	← Vector-specific differentiation (Colwell 1986)	
	Differentiation within populations	
Homogeneous	↔ Differentiation within populations (Werner and Gilliam 1984, Moran 1988, 1992)	Heterogeneous (stages, polymorphism, polyphenism)
Intense	↔ Competition between phenotypes (Roughgarden 1972)	Lax
Monomorphism	← Unstable protected polymorphisms (Getz and Kaitala 1989)	
	Competition	
Interspecific	↔ Intense competition (McNaughton and Wolf 1970, Bernstein 1979, Polis 1984, Taper and Case 1985)	Intraspecific
Contract patch use	↔ Proximate response to intense competition (MacArthur and Wilson 1967)	Expand diet

organisms becoming restricted in the range of resources that they can use. Thus, specialization is the best option ('making the best of it' according to Kolasa) once a species is caught in a restricted part of ecological hyperspace (*sensu* Hutchinson 1957).

Futuyma and Moreno (1988) noted that almost all mathematical and logical models of the evolution of specialization hinge on an antagonism in the efficiencies of using alternative resources. If efficiencies are determined as functions of many aspects (i.e., finding, handling, digestion, detoxification, etc.) then the simple antagonistic expectation may not be a valid one because of the interaction between factors. As Futuyma and Moreno (1988: 219) put it: 'A trade-off at a reductionist level may not operate at the organismal level, at which compensatory changes come into play'. In any event, there seems to be little empirical support for the antagonism assumption (Huey and Hertz 1984, Futuyma and Philippi 1987, although see Benkman 1988, 1993). In addition, there is some evidence suggesting that specialized organisms need not be competitively superior to generalized ones even within their domain of specialization (e.g., Jackson 1973), which undercuts models in which this is a key assumption (e.g., Cody 1974, Lynch and Gabriel 1987).

Whatever pressure there is to specialize for greater efficiency of resource use may depend upon how different resources are from one another. If alternative resources require very different organizations or capacities for efficient exploitation, then specialization may be favoured. However, if the resources differ little in this regard then more generalized exploitation may not be unfavoured (Levins 1968). Similar arguments have been used for the evolution of ubiquity (e.g., Rosenzweig 1991: Fig. 11).

McNaughton and Wolf (1970: 137) attributed differences in versatility (niche breadth in their terms) to dissimilar levels of genetic diversity. According to them, if two species initially utilize the same niche dimension, one is certain to be more abundant than the other. Greater abundance causes more intense intraspecific interactions through competition, hence high intraspecific pressure for genetic divergence. The less abundant species will not diverge as rapidly because of reduced intraspecific contact compared with the more abundant species. Genetic divergence in the less abundant species then slows and contracts because of competition with the more genetically diverse abundant species, causing relative specialization and subordination in the initially less abundant taxon. Thus, lower initial abundance on one niche dimension transforms into persistence only through specialization on another dimension. This idea is a qualitative argument and has not, as far as I am aware, been cast in a quantitative, testable form.

The correlation between genetic diversity and ecological versatility has proved to be a contentious point. Although some experiments have indicated that genetic diversity and ecological versatility may be correlated in some species of *Drosophila* (Powell 1971), Sabath (1974) found little evidence for this correspondence. Later claims of a similar relationship in species of *Drosophila* from Hawai'i (Steiner 1977) were contested by Hallett (1980). He noted that different measures of genetic variability yield different results – chromosomal variability does not map uniquely onto electrophoretic variability, for example. Thus, the relationship between genetic variability and ecological versatility remains unresolved but seems unlikely to be a general explanation for differences in versatility.

In three fields, tight coevolutionary coupling has been raised as a potentially important agent in the occurrence of specialization, namely, mutualism, parasitism (including parasitoids) and in insect herbivores (e.g., Ehrlich and Raven 1964). In mutualisms, the coupling is deemed to be mutually beneficial such as in pollination systems in which plants trade food resources for the conveyance of their pollen to other conspecific individuals. In parasitism and insect herbivores, the relationship is often regarded as an 'arms race' in which the resource species and exploiter enter into a coevolutionary sequence of innovation and response. Although a superficially appealing explanation for a vast amount of apparent specialization, much recent thought regards coevolution between species pairs as being unlikely (Schemske 1983, Bernays and Graham 1988). The conditions necessary for effective bilateral coevolution seem to be rather restrictive and improbable (Howe 1984, Thompson and Pellmyr 1992).

Many workers believe that there is a generalization of resource use as densities increase. Individual organisms are thought to occupy or use the highest quality or optimal resources first, but at some stage must use inferior resources. Thus, in this view, less specificity of resource use occurs as densities increase (assuming optimal choice, an accurate and immediate perception by individuals of the current density of the population, etc.). This mechanism underlies many models of the dynamic changes in versatility and ubiquity (e.g., Svärdson 1949, MacArthur and Wilson 1967, Fretwell 1972, Vandermeer 1972, Rosenzweig 1981, Glasser 1982).

Unfortunately, the direction of this causality is not well established (i.e., that density determines versatility rather than versatility determining density). Does high density cause generalization or greater diversity of resource use? Emlen (1968: 388) and others seem to believe that this is the case: 'A species may be specialized in habitat or resource selection as a result of being rare, and not necessarily rare as a result of over-specialization.' On the other hand, some ichthyologists believe that omnivorous feeding habits (i.e., greater generalization) foster greater population densities (e.g., Mann 1965, Persson 1983, Braband 1985). Niche apportionment models of the determination of distributions of densities in communities are underpinned by the assumption that versatilities determine densities rather than vice versa (Tokeshi 1993: 135–136). Therefore, perhaps by virtue of exploiting greater amounts of resources than specialists, populations of generalists are more numerous than those of specialists (see Chapter 6). Admittedly, this picture does not explain the origin of differential versatility as such, but this has not deterred speculation on the density–niche apportionment issue (see Tokeshi 1993: 132 ff.). If density does indeed determine versatility then this may be an important mechanism in dictating versatility. However, differences in densities then have to be explained, which requires additional mechanisms such as the effects of natural enemies (e.g., Emlen 1968).

How might one distinguish between these alternatives? Let the hypothesis that rarity *causes* specialization be denoted by the symbol $H_{r \rightarrow s}$, and the hypothesis that specialization *causes* rarity be indicated by the symbol $H_{s \rightarrow r}$. In principle, if one were to increase experimentally the density of a rare *and* specialized population while holding other factors constant (e.g., the densities of potential competitors and natural enemies), then there is an expectation that the resources used by the population should diversify under $H_{r \rightarrow s}$ but be unaffected if $H_{s \rightarrow r}$ were true. The former corresponds to the partial niche model developed by Svärdson (1949), Vandermeer (1972) and others, while the latter indicates obligatory specialization. On the other hand, if the resource upon which the specialty is based were supplemented (rather than densities being increased) while other extraneous factors were held constant, then densities should increase more or less monotonically with the amount of supplementation (up to a point), given sufficient time under either hypothesis. The degree of specialization under $H_{r \rightarrow s}$ may decrease as densities grow but this need not necessarily happen. And finally, if alternative resources were supplemented rather than the one upon which the speciality is based, then, holding other factors constant as usual, one might anticipate that both the density and diversity of resources should increase under $H_{r\rightarrow s}$, but remain unaffected under $H_{s\rightarrow r}$. Of course, a parallel series of experimental contrasts could be envisioned for abundant generalists to track their responses to artificially altered densities and resource availabilities. Under what circumstances could such experiments be contemplated? Relatively few I suspect, perhaps insular granivorous ants may be a possibility, or direct-developing, rocky-shore molluscs (either algal grazers or predators). One would need systems that are not prone to high mobilities of the target organisms and in which the resource, probably food, can be easily manipulated. Unfortunately, such systems are hardly typical ecological situations.

In any event, this set of experimental contrasts basically corresponds to regarding specialization as a facultative/dynamic feature ($H_{r\rightarrow s}$) or as fixed/obligatory characteristic ($H_{s\rightarrow r}$). Thus, the experimental program could only possibly identify the functional relationship between rarity and specialization in a contemporary sense and not the process leading to its evolution.

Thermal physiology might influence versatility according to some authorities. Schoener (1971) mentioned that owing to differences in energy requirements between poikilotherms and homeotherms, the former could often afford to be more selective in foods consumed. He illustrated this point by using a comparison between snakes and birds of the same body mass, the former being more stenophagous than the latter. However, this may be a specious argument for several reasons. First, birds must have greater opportunities to find a wide variety of prey owing to their greater mobility, so that the array of prey items to which birds are exposed is significantly more diverse. And second, by virtue of their greater energy requirements, birds are also more likely to appear to or actually track short-term changes in food availability (Zwarts and Esselink 1989). A snake may not need to feed for several days, whereas a similarly-sized bird needs to feed almost continually. Thus, thermal physiology *per se* may not be as crucial as factors that stem from it.

Position in food webs has been suggested to be a general determinant of versatility. High-order predators are often regarded as being more generalized than organisms at lower trophic levels (e.g., Glasser 1983, Pennings *et al.* 1993). Presumably, this idea means that top predators are *on average* more generalized in resource use than primary consumers, say, for there is well-documented evidence of substantial variation in versatility within trophic levels (e.g., Freeman 1979, du Toit and Owen-Smith 1989, Winemiller 1990, Hawkins *et al.* 1992). However, a trend to greater generalization at higher trophic levels can be decoupled easily, for instance, in systems in which feeding is by engulfment (e.g., plankton communities, Sprules and Bowerman 1988).

Environmental variation and predictability may be important in producing differences in ecological versatility. It has often been suggested that unpredictability, significant fluctuation and seasonality and heterogeneity in environments all favour generalization, while stability, homogeneity and predictability should foster specialized resource use (e.g., Levins 1968, MacArthur 1970, Emlen 1973, Roughgarden 1974b, Lynch and Gabriel 1987). One expression of this idea comes from Jaenike *et al.* (1980), who wrote: 'In a constant environment interclonal competition should result in the extinction of all but the most finely adapted clone in any particular niche. However, in a fluctuating environment ... successful clones should be relatively insensitive to environmental changes ... and probably characterized by ... "general purpose" genotypes, which permit long-term persistence in one area and dispersal through a heterogeneous environment to achieve broad distributions.' However, it is possible that some highly vagile species may be able to capitalize on the asynchronous availability of a particular resource in different habitats even though the availability at any one site is highly variable and unpredictable. Thus, a specialization can be supported even in the face of stochastic variation in resource availability (e.g., some nectarivorous, frugivorous and granivorous birds). In addition, recall from Chapter 7 that it may be difficult or inaccurate to characterize simultaneously the variability of an environment for all organisms because of resource-specific variation. Some organisms may 'perceive' an environment to be relatively constant because the resources that they use vary little in availability. Other syntopic organisms might construe the same environment to be much more labile because the availability of resources upon which they depend varies substantially through time. Thus, one has to classify levels of environmental variability in an organism-specific way, which means that 'environment' *per se* cannot be a consistent determinant of ecological versatility. An understanding of versatility requires analysis at the level of resources and not at the level of the entire environment.

In many organisms the pressure for dietary specialization is thought to be opposed by the need to acquire nutrients that only a diverse set of resources can provide. Westoby (1978) and others have used this potentially powerful argument to counter the uncritical application of optimal foraging models to some organisms, notably larger herbivores.

A general feeling is that these herbivores are more likely to be subject to nutrient-balancing demands than carnivores. At least in dietary terms, larger herbivores might be expected to be comparatively generalized compared with carnivores. This is precisely the reverse expectation from the trophic-level arguments referred to above, which engenders little faith in either hypothesis being an important, consistent factor in controlling versatility.

Nutrient balancing is akin to the idea of complementary resources – performance (e.g., growth rate, fecundity) is better on a mixture of foods than on a single food (e.g., Pennings *et al.* 1993). Thus, complementary foods enforce generalized diets (Futuyma and Moreno 1988). It is unclear whether substitutable foods necessarily lead to specialized diets because perfectly substitutable foods might be exploited in about the same proportions as their availabilities (i.e., like **Ch** or **Rh**). There may be the opportunity for specialization with substitutable foods but specialization need not inexorably follow.

The spatial distribution of resources may also affect ecological versatility (Futuyma and Moreno 1988). Resources that occur in clumps can be more efficiently harvested and defended, which may impose some pressure for specialization on such resources. In addition, if resources (especially foods) have to be gathered from distant locations, then specialization on relatively profitable resources may be expected (Schoener 1968, 1971). Almost all optimality arguments assume that there are energetic disadvantages to specializing on dispersed resources because of the wasted time and opportunities involved in searching and travelling (e.g., MacArthur and Pianka 1966, Schoener 1971, Fahrig and Merriam 1985). This is essentially the argument behind the compression hypothesis (MacArthur and Wilson 1967) – if competitors reduce the availabilities of resources then it pays to partition spatially rather than by food resources. This leads to reduced time and energy spent in searching for resources. As mentioned previously, empirical support for this idea is equivocal (see Dunham 1983).

By using arguments similar to the compression hypothesis, Emlen (1968) and others have suggested that animals should show dietary specialization if food is abundant but generalize when it is scarce. Again, patterns in natural communities fail to support this idea. For example, Herrera (1978a) reported from a study of an avifauna in southern Spain that resident and nonresident species exhibited opposite responses to changes in food availability. Resident species showed greatest versatility when food was freely available but reduced the diversity of foods they used when food was scarce (i.e., thus contradicting Emlen's hypothesis).

Herrera interpreted these observations as a retreat of each resident species into its own 'specialty' when food was harder to find. Nonresidents and migrants appearing when food was abundant specialized on particular foods. Migrant species that occurred during periods of food scarcity were opportunistic. Herrera's (1978a) results are intriguing and important because they demonstrate that two quite reasonable yet opposing expectations can occur in the one system. It certainly seems as logical that individuals might focus on the food types that they are most effective at exploiting when foods become scarce as it is that they might become less choosy. When foods are abundant, why should individuals not opportunistically feed upon any item that they encounter? Why should they necessarily restrict themselves to their specialties? The point is that food selectivity logically can be expected to increase *or* decrease (or even not change much) as availabilities change, which is what Herrera found. Once again this demonstrates that organisms almost always have a variety of ecological options open to them.

The versatility of resource use, particularly of food, must depend to some extent upon the discriminatory abilities of organisms (Soberón 1986). Some workers have contended that (apparent) specialization is merely a by-product of the limited neural capacities of some organisms (e.g., Futuyma 1983, Jermy 1988) or of their restricted sensory or biophysical capabilities (e.g., Barclay and Brigham 1991). In the former, the neural architecture is held to be insufficient to hold information for recognizing any but a few kinds of resource. In the latter, certain resources just do not 'register', given the sensory apparatuses with which an organism is equipped, or else the consumer cannot respond fast enough to utilize these resources. Thus, specialization from these perspectives is a passive reflection of limited sensory and behavioural capabilities. In classical optimality theory, organisms are often endowed with exceptional talents for discriminating between resources and perceiving the relative 'value' of alternatives (e.g., MacArthur and Pianka 1966, Charnov 1976; although see Westoby 1974, 1978). Specialization on the most rewarding foods is an axiom of optimal foraging. Thus, dietary specialization may arise from both active discrimination between alternative resources and also through passive perceptual mechanisms.

Specific ideas

Having recounted these general ideas on the factors affecting versatility, I now summarize briefly some of the theories and mechanisms that arise in particular contexts, such as in the various interspecific interactions (e.g., mutualism, parasitism). However, the overviews for each process provided in Chapter 4 are more definitive statements as are the chapter summaries for within-population differentiation (§5) and interspecific competition (§7). These should be consulted for more detail.

Herbivory

The distribution, abundance and apparency of hosts have been proposed as important factors affecting the versatility of herbivores, particularly small ones (e.g., Fox and Morrow 1981, Soberón 1986, Jaenike 1990, Andow 1991). The capacity to gauge the suitability of potential hosts and their relative abundances and distributions may be important factors (Levins and MacArthur 1969). Related to this are temporal changes in plant suitability or the heterogeneous quality of the parts of an individual plant, both of which might force herbivores to need to resample plants continually and thus increase their apparent dietary versatility (Westoby 1978). Grazing herbivores are more likely to be generalized than 'parasitic' ones (Thompson 1988a). This may be particularly evident in very large herbivores, such as most ungulates, because larger guts require higher throughput, which in turn can be satisfied only through comparatively nonselective feeding (Belovsky 1986, du Toit and Owen-Smith 1989). Thus, high ratios of digestion time to acquisition time may be associated with generalized herbivory, while small ratios permit greater selectivity (Verlinden and Wiley 1989).

Some workers believe that natural enemies play a signal role in constraining host use in small herbivores, particularly insects (Bernays 1989). Coping with toxins and other deterrents is another major theme (Bernays and Graham 1988; see coevolution above) although some small herbivores have turned allelochemicals to their defensive advantage (sequestering). Thus, rather than being limited to certain host plants because of the evolution of biochemical means to ameliorate the effects of plant-specific toxins, these small herbivores have restricted host ranges *because* their hosts produce these toxins and provide a means for the herbivore to become toxic or distasteful to potential predators and parasites (Hay *et al.* 1989, Denno *et al.* 1990).

Parasitism: parasitoids

The number of factors that affects the host specificity of insect parasitoids is large. They include, primarily, the nature of attack (koinobiosis, idiobiosis), specific adaptations of parasitoids, the abundance, feeding mode and location of hosts (endophytic, exophytic, gallers, leaf miners, etc.), the life form and densities of plants used by hosts (trees, shrubs) and the stage of host development (larvae, pupae). Koinobionts have been regarded as specialists and idiobionts as generalists, but this may be a relatively crude partition (Sheehan 1991). Mills (1993) provided a detailed assessment of the mechanisms involved in controlling the versatility of host use in insect parasitoids.

Parasitism: parasites

The mode of transmission, abundance, and vagility may be the major controls on the versatility of host use in parasites (Rohde 1979, Garnick 1992). On theoretical grounds, Garnick believed that of all transmission modes, only direct infection favours host specificity, while Rohde thought that the ability to locate mates is the best explanation for extreme specialization in marine parasites. Generalization might allow parasites greater security and more extensive geographic ranges, while the ecological specialization of the hosts themselves may limit the versatility displayed by parasites (Cameron 1964). One must exercise some caution when discussing the prevalence of pronounced host specificity in many parasites because a single species of parasite can develop quite distinct morphologies in different hosts (Downes 1990). Clearly, this feature could lead to spuriously high degrees of specificity unless workers employ means independent of parasite morphology to identify specific differentiation. Most of the ideas on parasitic versatility involve population densities and the ability to find resources (hosts), which are versions of common themes throughout versatility theory.

Predator–prey relationships

The ways in which prey are captured may affect versatility. Several workers have suggested that active pursuit is more likely to be associated with prey specialization than is more general searching (e.g., MacArthur and Pianka 1966). Presumably, this reflects the greater investment in time and energy involved with pursuit, which needs to give a comparatively high return for effort (Schoener 1971). The nature and distribution of prey alternatives have often been regarded as important because if prey alternatives yield similar returns, or are difficult to distinguish, then there is little reason to specialize (Westoby 1978, Hughes 1979). On the other hand, if returns can be estimated well and/or are substantially different, then specialization becomes more advantageous. Encounter rates with different prey types may be important, particularly in terms of prey recognition and handling experience (Hughes 1979). As with resources in general, if preferred foods are scarce there is an expectation that predators will generalize their use of food and vice versa.

Mutualism

Even Howe (1984), who was generally dubious about specificity in mutualisms, conceded that extreme specialization of symbiotic relationships may occur via an initially parasitic relationship. However, recent work demonstrates that examples of symbioses previously thought to be highly specific may not be so (e.g., mycorrhizae, Sanders 1993); Briand and Yodzis (1982) could find little evidence of obligatory mutualisms. There is some evidence suggesting that mutualisms increase the versatility of at least one partner, such as insects with ant protection (e.g., Pierce and Elgar 1985, Pierce 1989), but, generally speaking, there is an expectation that mutualisms will be 'diffuse' or nonspecific (Addicott 1984).

Ecologically differentiated populations

The perceived versatility of populations clearly depends upon the degree of ecological differentiation between components such as phenotypes, stages, sexes, and so on (Van Valen and Grant 1970, Roughgarden 1972). The potential for differences between individuals of the same phenotype or stage (e.g., Bryan and Larkin 1972, Real 1975, Ehlinger 1990, Werner 1992) and plastic exploitation by individuals (e.g., Glasser 1982) also are important factors in determining versatility.

The details of models for differentiated populations are too complex to consider at any length here (see Chapter 5). However, it is worth pointing out that there are some contradictory results in existing treatments. For example, although intense competition between phenotypes usually is thought to favour expanded versatility, Roughgarden (1972) found under some conditions that versatility might decline when competition is fierce and expand when it is lax. Thus, there is the need to continue to develop community models in which populations are treated as heterogeneous collections of individuals (see Werner and Gilliam 1984, Milligan 1986, Ebenman 1988, DeAngelis and Gross 1992). Secular changes in resource use, which are common in many organisms (e.g., Wilbur and Collins 1973, Moran 1988, Ebenman 1992), also need to be addressed more generally in interactive models of community dynamics. In many respects, the modelling of resource use in differentiated populations is at a rudimentary level. For example, we have no way of forecasting just how many ecological phenotypes should occur, and how ecologically different they should be from one another. The usual treatments are based on a specified number of initial phenotypes, with the dynamics being followed subsequently.

Interspecific competition

Having devoted an entire chapter to competition, it should be clear that one cannot briefly summarize how this process might affect ecological versatility. I would say that the sorts of expectations derived under the auspices of the 'MacArthurian paradigm' (Wiens 1989b: 256) do not seem to be all that compelling (see Chapter 7). The necessity for specialization implicit in or derived from many treatments (e.g., MacArthur and Levins 1967, MacArthur 1970), is probably an artefact of the assumptions and of the nature of the mathematical models used in competition theory.

The conditions under which coherent patterns of resource use might occur, perhaps mediated by morphological differentiation (e.g., Hutchinsonian series and character displacement), seem to be satisfied in only a rather limited set of circumstances. These occur where individual mobility and migratory fluxes are very restricted, recruitment arises only from within the local populations themselves and systems are relatively stable and isolated for many generations. Under these seemingly rare conditions, it might be possible to accumulate the necessary genetic differentiation for specialization based on interspecific competition. Perhaps a few situations, such as nonvagile organisms on islands, on mountain-tops, or in hydrologically isolated lakes or direct-developing molluscs on rocky shores, might be the best candidates for applying conventional competition theory. Roughgarden's (1983) vigorous defence of the reality of competitive coevolution of species of *Anolis* in the eastern Caribbean basin is indicative of a bias engendered by one's study organisms. The anoles are *constrained to occupy* essentially small areas of land surrounded by a hostile medium, so that fluxes are necessarily small. They also occur in high densities (typically ≥ 0.5 individuals m^{-2}) in warm conditions so that the potential for resource limitation is high. The point is that whether or not competitive coevolution governs communities of *Anolis*, these circumstances in which such coevolution might occur are *uncharacteristic* of ecological communities. I find it curious that so many of the leading proponents of competition theory should have been ornithologists (although see Ricklefs 1987), who often deal with populations showing strong seasonal fluxes in continental arenas, and in which philopatry is low or uncertain.

'Fine-tuning' between resource use and availability is predicted under the impact of interspecific competition (e.g., MacArthur 1969, Case 1980, Gatto 1982) and is a fundamental assertion of invasibility theory (MacArthur and Wilson 1967, Case 1990). However, it seems an unlikely expectation when environmental variation and unpredictability (Wiens 1977), recruitment from other populations (Gaines and Roughgarden 1985) and frequent local extinction and recolonization (DeAngelis and Waterhouse 1987) are taken into account.

In short, it is unlikely that interspecific competition will lead to coherent patterns of differentiated resource use in the majority of cases. This may be a view at odds with many who believe that features such as resource partitioning (e.g., Schoener 1974a, Toft 1985) are explained best as the results of competition, either currently or in the past (see Connell 1980, Jackson 1981). However, there are many observations inconsistent with conventional expectations, many cases in which resource partitioning is not observed and a general inability to rule out other explanations for the patterns (e.g., independent allopatric evolution and secondary contact). There is also the logical problem that it is difficult to envisage how tight coevolution between competitors might evolve given the rather restrictive conditions outlined above. Together, these points indicate that the effects of interspecific competition on ecological versatility are likely to be inconsistent, unsystematic and variable.

9.2 Questions and answers

In the first chapter, I posed three questions about ecological versatility to provide a focus for this book. The first two of these questions are considered next, but as the third (*3: what determines the ubiquities of species?*) has just been discussed in some detail, the reader should consult §8.6 for my concluding comments on this issue. As for the other questions, I will not answer them pat, but rather discuss them in general terms.

Question 1: How much is known about the patterns of versatility in natural communities, and how consistent are studies in terms of the sorts of quantities

that are monitored to summarize and explain variation in ecological versatility?
Given the diversity of field studies, topics and theory covered in Chapters 3 to 8, it is clear that a vast amount of data has accumulated about the versatility of resource use, and much of it has been treated theoretically. The information tends to be patchy because of the intense interest in the reasons for specialization in some groups of organisms (e.g., parasites and small herbivores) and explanations for generalization in others (e.g., large herbivores). One group for which a balance is achieved is insect parasitoids, where a picture is beginning to emerge of a complex interaction of effects leading to different degrees of versatility (Mills 1993).

Perhaps the most striking feature of the review presented in Chapter 3 is the variation in quality between both studies and subdisciplines. There is no need to repeat here in detail the unevenness in the quality of designs of field studies; this has been analysed in sufficient depth in Chapter 3. Nevertheless, it is worth mentioning that one would find it difficult to conduct a quantitative, statistical meta-analysis based on published information. This reflects partially the particularity of the questions addressed by community ecologists, but also an almost total lack of attention to general design (see Wiens 1989b: 257–264). Admittedly, no such design has been promulgated, but several factors seem to be both obvious and essential in field studies of resource use: Do individuals or components of populations differ from one another? Do individuals change through time? How does use conform with availability? How pronounced are seasonal changes? Are patterns repeated from year to year? Are they similar in different habitats? All of these questions are fairly basic ecological information that obviously should be incorporated into most studies. Yet, studies varied enormously in attention to these design issues, mostly erring on the side of inadequacy. One would like to be able to allocate levels of importance to each factor (e.g., differences between individuals, habitat-specific patterns, seasonal effects, changes between years, etc.) by reference to literature sources, which had been an original intention of this book, but it soon became evident to me that studies are just too inconsistent in detail to attempt such a meta-analysis. I offer more comments on the issue of sampling designs in §9.3 – Design.

The subdisciplines differ of course in their attention to aspects of design, further contributing to quality problems. For example, dynamic changes in host availability may be a fairly meaningless issue for a study of insect herbivores of trees at one location because of the comparatively slow rates of change of tree populations. But it becomes important if one wishes to consider different populations of a given herbivore in many habitat types, as shown by Fox and Morrow (1981). On the other hand, a knowledge of the dynamic conformance between resource use and availability is crucial in investigations of predation, competition and parasitism, even in localized studies (e.g., Fretwell 1972). Collecting enough information to satisfy the requirements outlined in Chapter 3 may seem like overkill in many cases but it would contribute to a systematic approach to studying resource use, and the data would hardly go astray.

Having said all this, what do we know about distributions of ecological versatility in natural communities? Despite some evidence for specificity in parasites and small herbivores, data in which resource use and availability can be compared seem to indicate a relatively uniform distribution of versatilities (Fig. 3.9). This result is very preliminary for the reasons outlined in §3.4. If this result turns out to be a general one then it suggests that speaking of specialists and generalists as such may not be profitable because they represent two ends of a continuous spectrum (i.e., a trivial distinction) rather than distinct alternative strategies, which has been the traditional view (e.g., Cody 1974, Garnick 1992). It may also indicate that levels of versatility are fashioned concurrently by a variety of more or less independent factors, leading to a numerical dominance of neither specialization nor generalization.

In conclusion, if one had to nominate those factors most influential in controlling ecological versatility, then the three most common themes are: (1) the abundance of consumer populations; (2) consumer body size; and (3) the ability to locate and/or process resources. Many think that abundance maps rather directly onto versatility (i.e., abundant species are generalized, rare species are specialized). Large organisms are thought to be more generalized than smaller ones for a variety of reasons (e.g., Glasser 1983). The degree of difficulty in finding a particular resource is often regarded as a key factor, as is the morphological, behavioural and physiological equipment needed to process or utilize the resource. I think that abundance, size and the location of resources are not sufficient in themselves to account for patterns of versatility, but they have arisen time and again as important elements in the views of many ecologists.

Question 2: How well does existing theory account for observations, what are the sources of contention between rival schools of thought, and what sorts of insights can be derived from modelling alternative strategies of resource use? Not surprisingly, a body of ideas peculiar to each subdiscipline has arisen to account for patterns of resource use. Some common ideas appear to have emerged in parallel between subdisciplines (e.g., arms races in small herbivores and parasites), but there are instances of contention. For example, interspecific competition has been promoted by some as a leading factor in constraining host use in both parasites and parasitoids (e.g., Askew 1980), drawing stinging criticism (e.g., Rohde 1979, Mills 1992, 1993).

Despite the existence of characteristic bodies of theory for each subdiscipline, is it possible that the specificity has been overemphasized at the expense of a more general qualitative framework? After all, resource use is concerned with several simple factors: (1) the spatial and temporal variation of resource availability; (2) the types of machinery needed to cope with the defences of resources (if they are foods); and (3) the impact of other syntopic populations (natural enemies and competitors). For example, much is made of the problems posed by the alternative feeding styles of the hosts of insect parasitoids, whether hosts are endophytic or exophytic, larvae or pupae, and so on (Mills 1993). It seems that these properties are not all that different from the characteristics of prey in conventional predator–prey theory – they just affect the comparative profitability of attacking various hosts. The differences between small and large herbivores may hinge on factors as simple as the size of the digestive tract and the associated relative times taken to procure and process food (Caughley and Lawton 1981, Belovsky 1986). It seems that optimal foraging theory developed for predators (e.g., Schoener 1971) is inappropriate for herbivores (Westoby 1974, 1978), not because of a fundamental inadequacy, but rather because digestion was deemed to be nonlimiting for predators and jettisoned from optimal foraging models (Verlinden and Wiley 1989). This is one of the main problems with much of the theory – components crucial to some processes are eliminated from theories for other processes at an early stage of development. This produces the appearance of a qualitative gulf rather than a quantitative one (i.e., the difference between excluding a term from a model completely and setting it to some small value).

9.3 Commentary

It would seem to be an appropriate way to conclude this book by considering the ways in which the study of ecological versatility might be improved, and also some of the problems that obstruct these aims. There are several ways that the study and modelling of ecological versatility might be progressed. It seems crucial that the examination of the reasons for variation in resource use becomes more systematic, with common procedures and semantic conventions being applied by investigators. Currently, each subdiscipline engenders its own subcultural methodologies, although even these are not necessarily applied consistently. Therefore, I discuss first some of the elements that may streamline and systematicize research into how individuals within populations use the variety of resources that they do. Once there is more consistency in the study of versatility, we may be in a much better position to answer more general questions as to the origins and maintenance of different degrees of specialization in natural communities. Next, I offer a few comments on the development of modelling methods, which, when used in conjunction with improved sampling designs, may produce a superior understanding. I think that it is important to raise the issues related to species' idiosyncrasies, historical contingency and the small numbers of cases with which ecologists typically have to deal. These are, of course, significant insofar as they present problems in generalizing models or theories to account for or predict patterns.

Improved sampling designs

I commented in §9.2 on the lamentable inconsistency of studies designed to quantify the ecological versatility of species. But what would constitute a good protocol for designing such studies? There seem to be several stages to the process and it may be worthwhile to discuss these now, not in any great depth but in sufficient detail to indicate the sorts of thought and planning that should go into a 'good' design for characterizing the ecological versatility of a population.

Stage 1: Pilot work and preliminary screening

As with any work in which one intends to invest a great deal of effort, it is initially important to collect basic natural history information on the designated species. This can indicate many things that may help to simplify or reduce the logistic demands of the design. For example, does there appear to be substantial gender- or stage-specific variation in resource use? Often, this can be determined from a relatively cursory examination of the species or of existing information, and if such differentiation is not evident, then the design can be simplified accordingly.

If the target species occupies several or many different types of habitat, then, given the inevitable logistic constraints that researchers face, one would have to choose a selection of habitats that reflects the possible habitat-specific variation in resource use that may occur in populations of the species. (This may also apply to spatial isolates of a species in which habitat differences are not especially important, see J. T. Emlen 1981, Emlen and DeJong 1981.) Thus, existing data bases, literature sources and experience should be combined to help with this aspect of the design. In so doing, we may be able to improve significantly our knowledge about the relationship between ecological versatility and ubiquity, about which we know almost nothing (§8.5).

Stage 2: Design

The limit on the number of distinct habitat types to be considered must in part be determined by having *replicate* examples of the habitat types that are selected (see Mac Nally 1994b, 1995a). Otherwise, we will not be able to distinguish between habitat-specific variation in resource use (about which we are interested) and a specific dependence on *particular* plots (which is of little systematic importance). So, if the choice were between selecting four plots each of a different habitat type and two examples of each of two distinct habitat types, then the latter is unquestionably the desirable option. Admittedly, the statistical power of such a design would not be great but at least one has grounds for having a measure of confidence (or otherwise) in the inferences drawn about habitat-specific dependencies.

I think that there are five critical aspects that have to be built into systematic studies of ecological versatility once the study locations have been decided. These aspects can be summarized by the acronym '*F–I–V–A–R*': (*F*)itness basis, (*I*)ndividual patterns, (*V*)ariation in space and time, (*A*)vailabilities of resources, and clear and justifiable (*R*)esource specifications.

As outlined in Chapter 2, the difficulties associated with ecological scale among many other factors mean that it is most unlikely that definitions for resource states (*R*) can be generalized to apply to all cases and all situations. The minimum acceptable standard is clear, justified statements of the bases used for estimating versatility. It is probable that if workers think about what is needed to justify their resource classification in the planning or design stage of a study then much better and more consistent results will ensue. It may not be a vain hope that standard protocols could emerge for the specification of resource categories for similar systems.

The fitness differentials (*F*) associated with the use of different resources are indicative of the degree to which the resources contribute to the 'success' of an organism. My operational definition of ecological versatility links these differentials to the relative availabilities of the resources (*A*). For example, although a food resource might be 75% of the bulk of the consumption of an organism, it may deliver a much lower fraction of energy or essential nutrients than that, or entail a heavy processing load in terms of detoxification or tooth wear. The fitness returns associated with the resource may end up being similar to a resource that forms only 10% of the ingested bulk. It is these fitness returns rather than the raw consumption fractions that must be related to relative availabilities to gauge the dietary versatility of organisms.

A strong (but inconsistent) dependence of versatility on time and place (*V*) is evident in most field studies. There is certainly considerable field evidence showing the variation in versatility throughout the annual cycle and also in the corresponding seasons of different years. There may also be diurnal variation, although none of the studies in the survey was organized in a suitable way to detect it. Spatial variation consists of two aspects: selecting habitats – as described in stage 1 – and the variation in microhabitat use or of resource use (e.g., food) within microhabitats (see Fig. 3.4). The way to regard microhabitats depends very much on the distinction between viewing microhabitats as resources in their own right or as integrals of resources and constraints. There may be circumstances in which either case is appropriate so that no general heuristic seems to be appropriate. The process of just *thinking* about microhabitats and how they relate to a particular population probably will be beneficial.

Individual patterns of resource use (*I*) seem to be crucial components of an analysis of versatility (Sherry 1990), although pooling of phenotypic or stage data may be necessary or unavoidable if sampling is destructive or if there is no way to distinguish consistently between individuals. However, the results of the work of Humphrey *et al.* (1983) on diets of surface-gleaning Panamanian bats provided a fascinating insight into how misguided some of the cherished notions of the consistency of resource use among conspecific individuals might be (i.e., guild structure, resource partitioning, ecological displacement). Other workers have discovered marked individualistic patterns in groups as different as marine sea slugs (Trowbridge 1991) and terrestrial birds (Werner and Sherry 1987). Much of the important information on resource use and population dynamics will be masked if the distinctions between individuals are obscured.

Changing directions in ecological modelling?

There are promising signs that theoretical ecology might eventually escape the straitjacket imposed by logistic and Lotka–Volterra models and stability criteria by recognizing that spatial dynamics (e.g., Danielson 1991, 1992, Czárán and Bartha 1992, Danielson and Stenseth 1992, Wilson 1992, Tilman 1994) and diverse strategies (e.g., Ebenman 1988, Werner and Gilliam 1984, Crowder *et al.* 1992, Werner 1992) are the appropriate bases for future models of ecological communities. For example, a quite different picture for the coexistence of potential competitors can be constructed by specifying exploitation strategies first and placing these in various resource environments (see Chapter 7). Complete generalists might potentially coexist for periods much longer than they are likely to remain isolated or under a similar environmental regime (e.g., communities consisting of **Cs**-like strategists). Outcomes are sensitive to so many factors that much existing theory is simply too naïve to provide much useful guidance or many testable predictions.

The integration of spatial dynamics with more complex exploitation strategies into general models should involve less attention being paid to the development of abstruse mathematical models and greater thought to designing computer-based, realistic simulations of communities. The real intelligence behind modelling in the future must be in asking pertinent questions and in developing appropriate experimental designs for simulations. Selecting the appropriate technologies for implementing models will also be important (e.g., cellular automata? Phipps 1992, Ermentrout and Edelstein-Keshet 1993; fractal landscapes on conventional lattices? Seagle and Shugart 1985, Palmer 1992; see Haefner 1992 and Villa 1992).

Rather than being a deterrent, I see a reduced reliance on complex mathematical models that have dubious applicability in favour of realistic simulation as an opportunity for greater participation by ecologists in general – at present many are deterred from participating in the development of ecological theory by a lack of mathematical skills. The development of algorithms also forces judgements to be made about ecological requirements in a fairly direct way (Taylor *et al.* 1989), and may provide new ecological insights and inspire important empirical tests of assumptions. As Hawkins and MacMahon (1989: 439) put it: 'creation of a [simulation] model has heuristic value in the sense that an investigator must develop a blueprint of the system, and this activity reveals voids in knowledge of system components and processes.'

However, it is absolutely crucial that these programs not be carried out in isolation from observations and measurements on natural communities. The potential sensitivity of outcomes to particular rate-values is clear from the results presented in Chapter 7. The use of observed values for the rates of resource renewal, resource utilization and dispersal will be crucial in a modelling era in which asymptotic stability no longer is the main theoretical criterion. Only by reference to measured values can relevant levels of resource stochasticity, geometric characteristics of landscapes and population fluxes be meaningfully applied to simulations. It is important not to stray into regions of parameter space that are irrelevant to any natural system.

Idiosyncrasy, contingency and small numbers

Based on the comments at the end of §9.2, there may in principle be much potential for building a general picture of the constraints on ecological versatility that is not so intimately connected with particular processes, but based on a small set of factors. But there is one daunting problem impeding the development of such a theory, namely, the widespread occurrence of *idiosyncratic* species. Idiosyncratic species show patterns of resource use or behaviour at odds with most other (related) species, which are usually found to conform with model expectations. They also demonstrate that novel (perhaps abnormal is a better adjective) solutions have arisen to circumvent or nullify constraints that seem both theoretically logical and also binding on most other species *under the same conditions.*

Many examples of idiosyncratic species have been encountered in previous chapters. Ehrlich and Murphy (1988) reported how some herbivorous insects feed upon xylem and plant embryos, which appears to allow them to use a far greater diversity of plant hosts than most other herbivorous insects. Only a limited number of herbivorous insects possess trenching behaviour to counteract canalized chemical defences, such as latex (Doussard 1993). Special anatomical and physiological adaptations allow tachinid parasitoids (Diptera) to use a greater range of hosts than koinobionts might normally be expected to do (Hawkins and Gross 1992). Although some species of birds expand into habitats left unoccupied by similar species on some islands in the Lesser Antilles, others do not (Terborgh *et al.* 1978). Several species of birds in the lower Colorado River valley displayed responses to seasonal changes in food availability inconsistent with Fretwell's (1972) model, although most species conformed (Rice *et al.* 1980). One is struck in Miller's (1992) analysis of herbivory in notodontid moths by the prevalence of idiosyncratic host use. This is just a small selection of instances of idiosyncrasy; no doubt, most ecologists have encountered idiosyncratic species at some time.

The existence of these relatively uncommon, idiosyncratic strategies presents two serious dilemmas: (1) why do more species not use the strategies? and (2) what does idiosyncrasy mean for models, theories and testing?

Why only a limited number of idiosyncratic strategists should occur in an ecological system might be due purely to chance or historical contingency, but this is both unsatisfying and untestable yet quite possibly true. In a similar vein, Cornell and Lawton (1992: 10) recognized the disquieting contingencies in the adaptive diversification of certain taxa and not others. They posed the conundrum: 'the interesting question about the Hawaiian drosophilid radiation may not be the actual radiation, but why drosophilids and not something else?' One can equally ask: why have the tachinid parasitoids evolved special characteristics for overcoming the koinobiont bind, or why is there only a small number of leaf-trenching insect taxa? Interspecific competition might be responsible (i.e., insufficient 'niche space' for any but a few idiosyncratic species) but there seems to be little reason to expect this process to lead to particular patterns of resource use. I can see no ready answers to the first dilemma at this stage beyond an appeal to historical contingency and taxon-specific, evolutionary potential.

The occurrence of idiosyncratic species undoubtedly threatens generality in ecological understanding. The scientific method involves continually probing existing theory to refute and/or refine ideas, with tests being conducted on particular systems or species. The problem with idiosyncratic species is that they just do not conform with most other species so that framing a general model may be intractable. Idiosyncratic species frequently show patterns that are inconsistent with theoretical

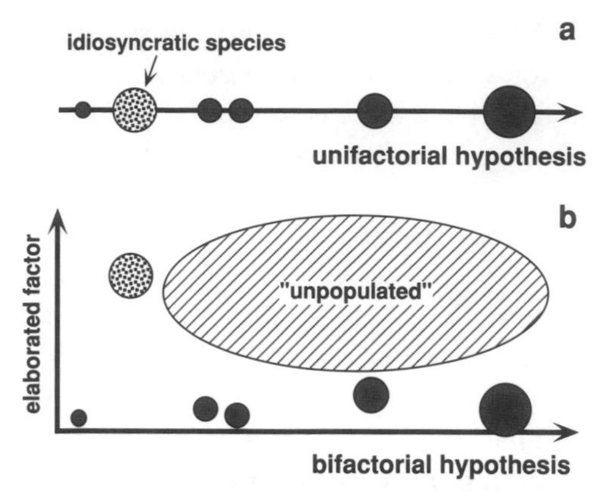

Fig. 9.1 (a) distribution of individual species with respect to a unifactorial hypothesis, which provides good explanatory (and possibly good predictive) power (indicated by the area of circles) apart from one idiosyncratic species. To 'explain' the deviance of the latter, further studies are conducted and a second factor is introduced (b). However, the elaboration still is unnecessary for most of the initial set of species and the bifactorial hypothesis now has a region that is 'unpopulated.' Although explanatory power has been improved because the idiosyncratic species has been incorporated into a more general framework, predictive power has been diminished – what do we expect to happen in the unpopulated zone? The problem is compounded with each elaboration (increased dimensionality).

expectations in a way that was not anticipated at the outset of a test. In other words, they not so much disprove models as side-step them by violating crucial and often apparently reasonable assumptions. I think that it is important to resist the temptation to 'sweep under the carpet' idiosyncratic species that fail to conform with a general picture. For example, Case and Bolger (1991) labelled a pair of unconventional anoles of the Greater Antilles (*Anolis eugenegrahami* and *A. vermiculatus*) as 'oddball', which seems unnecessarily disparaging and possibly designed to lessen the threat they might pose in theoretical terms.

In one sense, the idiosyncratic species 'drive' the research program by forcing an increasingly broad picture on a subdiscipline. The models escalate in complexity to accommodate the unusual characteristics of the idiosyncratic species with, perhaps, sets of rather restricted combinations of factors being necessary to do this. Were it not for the idiosyncratic species, the elaborations to the model may not be needed (Fig. 9.1 a, b). But the problem in this continual accretion of terms or factors into models is that much of the ecological hyperspace that is produced

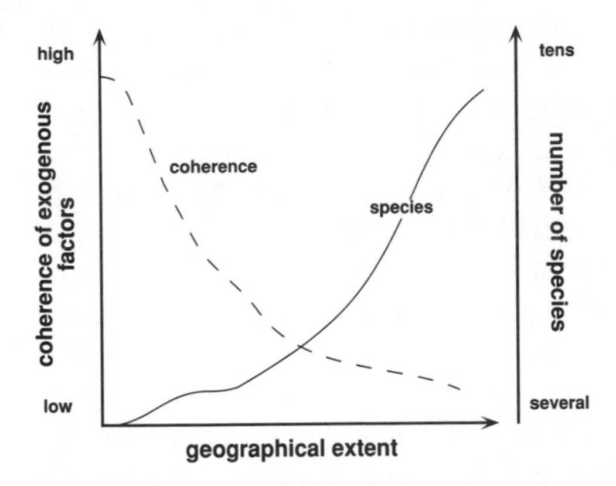

Fig. 9.2 The contention between the number of species for which an hypothesis is to be tested and the coherence of exogenous factors as functions of geographic extent. In most ecological studies, the number of species with respect to which hypotheses are framed and tested is small. To increase statistical power and reduce the impact of idiosyncratic species, workers seek similar species from further away for tests. However, as successively greater areas are considered, the coherence of exogenous factors decreases, which means that there is greater uncontrolled variation and actually diminished statistical resolution.

remains 'unpopulated' by natural species (Fig. 9.1 b). The additional complexity necessary to accommodate idiosyncratic species, although possibly leading to a more general picture, nevertheless reduces predictive capability because of the lack of constraints in the sparsely populated parameter space.

Almost all analyses of general patterns in community ecology also suffer from a bind that limits our capacity to generalize and to predict. This is the contention between the numbers of cases (typically populations) and the geographic realm over which there is a moderate to high coherence of exogenous factors (Fig. 9.2). In ecological analyses we typically consider 'middle-number' systems (at most) (Allen and Starr 1982) so that the opportunities for idiosyncratic behaviour and historical contingencies are rife while statistical power is low (small sample sizes). To overcome this difficulty, workers often try to extend their analyses to larger numbers of cases, which introduces heterogeneity in at least two forms.

First, the valid use of literature sources depends upon a high level of consistency in the collection, screening, interpretation and reporting of results by the original workers, which, as we have seen in Chapter 3, is a vain hope. Heterogeneous data sets have plagued many ambitious analytical programs, for example, food web theory (see Peters 1988, Martinez 1991, Polis 1991, Cohen *et al.* 1993, Hall and Raffaelli 1993).

And second, the laudable desire to increase sample sizes inevitably draws a cost, namely, the introduction of historical contingency and the influence of particular characteristics associated with different geographic and biogeographic regions. There are of course many examples of this effect, but a short discussion of one of these should suffice here. Steinberg and van Altena (1992) reported that three Australasian marine herbivores (the sea urchins *Tripneustes gratilla* and *Centrostephanus rodgersii*, and the gastropod *Turbo undulata*) showed little or no aversion to brown algae nor were there any demonstrable effects on fitness despite high concentrations of polyphenolic, 'deterrent' chemicals. These results differed markedly from studies conducted in ostensibly similar alga–herbivore systems from the northern Pacific, in which there was an avoidance of algae containing higher concentrations of polyphenols. It is noteworthy that the concentrations in the northern algae were at most one-third of the typical values recorded by Steinberg and van Altena (1992) for the Australasian algae. Thus, if one were to try to build up sample sizes for testing an hypothesis of the interaction between marine herbivores and algae derived from observations in the northern Pacific, then the extension into a distant region such as Australasia would prove to be counter-productive.

By moving into other biogeographic realms, other factors come into play. Thus, in the current example, Estes and Steinberg (1988) hypothesized that the absence of diving, predacious, marine mammals such as the sea otter (*Enhydra lutris*) from Australasian waters over the past several million years has permitted the occurrence of high densities of sublittoral marine herbivores. They argued that this factor would be associated with a more stringent coevolutionary coupling between the herbivores and marine algae, leading to two hypotheses: (1) concentrations of deterrent chemicals in Australasian algae should be significantly higher than in northern Pacific kelp; and (2) Australasian marine herbivores should have developed mechanisms to tolerate or ameliorate these high concentrations. Both hypotheses seem consistent with experimental results (Steinberg and van Altena 1992).

It seems clear that when one takes a broad overview – one not tied to particular taxa or ecological interactions – then a fundamental obstacle in coming to a systematic understanding of ecological versatility is simply the paucity of comparable cases in almost all systems. Much of the theoretical heritage of the 1960s and 1970s is essentially a uniformitarian picture of communities based on existing ecological processes that does not attempt to accommodate the evident historical contingency and potential for evolutionary novelty. The paucity of cases generally cannot be overcome by 'widening the net' to include taxa from farther afield. I suspect that contingency and novelty have, and have had, more influence than uniformitarian processes, giving rise to the apparent failure of community theory to account for and predict patterns in ecological communities.

Prospectus

Idiosyncratic species challenge what ecologists mean by having a 'good' model of nature and using such a model to 'predict'. How many special cases have to be accommodated before a general model is no longer of much use (Murray 1986, Wiens 1989b)? Is the conventional scientific 'propose–test–refine–predict–test . . .' paradigm completely appropriate for community ecology? Is prediction in the usual scientific sense applicable to community ecology when so many characteristics or variables have to be measured before a prediction can be made? As we have seen, we generally cannot increase the number of cases without incurring unacceptable increases in uncontrolled variation. With typically less than several dozen cases with which to work in any problem, and very often with fewer than ten, there seems to be much opportunity for evolutionary novelties to arise in most systems that will cause generalizations and predictions to founder. This indicates, I think, that ecologists may need to develop a conceptual basis for gauging the applicability of models in a way that may not be acceptable to the Popperian purist, but one that nevertheless reflects the reality that we are dealing with systems endowed with many degrees of freedom but without having the luxury of the laws of large numbers.

At the beginning of Chapter 1, I wrote of the main objective of community ecology – to provide a general picture for the reasons for patterns of resource use among individuals and populations in communities. We are still a long way from achieving this aim despite the efforts of many ecologists for over a century. Readers will judge for themselves whether the diverse set of explanations suggested for various ecological processes might not be expressions of more general processes that are based on a simple set of characteristics, with particularity emerging from different emphases on one factor or another. This is where specific information is so crucial – it tells us which factors to downplay and which to focus on in particular situations. In a qualitative sense, I think a general understanding might well be achieved. But, successful or even adequate quantitative models are unlikely in the foreseeable future because of the variety of strategies and tactics available to organisms. Some features such as soft/hard exploitation were modelled in Chapters 6 and 7, but these represent a very basic set of possible responses. The potential for hybrid strategies, for example, makes successful quantitative modelling of natural communities rather challenging (see §7.6).

One of the difficulties in understanding the evolution and maintenance of different patterns of ecological versatility and ubiquity (and community ecology in general) is that the determinants are inherently multifactorial. Many of the factors are the endogenous characteristics of particular organisms while others are exogenous factors affecting those individuals. The relative importance of factors can be shown to differ spatially and temporally at contemporary time scales, notwithstanding the variation that is forever hidden from us in the community dynamics of the past (see Fig. 4.10). The main technique by which causation is inferred is experimental manipulation, which is inherently more suited to unifactorial (or problems involving small numbers of factors) than multifactorial situations, and also is rooted in linear models. For numerous reasons, such experiments often have the aura rather than the substance of understanding the processes involved. I think that it is unwise to stake everything on conventional experimental designs, and that there needs to be a concurrent and complementary development of techniques or methodologies by which multifactorial inference can be conducted. I believe that this should be one of the objectives of community ecologists in the near- to middle-term future.

Glossary of terms

The specialized nature of this book involves the introduction and use of many terms that may not be familiar to readers. Therefore, this glossary should be a helpful quick-reference guide to many of the technical terms arising in the text. Where possible, I cite a primary reference for a term used in the literature. Each term has a chapter reference (e.g., §7) either where that term first is used or else where it is defined or discussed in greater detail.

TERM	DEFINITION
aliasing	artefacts in estimating versatility produced by definition of resource states in either a too coarse (lumped) or too fine (split) fashion than is biologically meaningful (§2)
Allee effect	density in sexual populations in which individuals are so unlikely to find mates that the population cannot be sustained (Hopf and Hopf 1985) (§6)
asymptotic equilibrial stability	the stable equilibrium state to which model communities converge, normally as solutions to equilibrium analyses of Lotka–Volterra models (§7)
Ch	a strategy of hard coherent exploitation used in modelling (§6)
coenocline	the distribution of species along environmental gradients, often represented in terms of density responses (Whittaker 1975)
coherent versatility	no significant differences between individuals in exploiting resources in the proportions in which they are available (§6)
complementarity	an inverse correlation between versatility and ubiquity (Cody 1974) (§8)
complementary resources	resources that must be gathered together by a consumer because they fulfil distinct needs – the resource acquired

Term	Definition
	(and assimilated) at the slowest rate effectively limits the per capita growth rate (León and Tumpson 1975) (§2)
component (environmental)	anything impinging on an individual organism, including food, refuges, nesting sites, ecological gradients or patches such as salinity, light or trace elements, predators, conspecific individuals, etc. (§2)
compression hypothesis	constriction of habitat use rather than food use should occur when a residential species encounters an invading species, at least at proximal time scales (MacArthur and Wilson 1967) (§8)
conformance index	a niche index in which either utilization or exploitation of resources is compared with availability (§2)
constraint	any environmental component limiting or impeding free use of resources (e.g., limiting factors, malentities, §2)
cosmopolitan	occupying an extensive geographic range, though not necessarily every habitat within that range (Hesse *et al.* 1951) (§2)
Cs	a strategy of soft coherent exploitation used in modelling (§6)
disjunctive metamorphosis	extremely pronounced change in somatic organization during development, equivalent to complex life cycle (Wilbur 1980) (§5)
dominance	the degree to which species exert influence on the occurrence of other species – dominant species exert a strong influence and subordinate species are greatly influenced (McNaughton and Wolf 1970) (§8)
DRM	digestive rate model of Verlinden and Wiley (1989) (§4)
ecological release	expansion of densities and many other ecological attributes when competitors are absent (§8)
efficiency	efficiency of conversion of utilized resources into realized fitness gain (§6)
euryphagy	consumption of many foods (§4)
exploitation	the fitness return derived from utilizing a resource (§1)
exploitation, hard	see *hard exploitation*
exploitation, soft	see *soft exploitation*

TERM	DEFINITION
extraneous resources	nonlimiting resources of which individuals can utilize as much or as little as they need to without being forced into competitive situations or optimization decisions (Real 1975) (§2)
feasibility	model communities in which there are 'stable' equilibrium solutions with all population densities being positive (Pimm 1984) (§7)
functional response	changes in utilization rate of a resource in response to levels of utilization pressure and availability of the resource (soft strategists only) (§6)
general versatility	see *coherent versatility*
glyph	graphical device for displaying multidimensional data (Crawford and Falls 1990) (§7)
habitat versatility	see *ubiquity*
hard exploitation	an inability to adjust utilization rates of resources in response to levels of utilization pressure and the availability of the resource (§6)
homologous strategies	pairs of versatility strategies that differ only in that one involves fixed utilization rates (the hard strategist) and the other variable utilization rates (the soft strategist), such as **Ch** and **Cs** (§6)
ideal free distribution	individuals within populations should distribute themselves amongst habitats or microhabitats to maximize their individual fitnesses (Fretwell and Lucas 1970) (§8)
idiobiosis	parasitoids arresting development of the host individual upon infection, hence idiobiont (Hawkins and Gross 1992) (§4)
incoherent versatility	significant differences occur between individuals and exploitation summed over the population differs significantly from availability (§6)
invasibility analysis	method of determining solutions to model communities in which combinations of populations are sought whose joint resource use prevents the establishment of invading populations (§7)
juvenile bottleneck problem	contention between the demands for efficient exploitation of resources by different stages (Lande 1982) (§5)

Term	Definition
koinobiosis	parasitoids allowing host development to continue following infection, hence koinobiont (Hawkins and Gross 1992) (§4)
landscape	spatial scale over which populations of a species can influence each others' dynamics via dispersal and migration (Danielson 1992) (§8)
localized	occupying a small geographic range, although not necessarily restricted to few habitats within that range (Hesse *et al.* 1951) (§2)
LVM	Lotka–Volterra models of dynamics of interacting populations (§7)
marginal utilization rate	utilization rates below which resources are no longer sufficient to use economically (§6)
maximum utilization rate	maximum utilization rate of a (currently) unlimited resource (§6)
metamorphosis, disjunctive	see *disjunctive metamorphosis*
metamorphosis, nondisjunctive	see *nondisjunctive metamorphosis*
niche pattern	the relationship between niche position, ubiquity, and mean density (Shugart and Patten 1972) (§8)
niche position	in an abstract space derived from ordinations, the absolute distance from the mean location of habitats occupied by a species to the mean of all habitats in a sample (Shugart and Patten 1972) (§8)
nonconformance index	a niche index in which availability of resources is not explicitly considered (§2)
nondisjunctive metamorphosis	small or moderate change in somatic organization during development (§5)
numerical response	changes in population density in response to levels of utilization pressure and availability of the resource (§6)
oligophagy	eating few types of food (Fox and Morrow 1981) (§4)
phenomenological models	models that rely on empirical findings to account for other observed results, i.e., such models are incomplete representations of natural communities (§8)
physiognomy	structural characteristics of habitats (Whittaker 1975) (§8)

TERM	DEFINITION
polyphagy	eating many types of food (Fox and Morrow 1981) (§4)
polyphenism (adaptive)	the capacity of organisms to follow alternative developmental pathways according to ecological cues perceived or detected during development (Moran 1992) (§5)
range (geographic)	the total area of a polygon enclosing the extreme recorded locations of all populations of a species (Rapoport 1982) (§2)
resilience	model communities in which systems quickly return to 'stable' equilibria following disturbance (Pimm 1984) (§7)
resource (true)	any ecological factor to which an organism can gain exclusive access for some period of time, and that by its use increases the organism's fitness (§2)
resource, complementary	see *complementary resources*
resource, extraneous	see *extraneous resources*
resource, substitutable	see *substitutable resources*
resource-like versatility	exploitation of resources, when summed over the population, is in the same proportions as the availability of the resources, but individuals differ significantly in the resources they exploit (§6)
response, functional	see *functional response*
response, numerical	see *numerical response*
restricted	population densities differ according to habitat; at the most extreme, species have zero density everywhere except in one habitat type (Hesse *et al.* 1951) (§2)
Rh	a strategy of hard resource-like exploitation used in modelling (§6)
Rs	a strategy of soft resource-like exploitation used in modelling (§6)
Sh	a strategy of hard specialized exploitation used in modelling (§6)
sink habitat	habitat in which local recruitment is insufficient to maintain population densities, and immigrants from source habitats can be accommodated (Pulliam 1988) (§8)

Term	Definition
soft exploitation	the capacity to adjust utilization rates of resources in response to levels of utilization pressure and availability of the resource (§6)
source habitat	habitats in which recruitment exceeds numbers required for stable population sizes, so that these habitats act as centres of emigration (Pulliam 1988) (§8)
span	a representation of modelling results spanning parameter space (§7)
specialized versatility	all individuals exploit resources in the same way but not in direct proportion to resource availability (§6)
Ss	a strategy of soft specialized exploitation used in modelling (§6)
stenophagy	use of one (monophagy) or few (oligophagy) species as food (§4)
substitutable resources	resources satisfying the same requisite needs, and that can be freely interchanged. The net result of perfect substitutability is that the *total* amount of all substitutable resources limits the per capita growth rate of the population (León and Tumpson 1975) (§2)
supplementation	a positive correlation between versatility and ubiquity (Cody 1974) (§8)
ubiquitous	similar population densities irrespective of habitat type (Hesse *et al.* 1951) (§2)
ubiquity	the evenness of densities of populations of a species with respect to different types of habitat (Hesse *et al.* 1951) (§2)
utilization	consumption or occupation of a resource (§§1, 2)
utilization pressure	total pressure exerted on a resource by a population – the product of population density and average utilization rate (§6)
versatility (ecological)	the degree to which fitness derived from utilizing various resources matches the availability of those resource (§1)
versatility, coherent	see *coherent versatility*
versatility, general	see *coherent versatility*
versatility, habitat	see *ubiquity*

Term	Definition
versatility, incoherent	see *incoherent versatility*
versatility, resource-like	see *resource-like versatility*
versatility, specialized	see *specialized versatility*
viability floor	density at which a population is effectively extinct (§6)

Appendix A

In this appendix, I provide details related to the survey of studies described in Chapter 3. The complete list of papers surveyed for the analysis in Chapter 3 is presented in §A.1 (see the reference list for complete entries). The reasons why these studies were selected for analysis are described in §3.2. The other section to this appendix (§A.2) is an overview of the breadth and overlap measures that were encountered in the survey, including formulae (where appropriate), numbers of instances of use, and comments on the usefulness of each measure.

A.1 List of studies used in the survey

Abbott, I., Abbott, L. K. and Grant, P. R. (1977) *Ecological Monographs*, **47**, 151–184.
Addicott, J. F. (1978) *Canadian Journal of Zoology*, **56**, 1837–1841.
Adolph, S. C. (1990) *Ecology*, **71**, 315–327.
Aldridge, H. D. J. N. and Rautenbach, I. L. (1987) *Journal of Animal Ecology*, **56**, 763–778.
Alford, R. A. (1989) *Ecology*, **70**, 206–219.
Anderson, S. H., Shugart, H. H., Jr and Smith, T. M. (1979) In *The role of insectivorous birds in forest ecosystems.*, ed. Dickson, J.G., Conner, R.N., Fleet, R.R., Kroll, J.C. and Jackson, J.A., pp. 261–268. Academic Press: New York, NY, USA.
Barker, K. M. and Chapman, A. R. O. (1990) *Marine Biology*, **106**, 113–118.
Bell, H. L. and Ford, H. A. (1990) *Advances in Avian Biology*, **13**, 381–388.
Bell, G. P. (1980) *Canadian Journal of Zoology*, **58**, 1876–1883.
Bennett, B. A. and Branch, G. A. (1990) *Estuarine, Coastal and Shelf Science*, **31**, 139–155.
Bergman, E. (1988) *Journal of Animal Ecology*, **57**, 443–453.
Black, R. (1979) *Journal of Animal Ecology*, **48**, 401–411.
Blake, J. G. and Hoppes, W. G. (1986) *Auk*, **103**, 328–340.
Bloom, S. A. (1981) *Oecologia (Berl.)*, **49**, 305–315.
Bouchon-Navaro, Y. (1986) *Journal of Experimental Marine Biology and Ecology*, **103**, 21–40.
Braithwaite, R. W., Cockburn, A. and Lee, A. K. (1978) *Australian Journal of Ecology*, **3**, 423–445.
Brakefield, P. M. and Reitsma, N. (1991) *Ecological Entomology*, **16**, 291–303.
Bratton, S. P. (1976) *American Naturalist*, **110**, 679–698.
Brown, J. H. and Kodric-Brown, A. (1979) *Ecology*, **60**, 1022–1035.

Carroll, C. R. (1979) *American Naturalist*, **113**, 551–561.
Chew, F. S. (1981) *American Naturalist*, **118**, 655–672.
Clarke, R. D. (1977) *Marine Biology*, **40**, 277–289.
Coley, P. D. (1983) *Ecological Monographs*, **53**, 209–233.
Cox, P. A. (1981) *American Naturalist*, **117**, 295–307.
Crowley, P. H. and Johnson, D. M. (1982) *Ecology*, **63**, 1064–1077.
Davies, R. W., Wrona, F. J., Linton, L. and Wilkialis, J. (1981) *Oikos*, **37**, 105–111.
Delbeek, J. C. and Williams, D. D. (1987) *Journal of Animal Ecology*, **56**, 949–967.
Denno, R. F. (1980) *Ecology*, **61**, 702–714.
Denno, R. F., Larsson, S. and Olmstead, K. L. (1990) *Ecology*, **71**, 124–137.
Diaz, N. F. and Valencia, J. (1985) *Oecologia (Berl.)*, **66**, 353–357.
Downes, B. J. (1986) *Oecologia (Berl.)*, **70**, 457–465.
DuBowy, P. J. (1988) *Ecology*, **69**, 1439–1453.
Ebersole, J. P. (1985) *Ecology*, **66**, 14–20.
Ehlinger, T. J. (1990) *Ecology*, **71**, 886–896.
Feinsinger, P., Swarm, L. A. and Wolfe, J. A. (1985) *Ecological Monographs*, **55**, 1–28.
Findley, J. S. and Black, H. (1983) *Ecology*, **64**, 625–630.
Fleming, T. H. (1985) *Ecology*, **66**, 688–700.
Flint, R. W. and Kalke, R. D. (1986) *Estuarine, Coastal and Shelf Science*, **22**, 657–674.
Ford, H. A. and Paton, D. C. (1977) *Australian Journal of Ecology*, **2**, 399–407.
Ford, H. A. and Paton, D. C. (1982) *Australian Journal of Ecology*, **7**, 149–159.
Fraser, D. F. (1976) *Ecology*, **57**, 238–251.
Gascon, C. (1991) *Ecology*, **72**, 1731–1746.
Gerard, V. A. (1990) *Marine Biology*, **107**, 519–528.
Gladfelter, W. B. and Johnson, W. S. (1983) *Ecology*, **64**, 552–563.
Hairston, N. G., Sr (1980) *American Naturalist*, **115**, 354–366.
Hairston, N. G., Sr, Nishikawa, K. C. and Stenhouse, S. L. (1987) *Evolutionary Ecology*, **1**, 247–262.
Hallett, J. G. (1982) *Ecology*, **63**, 1400–1410.
Harder, L. D. (1985) *Ecology*, **66**, 198–210.
Heithaus, E. R. (1979) *Ecology*, **60**, 190–202.
Herbold, B. (1984) *American Naturalist*, **124**, 561–572.
Herrera, C. M. (1978) *Journal of Animal Ecology*, **47**, 871–890.
Hines, A. H. (1982) *Ecological Monographs*, **52**, 179–198.
Hoffmaster, D. K. (1985) *Ecology*, **66**, 626–629.
Holmes, R. T. and Schultz, J. C. (1988) *Canadian Journal of Zoology*, **66**, 720–728.
Humphrey, S. R., Bonaccorso, F. J. and Zinn, T. L. (1983) *Ecology*, **64**, 284–294.
Jaenike, J. (1978) *Evolution*, **32**, 676–678.
Jaenike, J., Parker, E. D., Jr and Selander, R. K. (1980) *American Naturalist*, **116**, 196–205.
Jaenike, J. and Selander, R. K. (1979) *Evolution*, **33**, 741–748.
Jaksić, F. M. and Braker, H. E. (1983) *Canadian Journal of Zoology*, **61**, 2230–2241.
James, S. D. (1991) *Oecologia (Berl.)*, **85**, 553–561.
Joern, A. and Lawlor, L. R. (1980) *Ecology*, **61**, 591–599.
Joern, A. and Lawlor, L. R. (1981) *Oikos*, **37**, 93–104.
Johnson, R. A. (1986) *Ecology*, **67**, 133–138.
Kephart, S. R. (1983) *Ecology*, **64**, 120–133.
Kochmer, J. P. and Handel, S. N. (1986) *Ecology*, **56**, 303–325.

Kodric-Brown, A., Brown, J. H., Byers, G. S. and Gori, D. F. (1984) *Ecology*, **65**, 1358–1368.
Kohn, A. J. and Nybakken, J. W. (1975) *Marine Biology*, **29**, 211–234.
Lacy, R. C. (1984) *Ecological Entomology*, **9**, 43–54.
Landres, P. B. and MacMahon, J. A. (1983) *Ecological Monographs*, **53**, 183–208.
Lechowicz, M., Schoen, D. J. and Bell, G. (1988) *Journal of Ecology*, **76**, 1043–1054.
Lemen, C. A. (1978) *Oecologia (Berl.)*, **35**, 13–19.
Levey, D. J. (1988) *Ecological Monographs*, **58**, 251–269.
Leviten, P. J. and Kohn, A. J. (1980) *Ecological Monographs*, **50**, 55–75.
Lewis, W. M. J. (1978) *Ecology*, **59**, 666–671.
Lishman, G. S. (1985) *Journal of Zoology (Lond.)*, **205**, 245–263.
Lister, B. C. (1981) *Ecology*, **62**, 1548–1560.
Llewellyn, J. B. and Jenkins, S. H. (1987) *American Naturalist*, **129**, 365–381.
Lowry, L. F., McElroy, A. J. and Pearse, J. S. (1974) *Biological Bulletin*, **147**, 386–396.
M'Closkey, R. T. (1978) *American Naturalist*, **112**, 683–694.
Mac Nally, R. C. (1983) *Herpetologica*, **39**, 130–140.
Mac Nally, R. C. (1985) *Australian Journal of Zoology*, **33**, 329–338.
Mahdi, A., Law, R. and Willis, A. J. (1989) *Journal of Ecology*, **77**, 386–400.
Maiorana, V. C. (1978) *Canadian Journal of Zoology*, **56**, 1017–1025.
McEvoy, P. B. (1986) *Ecology*, **67**, 465–478.
McIvor, C. C. and Odum, W. E. (1988) *Ecology*, **69**, 1341–1351.
Mercurio, K. S., Palmer, A. R. and Lowell, R. B. (1985) *Ecology*, **66**, 1417–1425.
Meserve, P. L. (1976) *Journal of Animal Ecology*, **45**, 647–666.
Meserve, P. L. (1981) *Journal of Animal Ecology*, **50**, 745–757.
Miller, J. C. (1980) *Ecology*, **61**, 270–275.
Mitchell, J. C. (1979) *Canadian Journal of Zoology*, **57**, 1487–1499.
Mitchell, M. J. (1978) *Ecology*, **59**, 516–525.
Mountainspring, S. and Scott, J. M. (1985) *Ecological Monographs*, **55**, 219–239.
Moyle, P. B. and Vondracek, B. (1985) *Ecology*, **66**, 1–13.
Muotka, T. (1990) *Oecologia (Berl.)*, **85**, 281–292.
Mushinsky, H. R. and Hebrard, J. J. (1977) *Canadian Journal of Zoology*, **55**, 1545–1550.
Mushinsky, H. R., Hebrard, J. J. and Vodopich, D. S. (1982) *Ecology*, **63**, 1624–1629.
Nudds, T. D. (1983) *Ecology*, **64**, 319–330.
Parrish, J. A. D. and Bazzaz, F. A. (1979) *Ecology*, **60**, 597–610.
Parrish, J. A. D. and Bazzaz, F. A. (1985) *Ecology*, **66**, 1296–1302.
Parsons, P. A. (1978) *American Naturalist*, **112**, 1063–1074.
Pimm, S. L. and Pimm, J. W. (1982) *Ecology*, **63**, 1468–1480.
Platt, W. J. and Weis, I. M. (1977) *American Naturalist*, **111**, 479–513.
Polis, G. A. (1984) *American Naturalist*, **123**, 541–564.
Polis, G. A. and McCormick, S. J. (1987) *Ecology*, **68**, 332–343.
Preston, C. R. (1990) *The Condor*, **92**, 107–112.
Pulliam, H. R. (1985) *Ecology*, **60**, 1829–1836.
Raley, C. M. and Anderson, S. H. (1990) *The Condor*, **92**, 141–150.
Riechert, S. E. (1991) *Evolutionary Ecology*, **5**, 327–338.
Riechert, S. E. and Cady, A. B. (1983) *Ecology*, **64**, 899–913.
Ruiz, A. and Heed, W. B. (1988) *Journal of Animal Ecology*, **57**, 237–249.
Rusterholtz, K. A. (1981) *American Naturalist*, **118**, 173–190.

Sano, M. (1990) *Journal of Experimental Marine Biology and Ecology*, **140**, 209–223.
Schemske, D. W. and Brokaw, N. (1981) *Ecology*, **62**, 938–945.
Schluter, D. (1988) *Ecological Monographs*, **58**, 229–249.
Schoener, T. W. (1970) *Ecology*, **51**, 408–418.
Shelly, T. E. (1985) *Oecologia (Berl.)*, **67**, 57–70.
Sherry, T. W. (1984) *Ecological Monographs*, **54**, 313–338.
Shine, R. (1977) *Canadian Journal of Zoology*, **55**, 1118–1128.
Skeate, S. T. (1987) *Ecology*, **68**, 297–309.
Smith, J. N. M., Grant, P. R., Grant, B. R., Abbott, I. J. and Abbott, L. K. (1978) *Ecology*, **59**, 1137–1150.
Snow, B. K. and Snow, D. W. (1972) *Journal of Animal Ecology*, **41**, 471–485.
Sogard, S. M., Powell, G. V. N. and Holmquist, J. G. (1989) *Bulletin of Marine Science*, **44**, 179–199.
Spence, J. A. (1983) *Journal of Animal Ecology*, **52**, 497–511.
Stamp, N. E. and Ohmart, R. D. (1978) *Ecology*, **59**, 700–707.
Steenhof, K. and Kochert, M. N. (1988) *Journal of Animal Ecology*, **57**, 37–48.
Stiling, P. D. (1980) *Journal of Animal Ecology*, **49**, 793–805.
Stiling, P. D. and Strong, D. R. (1983) *Ecology*, **64**, 770–778.
Talbot, J. J. (1979) *Copeia*, **1979**, 472–481.
Tanaka, K. (1991) *Oecologia (Berl.)*, **86**, 8–15.
Thomas, C. D., Vasco, D., Singer, M. C., Ng, D., White, R. R. and Hinkley, D. (1990) *Evolutionary Ecology*, **4**, 67–74.
Thompson, J. N. and Willson, M. F. (1979) *Evolution*, **33**, 973–982.
Thorman, S. and Wiederholm, A-M. (1986) *Journal of Experimental Marine Biology and Ecology*, **95**, 67–86.
Trivelpiece, W. Z., Trivelpiece, S. G. and Volkman, N. J. (1987) *Ecology*, **68**, 351–361.
Trowbridge, C. D. (1991) *Ecology*, **72**, 1880–1888.
Via, S. (1991) *Ecology*, **72**, 1420–1427.
Vitt, L. J., van Loben Sels, R. C. and Ohmart, R. D. (1981) *Ecology*, **62**, 398–410.
Wagner, J. L. (1981) *Ecology*, **62**, 973–981.
Waldorf, E. S. (1976) *American Midland Naturalist*, **96**, 76–87.
Ward, M. A. and Thorpe, J. P. (1989) *Marine Biology*, **103**, 215–224.
Werner, P. A. and Platt, W. J. (1976) *American Naturalist*, **110**, 959–971.
Wheelwright, N. T. (1985) *Ecology*, **66**, 808–818.
Whiteside, M. C., Williams, J. B. and White, C. P. (1978) *Ecology*, **59**, 1177–1188.
Whittam, T. S. and Siegel-Causey, D. (1981) *Ecology*, **62**, 1515–1524.
Wolda, H. and Roubik, D. W. (1986) *Ecology*, **67**, 426–433.
Yu, D. S., Luck, R. F. and Murdoch, W. W. (1990) *Ecological Entomology*, **15**, 469–480.

A.2 Breadth and overlap measures

In this section are listed the niche breadth and overlap measures that were used in one or more of the studies consulted for the survey. In addition to the literature source of each measure, the usage in terms of numbers of studies (of a total of 145 studies), formulae, and general remarks (where appropriate) also are provided. The 'class' column is used to indicate whether a measure is standardized (*S*, usually standardized to the interval [0...1]) or not (*U*), whether the measure is a niche breadth (*B*) or niche overlap statistic (*O*), and whether the measure is a conformance (*C*) or nonconformance measure (*N*). Whence, a standardized conformance breadth measure carries the moniker *S–B–C*. In all formulae, Q indicates the number of resource categories, π_{ij} is the proportion of utilization associated with category i (by species j), n_{ij} is the number of items or events associated with category i (by species j), and α_i is the proportionate availability (more usually abundance, see Chapter 2) of category i. Breadth measures are listed first in order of popularity, followed by overlap indices similarly ordered.

Name/authority	Usage	Class	Formula/ description	Remarks
Levins $\beta_{1/\lambda}$ (Levins 1968)	11	*U–B–N*	$1/\sum_{i=1}^{Q}\pi_{ij}^2$	avoid
Shannon–Wiener $\beta_{H'}$ (Shannon and Weaver 1949)	10	*U–B–N*	$-\sum_{i=1}^{Q}\pi_{ij}\log_e\pi_{ij}$	avoid
Standard Levins $\beta_{1/\lambda}$ (Levins 1968)	5	*S–B–N*	$1/Q\sum_{i=1}^{Q}\pi_{ij}^2$	avoid
Colwell–Futuyma (Colwell and Futuyma 1971)	2	*S–B–C*	consult original paper	usable
Standard Shannon–Wiener $\beta_{H'}$ (Shannon and Weaver 1949)	2	*S–B–N*	$-\sum_{i=1}^{Q}\frac{\pi_{ij}\log_e\pi_{ij}}{\log_e Q}$	avoid
Adapted Schoener *PS* (Feinsinger *et al.* 1981)	2	*S–B–C*	$1-\frac{1}{2}\sum_{i=1}^{Q}\lvert\pi_{ij}-\alpha_i\rvert$	usable and simple, recommended
Ivlev electricity (Ivlev 1961)	1	*S–B–C*	$\sum_{i=1}^{Q}\frac{\lvert\pi_{ij}-\alpha_i\rvert}{(\pi_{ij}+\alpha_i)}$	summation of absolute values of Ivlev electivities is equivalent to Canberra metric (Lance and Williams 1967) – recommended

Name/authority	Usage	Class	Formula/ description	Remarks
Petraitis *W* (Petraitis 1979)	1	*S–B–C*	$(\prod_{i=1}^{Q}[\alpha_i/\pi_{ij}]^{n_{ij}})^{(1/N_i)}$ where $N_i=\sum_{i=1}^{Q}n_{ij}$	sample size dependent, and usable only with resource states that can be 'counted'
Dueser–Shugart *V* (Dueser and Shugart 1979)	1	*U–B–N*	coefficient of variation of values in a derived multivariate space	avoid
Pielou (1975) method	1	*U–B–N*	a sampling-based method using the Brillouin diversity index	avoid
Hurlbert's adaptation of mean crowding (Hurlbert 1978)	1	*U–B–C*	$\sum_{i=1}^{Q}(\pi_{ij}^2/\alpha_i)$	usable
Standard deviation of principal components (Hoffmaster 1985)	1	*U–B–N*	—	avoid
Standard deviation of physical gradient (Werner and Platt 1976)	1	*U–B–N*	—	avoid
Schoener *PS* (Schoener 1970)	30	*S–O–N*	$1-\frac{1}{2}\sum_{i=1}^{Q}\lvert\pi_{ij}-\pi_{ik}\rvert$, or $\sum_{i=1}^{Q}\min(\pi_{ij},\pi_{ik})$	avoid
Pianka–May (May 1975, Pianka 1973)	10	*U–O–N*	$\dfrac{\sum_{i=1}^{Q}\pi_{ij}\bullet\pi_{ik}}{\sqrt{\sum_{i=1}^{Q}\pi_{ij}^2\bullet\sum_{i=1}^{Q}\pi_{ik}^2}}$	avoid
Hurlbert *L* (Hurlbert 1978)	3	*U–O–C*	$\sum_{i=1}^{Q}(\pi_{ij}\pi_{ik}/\alpha_i)$	usable
Asymmetric MacArthur–Levins (MacArthur and Levins 1967)	2	*U–O–N*	$\dfrac{\sum_{i=1}^{Q}\pi_{ij}\pi_{ik}}{\sum_{i=1}^{Q}\pi_{ij}^2}$	avoid
Morisita (Horn 1966)	2	*S–O–N*	$\dfrac{2\sum_{i=1}^{Q}\pi_{ij}\pi_{ik}}{\sum_{i=1}^{Q}\pi_{ij}^2+\sum_{i=1}^{Q}\pi_{ik}^2}$	avoid

Name/authority	Usage	Class	Formula/ description	Remarks
Horn R_0 (Horn 1966)	2	*S–O–N*	see Horn (1966)	avoid
Colwell–Futuyma (Colwell and Futuyma 1971)	2	*S–O–C*	consult original paper	usable
Adapted Schoener (Schoener 1974b)	1	*U–O–C*	$\frac{\sum_{i=1}^{Q} \pi_{ij}^2/\alpha_i \bullet \pi_{ik}/\alpha_i}{\sum_{i=1}^{Q} \pi_{ij}^3/\alpha_i^2}$	usable, but unstandardized
Levins β-weighted (Levins 1968)	1	*U–O–N*	$\sum_{i=1}^{Q} \pi_{ij}\pi_{ik}\beta_{j(i);1/\lambda}$	avoid: asymmetric and compiled over different habitat types

Appendix B

This technical appendix contains information on several issues related to the modelling described in Chapters 6 and 7. None of this material is essential reading for understanding the flow of results in these chapters, but it is provided for completeness. The first few sections describe the structure of the diagnostic statistics that are used to characterize the behaviour of model populations. These statistics measure the conformance between the utilization and the availability of resources, the specification of the variation in resource availability, and the degree of heterogeneity of resource use among phenotypes within populations. I also provide largely verbal descriptions of the algorithms for each of the six exploitation strategies (i.e., **Ch**, **Rh**, **Sh**, etc.), with comments on implementation issues. These descriptions include notes on the assumptions that are specific to each algorithm. I believe that most decisions are ecologically reasonable choices that offer the most natural way of selecting between alternatives when contentious situations arise. Unfortunately, the verbal description of relatively complex algorithms makes for rather dry reading, but short of providing a program listing consisting of 7500 lines of C or 'pseudocode', this is the best way to describe how the population dynamics of the different strategies function. The last part of this appendix recapitulates the main assumptions and limitations involved in the modelling techniques.

B.1 Diagnostic statistics

There are many criteria that might be used to characterize the behaviour of a modelling suite, but the key issue from a versatility standpoint is the pattern of conformance between utilization and availability through time. Other points of interest are the temporal changes in density and the rates of resource utilization.

Conformance indices

There are two main facets of conformance, namely: (1) the overall conformance between utilization and availability in a population as a whole; and (2) the variation between the components of a population. Given that phenotypes represent the unit of analysis in the models, the second aspect relates to between-phenotype variation in utilization. I briefly address how these facets were evaluated.

Consider a population consisting of M phenotypes, with current densities $[\eta_1, \ldots, \eta_M]$. Assume that there are R resources, with current availabilities $[A_1, \ldots, A_R]$. For a given phenotype, p say, there will be R utilization rate coefficients, some of which may be zero: $[\tau_{p1}, \ldots, \tau_{pR}]$. The population-wide conformance index, B_{pop}, used here is a simple extension of the well-known proportional similarity index (Schoener 1970), except that the role of utilization in that statistic is replaced here by a scaled, density-weighted average of the utilization rates of the phenotypes of the population. By setting

$$\sum_{p=1}^{M} \sum_{r=1}^{R} \eta_p \tau_{pr} = (\eta\tau)_{\bullet\bullet}, \tag{B.1}$$

the weighted utilization value for resource r is

$$U_r = \textstyle\sum_{p=1}^{M} \eta_p \tau_{pr} / (\eta\tau)_{\bullet\bullet}. \tag{B.2}$$

(see Siegismund *et al.* 1990). Note the use of the dot-summation convention, where the dot subscript implies the summation over the index in the corresponding subscript position. Thus,

$$B_{pop} = 1 - \tfrac{1}{2} \textstyle\sum_{r=1}^{R} |U_r - A_r|. \tag{B.3}$$

Strictly speaking, the equations (B.1) and (B.2) should include the efficiencies of translation of resources into fitness gains.

A way of monitoring the between-phenotype variation in resource utilization also is needed. A straightforward method for doing this is to use the Czekanowski similarity measure, averaged over all pairwise combinations of phenotypes. For two phenotypes, the Czekanowski similarity is:

$$C_{pq} = \frac{2 \times \sum_{r=1}^{R} \min(\tau_{pr}, \tau_{qr})}{\sum_{r=1}^{R} (\tau_{pr} + \tau_{qr})}, \tag{B.4}$$

where 'min' indicates the minimum of the two values (see Greig-Smith 1983: 192 ff., for a detailed discussion of similarity and dissimilarity measures). Thus, B_{pop} corresponds to the population-wide conformance between utilization and availability, while the mean Czekanowski similarity represents the degree of similarity between phenotypes in the utilization of resources. Clearly, the latter is not relevant to populations consisting of a single phenotype (e.g., **Ch**, **Sh**), and, therefore, is undefined.

B.2 Variation in resource availability

The availabilities of many natural resources fluctuate markedly through time, often with both deterministic or systematic (e.g., diurnal, seasonal) and seemingly random components. The variation in resource availability in the models considered here was generated by using elements corresponding to systematic changes (sinusoidal variation) and also to random disturbances. The general form for producing variation in resource availability is:

$$A(t) = A(0) + W \sin\left(\frac{t - L}{P}\right) + G(0, \gamma), \qquad \text{(B.5)}$$

where: $A(t)$ is the current availability (i.e., at time t); $A(0)$ is the base level availability; W is the amplitude of the systematic variation; L is a temporal lag or offset, permitting the asynchronous variation between resources; P is the periodicity of the systematic variation; and $G(0, \gamma)$ is a Gaussian random disturbance with zero mean and a standard deviation of γ.

The ways in which the variation in availability is produced need not concern us greatly, but several variants have been used in the past (e.g., self-renewing resources, chemostats; see Wilson and Turelli 1986). But the main purpose for introducing this form of explicit variation in resource availability is to gauge the direct impact of resource variation on population dynamics. Stochasticity of resource availability has been studied (theoretically) for many years (May 1973). In general, such stochasticity has been modelled in an obscure way that prevents the study of the causal relationships between specific resources and dynamics (e.g., Chesson 1988).

B.3 Resource availability

'Perceived' and actual availabilities

All algorithms distinguish between the *perceived* and the *actual* availability of resources at any time. The latter is defined precisely for all resources by equation (B.5), and these values are used in calculations of the conformance statistics for each timestep (equation (B.3)). However, the utilization behaviour of any phenotype or population cannot be governed by these values because resource-tracking by organisms is imperfect, and/or lagged (Boyce and Daley 1980, Glasser 1983, Wiens 1984, Stephens 1987). Therefore, the algorithms for determining how different exploitation strategies work must depend on the perceived availability of resources by organisms. An obvious candidate for characterizing the perceived availability is the previous value (i.e., the availability in the previous timestep of the model), which often will be a good estimate of the current availabilities providing resources vary systematically or deterministically. However, the immediate past value will be a poor guide when there is severe stochastic variation in the availability of resources. In general, resource tracking under these conditions must be inaccurate (Fig. B.1). An alternative approximation is to extend the 'memory' of fluctuations further into the past. This approach would weaken the conformance when resources vary systematically (the last value probably is the best estimate), but serves to smooth out large stochastic fluctuations.

A compromise solution is to use differential weighting such that the more recent values more heavily influence the perceived availability. The exponentially-weighted moving average (EWMA) fits this prescription well (Pindyck and Rubinfeld 1981: 484–487), and is used here to estimate the perceived availability. Note that it is reasonable to expect that individuals within populations will have the capacity to distinguish between volatile and smoothly varying resources, and, therefore, to generate perceived availabilities differently for each resource.

EWMAs depend upon just a single parameter, α, which, when set to unity, discounts all previous values except the last (Fig. B.1 a). Smaller values of α (>0.0) correspond to greater significance being attached to older values. Very small values also have the unfortunate side-effect of reducing the overall estimate too greatly (Fig. B.1 c), so α normally is set to at least 0.5. Resources fluctuating in an extremely stochastic fashion

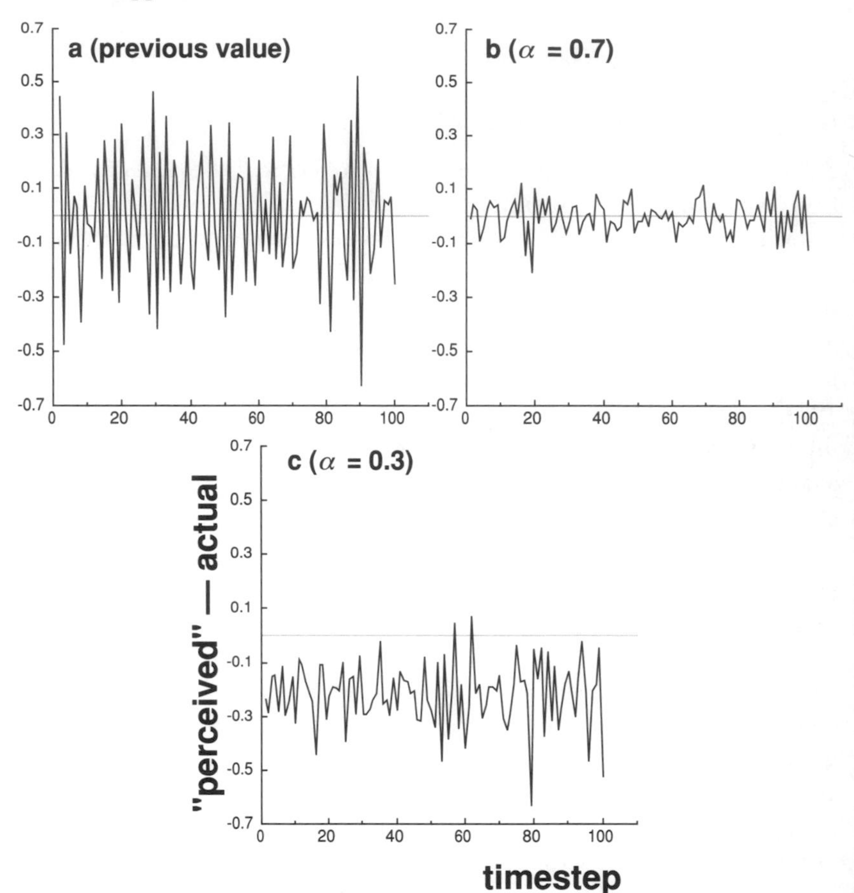

Fig. B.1 Differences between the 'perceived' resource availability based on EWMA and the actual availabilities. The effect of three values of α are shown: (a) $\alpha = 1$, which discounts all prior values except the penultimate one; (b) $\alpha = 0.7$; and (c) $\alpha = 0.3$. The faint horizontal lines represent no difference between actual and perceived resource availability. Note the small deviations for $\alpha = 0.7$ compared with $\alpha = 1$, and also, that much of the available resource will not be exploited if values of α are too low ($\alpha = 0.3$). In other words, the perceived availability will consistently underestimate the actual availability by a large amount. All data are derived from the stochastic environment.

therefore have α near to 0.5 (Fig. B.1 b), whereas deterministically varying resources have α set to 1.0.

In all of the discussions of the versatility algorithms presented below, availability refers to the perceived availability. Thus, the degree to which a resource is overtaxed is a measure of the demand and availability of the previous timestep of the simulation. However, once all functional and numerical responses have been calculated, the reported conformance statistics refer to the actual availability. This merely means that algorithms for satisfying all constraints use information existing prior to the calculation of the current or actual availabilities.

B.4 Exploitation algorithms

In this section the six algorithms for the main exploitation strategies are described. These correspond, in order, to **Ch**, **Cs**, **Rh**, **Rs** (all generalized), **Sh**, and **Ss**. As mentioned previously, the incoherent strategies are effectively unconstrained and cannot be modelled in a general way.

Hard coherency algorithm

The quintessential characteristics of hard coherency are: (1) the total utilization rate over all resources is constant; and (2) the utilization rate of a given resource is proportional to its relative availability. These are the main constraints on the algorithm. The density of a population of **Ch** represents the only adjustable quantity to satisfy reductions of utilization pressure if one or more resources are overutilized (i.e., a purely numerical response). The information that needs to be supplied for a hard coherent population is the initial density, the total utilization rate, and the marginal total utilization rate, τ_c.

The algorithm is reasonably straightforward. First, the ratios of availability to total utilization (summed over all phenotypes and populations involved in the model) for each resource (i.e., ψ_r for the rth resource) are ranked from smallest to largest. The ordered list of ψ_r corresponds to the most heavily utilized resource through to the least utilized resource. Unavailable resources are excluded from the list.

If no resource is overutilized, then fitness accumulated during the previous timestep increases the current density, leading to short-term, exponential growth in population density. The utilization rate coefficients then are adjusted so that the resource-specific rates remain proportional to relative availabilities. A constant total utilization rate is propagated through time, as is an almost complete conformance between utilization and availability of resources (i.e., $B_{pop} = 1.0$).

Why is there imperfect conformance? The reason is that some resources, although available at low levels, become uneconomic to use at low abundances, depending on the τ_c of the population. This means that the utilization pressure on other, economically viable resources is slightly higher than the relative availability of those resources to compensate for the lack of use of uneconomic resources. Hence, conformance will not be perfect. (Of course, some amount of mismatching is to be expected due to differences between the perceived availability and the actual availability because exploitation strategies must be based on some measure of the recent behaviour of resources and not on the current availability. This source of nonconformance is more pronounced when stochastic variation is relatively high.)

If some resources are overtaxed (i.e., $\psi_{min} < 1.0$), then the population density must be reduced so that the excess utilization pressure is relieved. This condition requires finding the minimum density that will cause no resource to be overutilized when the updated utilization rates are found. That is, the adjusted density times the adjusted utilization rates for each resource must lead to $\psi_{min} \geq 1.0$.

Soft coherency algorithm

Soft coherency maintains the almost strict conformance between availability and utilization of resources. However, rather than fixing the total utilization rate, the rate is free to vary between the maximum figure and the marginal utilization rate, below which any further reduction in rate cannot occur (i.e., no further functional response). A **Cs** population then must reduce its utilization pressure by numerical responses such as death or emigration. The population will be forced to local extinction if the numbers become too small (see §B.5 – A viability floor).

At least some fitness increments derived from resource utilization must be used to increase the total utilization rate if that rate is less than the maximum rate. That is, consider a situation in which resources suddenly become abundant after a lean period in which the utilization rate was forced close to the marginal rate. The utilization rate cannot suddenly rise to the maximum value in a cost-free way to take advantage of the bloom to increase population density rapidly. Some of the fitness increments gained by using the resources must be channelled into increasing the total utilization rate. Biologically, this effect might consist of the restoration of somatic condition and energy reserves, or physiological or biochemical reorganization, that must precede reproduction itself.

Again, the availability to utilization ratios are computed for each resource and then ordered. If $\psi_{min} \geq 1.0$, no resource currently is overtaxed and the accrued fitness can be converted into increases in the total utilization rate, if necessary, population density, or both. The mix of increases of total utilization rate and density provides a quandary if the fitness increment derived from exploitation in the current timestep of the model is insufficient to elevate the total utilization rate to the maximum value. One could choose to divide the increment in some way, but I decided to allocate all of the gains to increasing total utilization rate. This choice effectively means that organisms deploy fitness increments in such a way as to put themselves in a better position to exploit resources rather than compromising themselves by premature reproduction (see Murphy 1968).

Hard resource-like algorithm

A key issue affecting resource-like exploitation is that of the relationships between phenotypes. Unless some action is taken, heterogeneous populations may tend to act like several independent populations rather than as a single population. The end-product of such independence is that differentiated populations will collapse quickly so that only one phenotype, which happens to gain a significant numerical advantage over the others, remains. Therefore, there must be 'communication' between phenotypes, particularly in terms of the relationships between utilization rates of each resource by phenotypes.

To produce several ecologically differentiated phenotypes in resource-like populations, the initial distributions of utilization rates were drawn from the log-normal distribution (MacArthur 1957). These utilization rates then were shuffled between resources within phenotypes by using a 'splaying' algorithm. The splaying algorithm is designed to maximize the between-phenotype variance in utilization rates over all resources. This approach ensures that the high conformance between utilization and availability in a resource-like population initially is due to the maximum differentiation of resource use among phenotypes. The log-normal distribution seems to be a suitable one to use because of its ecological importance (Preston 1962, Sugihara 1980). Thus, the maximization of between-phenotype variance in utilization rates corresponds to a phenotypic divergence due to pressure from within the population.

The algorithm for **Rh** strategists proceeds in the following way. The ratios of availability to utilization pressure are computed and ranked from least to largest, as for any of the exploitation strategies. If all resources are underused, then the densities of phenotypes can be adjusted by translating the fitness gains into increased densities. If at least one resource is overtaxed, then densities of some phenotypes may have to be reduced to alleviate excess pressure (usually the phenotypes that have highest demands for the overutilized resources).

The constraints inherent in resource-like exploitation mean that both densities and utilization rates vary for all phenotypes from timestep to timestep. In general, there are infinitely many ways of accomplishing this and still meeting the conditions for resource-like exploitation.

The algorithms described here address this mathematical underdetermination in two ways, depending upon whether any of the resources is overutilized. If none are, then densities are adjusted in accordance with the fitness gains accumulated previously, and then revised utilization rates are computed (see below). Calculations therefore follow the sequence – increase densities → modify utilization rates. On the other hand, if any resource is overtaxed, the updated utilization rates are computed first. These changes may be sufficient to alleviate pressures, but generally will not do so. Thus, the densities of each phenotype are adjusted so that, at worst, the utilization pressure on any single resource due to the resource-like population does not exceed the availability of that resource. The algorithm for the overutilization case follows the prescription – modify utilization rates → adjust densities (some may be decreased to reduce pressures, while others may be increased according to fitness gains). More specific details are provided below.

If no resources are overused, then phenotypic densities are adjusted in accordance with the fitness gains of the most recent timestep of the simulation. Revised utilization rates then are generated by using the algorithm described below with updated densities (see 'Utilization rate adjustment'). Otherwise, if some resources are overtaxed, then it seems prudent to adjust utilization rates first and then to address densities. So, altered utilization rates are generated, and then the densities are changed to ensure that no resource is overused relative to availabilities. The adjustments of densities are specific to phenotypes. If the 'new' exploitation pattern of a phenotype does not cause the overuse of any resource, then the usual increase in density of that phenotype can occur. However, if the modified rates of a phenotype in conjunction with its current density do not relieve its share of the excess utilization pressure on every resource, then the density must be reduced.

Utilization rate adjustment

The method by which utilization rates are updated for resource-like populations involves two main actions. First, the resources that remain economically useable are determined. This set depends upon the relative availabilities of resources and the marginal utilization rate of the population. Some resources may become so rare relative to total availability that to continue to use them would require utilization rates to fall below the marginal rate. Such resources are 'dropped' from the repertoire of the population, and do not subsequently affect the population unless they increase in availability again at a later stage. If that happens, the use of these resources resumes.

The second phase involves calculating and rescaling the utilization rates so that utilization mirrors the availability of used resources. This phase of the adjustment employs both phenotypic densities and the perceived availabilities. However, no account of whether a resource is overtaxed is taken – overutilization is handled explicitly as described above. The general requirement is

$$\sum_{p=1}^{M} \eta_p \tau'_{pr} = A_r, \tag{B.6}$$

where τ'_{pr} are the desired utilization rates for the next timestep. The densities, η_p, may or may not have been changed already depending on whether any resources were overused. A general weighting solution to (B.6) is

$$\tau'_{pr} = \frac{\omega_p A_r}{\sum_{q=1}^{M} \eta_q \omega_q}, \tag{B.7}$$

where ω are weights that are specific to phenotypes. The raw τ'_{pr} usually have to be rescaled so that the total utilization rate summed over all phenotypes is maintained at a constant level (i.e., **Rh** is a hard strategy), which is quite straightforward to do.

Soft resource-like algorithm

The almost complete conformance between availability and utilization of resources also is manifested by soft resource-like populations, **Rs**. However, the total utilization rate over all resources is not constant (cf. **Rh**), being able to vary between the maximum rate and the marginal utilization rate. As usual, the marginal rate represents the minimum below which any further reduction in total utilization rate cannot occur, and the **Rs** population must then enact a numerical response to alleviate the overutilization. There must be significant differentiation between phenotypes in the utilization rates of different resources, but the total use integrated over all phenotypes conforms closely with resource availability.

The **Rs** algorithm, naturally enough, is like a hybrid between the **Rh** (hard resource-like) and **Cs** (soft coherency) algorithms. If no resources are overtaxed, then fitness increments of the previous timestep can be converted into either an elevated total utilization rate, if necessary, and/or into increased population densities. Comments about this phenomenon provided in the section on **Cs** also apply here. The same method of utilization rate adjustment used for the **Rh** algorithm (see above) is employed to quantify the resource-specific utilization rates for each phenotype based on the updated densities and total utilization rates.

If any resource is overutilized, then, as is the case for **Cs**, the total utilization rate is decreased as far as possible (i.e., to the marginal rate) before numerical responses occur. Note that the total utilization rate and the density of each phenotype are alterable separately, provided that the conformance between availability and use ultimately occurs. The method by which this is accomplished is somewhat convoluted, but basically follows the path – adjust the utilization rates on the basis of the existing values and densities (using the **Rh** method) → check to see whether any resource remains overused given the adjustments → if one or more are, identify the phenotypes that need to be treated to eliminate the overutilization, and reduce their total utilization rates and/or their densities accordingly. If a phenotype does not cause any of the resources to be overburdened, then the total utilization rates and/or densities of that phenotype can be increased. If the total utilization rate of any phenotype is altered (up or down), then a rescaling of the resource-specific rates will be necessary.

Hard specialization algorithm

The algorithm for hard specialization is particularly simple. The utilization rates for each resource are fixed, independent of the availability of those resources. The ratios of the availability to the use of each resource *that the* **Sh** *population uses* are computed in a similar way to the hard coherent case, and the ratios subsequently ordered from least to greatest (i.e., from the most overused to the least overused).

If the availability exceeds the current use for all resources, then the fitness gains are converted directly into population growth. However, if any resource used by **Sh** individuals is overutilized, then the population size must be decreased because the utilization coefficients are fixed. That is, individuals within the population are incapable of relaxing their utilization rates in any measure, so that some mortality or emigration is required to release the excess utilization pressure. The degree to which the population is decreased is governed by the resource that is most overused (i.e., the smallest ψ). This is clear because any reduction of population density that is sufficient to abolish the overuse of the most heavily used resource must be enough to relax the pressure on any resource that is less overtaxed than that resource.

A strict adherence to these conditions implies that a **Sh** population can exploit only those resources that do not vary wildly or significantly in availability. Resources exhibiting high stochastic fluctuations certainly would be a threat to **Sh** populations, as would deterministically variable resources reaching low availabilities. Of course, a numerical response to the latter case may be mass emigration rather than mortality *per se*, but the local effect is the same – the **Sh** population would become locally extinct.

Soft specialization algorithm

The utilization rates in soft specialization can vary between the maximum rate for a given resource (that the population uses) and a value below which utilization no longer is economic, namely, the marginal rate τ_c. Additional flexibility can be added by permitting resource classes to be dropped from the repertoire of an **Ss** population if the resource is too heavily used. The algorithm for **Ss** permits the omission of resources that decline precipitously at relatively high levels of availability. Thus, resources such as *A* and *C* of the variable environment would be dropped from repertoires before they greatly affect densities. I used the heuristic that any resource in which the amplitude of variation exceeded the base level was dropped when the perceived availability fell below the base level to satisfy this assumption.

Of course, at least one resource must be retained for the population to persist. If dropped resources become relatively more abundant at some later time, then the use of these resources can occur again by a suitable allocation of accumulated fitness. This can be regarded as 'gearing up' for the utilization of previously dropped resources.

The ratios of availability to utilization for the resources used by a specialized population are computed again, and then ordered from the smallest to largest. Two possible avenues follow, one corresponding to the case in which no resources are overextended (no overutilization), and the situation in which one or more resources are overtaxed (some overutilization).

No overutilization

The first action is to determine how many, and the degree to which, dropped resources can be returned to the repertoire of individuals of the population. Prudence in reestablishing utilization rates dictates that the current level of use by other populations relative to availability be taken into account. That is, if the resource is not being used at all, then the specialized population should start to utilize the resource at the maximum rate if possible. At the other extreme, if the difference between utilization and availability were relatively small, so that the appropriate utilization rate for a reestablished resource was below the marginal rate, then there would be little point in individuals of the **Ss** population picking up that resource. The other proviso is that the reestablishment of use is not cost-free, but must be financed from the overall gains in fitness derived from the utilization of other resources. Thus, increases in population density or increases in utilization rates of resources currently being used must be sacrificed to reestablish the use of dropped resources.

Assuming that these adjustments are made, and that whichever dropped resources that can be picked up and supported by other fitness gains are, the algorithm then proceeds on the following basis. Each resource that the population currently uses is treated in turn, beginning with the resource that is closest to being overtaxed and ending with the one with the greatest oversupply relative to demand.

If the utilization rate of the current resource is less than the maximum rate, then the fitness increment associated with the resource is devoted to increasing the utilization rate up to the maximum, with any surplus applied to increases in population density. If the rate is already at the maximum level, then the entire fitness increment is directed to increased reproduction (i.e., increased density through local recruitment).

So, fitness gains in the case in which there is no overuse are translated in three ways: (1) the reestablishment of the use of resources previously dropped under extreme pressure; (2) an increase in the utilization rates that previously had been relaxed, up to the maximum rate for that resource; and (3) increases in population density.

Some overutilization

Each resource currently being used is considered in turn, one or more of which is overtaxed. The resources being overused relative to their availabilities have to be treated differently to underutilized resources.

Consider an overburdened resource. The pressure on that resource can be relaxed in two ways. In soft specialization, relaxation preferentially is accomplished by reducing the rates of utilization rather than reducing densities, which are lessened only as a last resort. Of course, the most drastic way of reducing the utilization rate is to drop a resource altogether, but at least one resource must be retained, particularly resources that are deemed to be significant to the soft specialized population. 'Significant' is interpreted as being any resource in which the maximum utilization rate exceeds the average maximum utilization rate calculated over all resources. Significant resources cannot be dropped – a relaxation of the utilization pressure on them can be accomplished only by reducing the utilization rate to the marginal rate, and, thereafter, by diminishing the population density.

If the utilization pressure can be relieved by reducing the utilization rate alone, and this does not cause the utilization rate to fall below the marginal rate, then this occurs. If such a reduction leads to the utilization rate falling beneath the marginal rate, then either the resource can be dropped, given the conditions outlined above, or else the utilization rate is reduced to the marginal rate and the population density reduced.

The prescription for underused 'current' resources is simple, mainly because the population density cannot be increased due to the constraints stemming from the overutilized resources. This means that fitness increments associated with undertaxed resources must be channelled into increasing the utilization rates. If these rates are less than the maximum, then the fitness increments are used to elevate them, but only up to the maximum rate exhibited by the population for that particular resource. Of course, some of the fitness returns also may be funnelled into picking up dropped resources.

B.5 More on assumptions and limitations

Some of the specific assumptions of models are referred to in the descriptions of the algorithms. However, there are several general assumptions and limitations that affect the conclusions of Chapters 6 and 7. The purpose of this subsection is to describe briefly these assumptions and limitations.

A viability floor

It is well known that the viability of a population drops dramatically below a particular density (Simberloff and Abele 1982, Pimm *et al.* 1988, Bolger *et al.* 1991, Tracy and George 1992, Lande 1993). This idea is

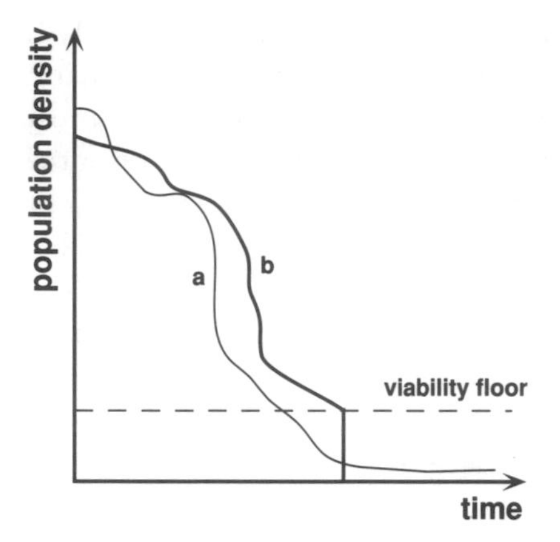

Fig. B.2 The effect of imposing a viability floor on a population. In (a), the declining model population is permitted to continue to persist at densities that are likely to be ecologically unsustainable. In (b) however, the population is regarded as extinct once the density reaches the minimum viability level.

particularly crucial in conservation biology. For example, Shaffer and Samson (1985) employed simulation models to estimate the minimum viable population sizes of the grizzly bear (*Ursus arctos*). There are at least two reasons for the lowered viability of low-density populations. The first is a deterministic mechanism, the Allee effect, which is a monotone decrease in the birth rate in (sexually reproducing) populations as a consequence of rarity (Asmussen 1979, Hopf and Hopf 1985). A second mechanism is stochastic in nature – chance effects are more likely to be deleterious to rare populations than to abundant ones (Shaffer 1981, Loeschcke 1985, Strebel 1985, Wissel and Stöcker 1991). Both the deterministic and stochastic mechanisms justify the introduction of a minimum model density below which the population (or phenotype) becomes locally extinct. The concept of a minimum population (or phenotype) density is referred to here as the *viability floor* (Fig. B.2).

Spatial isotropy
Resources are assumed to be distributed isotropically throughout space, which means variations in resource availability are independent of location. Neither resource nor population clumping is accommodated (see Paine 1992 for some effects of clumping in populations of consumers), and the resource environment is fine grained (see Emlen 1973). A closely related general assumption is that all individuals sense or perceive resource variation in the same way and at the same time.

Clearly, if some individuals perceive changes in resource availability more acutely than others, then the dynamics described in Chapter 6 may be affected. Lags in perception of change also are not modelled.

For some models of predator–prey dynamics, the physical distribution of prey seems to have a significant impact on the nature of interactions. On both theoretical and experimental grounds, Arditi and Saïah (1992) showed that if prey were distributed in an isotropic or homogeneous fashion, then dynamics were best understood in terms of prey dependence. Mobile prey are more likely to be distributed homogeneously, while more sedentary prey would be heterogeneously arranged. Prey dependence implies a principle of mass-action, so that the response of populations is related to the product of their biomass densities (Berryman 1992: 1531). Anisotropically distributed prey appears to force ratio-dependent behaviour, where ratio dependence means that predator–prey dynamics are couched in terms of ratios of prey-to-predator densities. Some regard interference between individual predators as a source of ratio-dependent behaviour (Ginzburg and Akçakaya 1992), while others see homogeneity–heterogeneity of prey distributions as being the root cause (Arditi and Saïah 1992). Arditi and Saïah (1992) emphasized that intermediate dynamics (i.e., neither strict prey dependence nor strict ratio dependence) also may occur. The implicit assumption, used in this book, of spatial isotropy in the distributions of resources therefore implies prey-dependent or mass-action dynamics.

Uniformity of exploitation rates

Models assume that all individuals of a phenotype of a soft strategy adjust utilization rates in the same way under the pressure exerted by intraspecific and interspecific competition. This also means that phenotypes themselves are relatively homogeneous (i.e., no significant differences in attributes, size say; see Gutierrez 1992: 1553). Recruits to a phenotype therefore adopt the same set of characteristics (utilization rates, marginal rates, etc.) that existing members of the phenotype currently display.

Phenotypic integrity

Reproduction of each phenotype is assumed to yield the same phenotype, where phenotypes are ecological phenotypes. Thus, phenotypes breed truly. More general models, particularly explicit models of sexual populations, need to address situations in which there are interphenotypic transfers of fitness (Travis *et al.* 1980) and cross-breeding between phenotypes (Roughgarden 1972, Getz and Kaitala 1989). Fitness increments of adults can appear as recruitment to juvenile phenotypes, while maturation of juveniles, which involves fitness-related quantities such as growth, increases densities of adults. Moreover, effects on preadult fitness may influence adult fertility, which in turn may have important effects on population dynamics (e.g., Prout 1986). These points clearly apply to species with multiple juvenile stages.

Isolated populations, no recolonization

Models do not permit recolonization or recruitment of individuals from external sources. This approach was adopted to allow analysis of the behaviour of isolated populations without interference from extraneous sources.

Pure strategies, no hybrids

The models do not permit switching of strategies under any circumstances, and particularly not in response to different regimes of resource availability (for example, no adaptive polyphenism, Moran 1992). Hybrid strategies probably are common in natural communities, but an understanding of pure strategies is required before addressing hybrid strategists.

Reproduction is possible during every timestep

Recruitment by local reproduction is assumed to be possible during each timestep. Population dynamics will be manifestly different if reproduction is constrained to occur at specific times or intervals. I regard my approach (i.e., nonpulsed reproduction) as being the most general case.

I also assume that there will be sufficient recruits available to satisfy the dynamic requirements at any time. This assumption may be inappropriate in numerous natural circumstances (see Gaines *et al.* 1985, Gaines and Roughgarden 1985, Underwood and Fairweather 1989, Downes 1995).

Exploitation competition

The modelling described here clearly is exploitative in nature. Competition for resources is mediated directly by utilization of resources and not by interference. This implies that competition coefficients of LVM and other models, the form of which is controversial (Abrams 1975), need not be specified to study system behaviour. Competition coefficients generally do not directly refer to the presumed resources through which the competitive process operates (see Getz 1984: 634). Interaction terms between populations are essential if competition is largely dominated by interference mechanisms (MacArthur 1970, Morse 1974, Getz 1984, Murray and Gerrard 1985). The exploitative nature means that all modelling is based on the premise that reductions of utilization pressure, where necessary, are proportional to the degree to which a population (or phenotype within a population) contributes to the excess pressure on resources. This is characteristic of exploitation competition. Disproportionate reductions are necessary to represent interference competition for limited resources, but this is beyond the scope of the current models. Prey-dependent dynamics derived from the application of spatial isotropy also are consistent with exploitation competition rather that interference competition.

Noninteractive resources

Chesson (1990) noted that a weakness in the consumer-resource models proposed by MacArthur (1969) is that these models don't account for interactions between 'resources', where Chesson viewed resources as prey in a predator–prey context. Thus, in Chesson's view, resources may be similar prey species between which interspecific interactions might occur. Resources in the current models are regarded in general terms, and, therefore, have availabilities specified at any time by using equation (A5). To this extent, resources vary independently of one-another.

B.6 Implementation, hardware, etc.

Algorithms described in preceding paragraphs were implemented in C in an *Irix*® 4.0.5 environment (Silicon Graphics, Inc.; SGI). The compiler was the ANSI C version 4.10 for this operating system. Simulations were run on several bit-compatible SGI workstations (*Iris*® 340 series and *Indigo*®). CPUs were MIPS R2000A/R3000 Processor Chips running at 25 or 33 MHz, or MIPS R4000 Processor Chips operating at 50 MHz. FPUs were either MIPS R2010A/R3010 VLSI Floating Point Chips or MIPS R4010 Floating Point Chips.

® Registered trademark of Silicon Graphics, Inc., Mountain View, CA, USA

References

Aarssen, L. W. (1983) Ecological combining ability and competitive combining ability in plants: toward a general evolutionary theory of coexistence in systems of competition. *American Naturalist*, **122**, 707–731.

Aarssen, L. W. (1989) Competitive ability and species coexistence: 'a plant's-eye view'. *Oikos*, **56**, 386–401.

Abbott, I., Abbott, L. K. and Grant, P. R. (1977) Comparative ecology of Galapágos ground finches (*Geospiza* Gould): evaluation of the importance of floristic diversity and interspecific competition. *Ecological Monographs*, **47**, 151–184.

Abrams, P. A. (1975) Limiting similarity and the form of the competition coefficient. *Theoretical Population Biology*, **8**, 356–375.

Abrams, P. A. (1980a) Are competition coefficients constant? Inductive versus deductive approaches. *American Naturalist*, **116**, 730–735.

Abrams, P. A. (1980b) Consumer functional response and competition in consumer-resource systems. *Theoretical Population Biology*, **17**, 80–102.

Abrams, P. A. (1980c) Some comments on measuring niche overlap. *Ecology*, **61**, 44–49.

Abrams, P. A. (1983) The theory of limiting similarity. *Annual Review of Ecology and Systematics*, **14**, 359–376.

Abrams, P. A. (1984) Variability in resource consumption rates and the coexistence of competing species. *Theoretical Population Biology*, **25**, 106–124.

Abrams, P. A. (1987a) The functional responses of adaptive consumers on two resources. *Theoretical Population Biology*, **32**, 262–288.

Abrams, P. A. (1987b) The nonlinearity of competitive effects in models of competition for essential resources. *Theoretical Population Biology*, **32**, 50–65.

Abrams, P. A. (1988a) How should resources be counted? *Theoretical Population Biology*, **33**, 226–242.

Abrams, P. A. (1988b) Resource productivity–consumer species diversity: simple models of competition in spatially heterogeneous environments. *Ecology*, **69**, 1418–1433.

Abrams, P. A. (1989) Decreasing functional responses as as result of adaptive consumer behavior. *Evolutionary Ecology*, **3**, 95–114.

Abrams, P. A. (1990) Adaptive responses of generalist herbivore to competition: convergence or divergence? *Evolutionary Ecology*, **4**, 103–114.

Abrams, P. A. (1993) Effect of increased productivity on the abundances of trophic levels. *American Naturalist*, **141**, 351–371.

Abrams, P. A. and Allison, T. D. (1982) Complexity, stability, and functional response. *American Naturalist*, **119**, 240–249.

Addicott, J. F. (1978) Niche relationships among species of aphids feeding on fireweed. *Canadian Journal of Zoology*, **56**, 1837–1841.

Addicott, J. F. (1984) Mutualistic interactions in population and community processes. In *A new ecology: novel approaches to interactive systems*, ed. Price, P. W., Slobodchikoff, C. N. and Gaud, W. S., pp. 437–455. Wiley: New York, NY, USA.

Addicott, J. F. (1986) On the population consequences of mutualism. In *Community ecology*, ed. Diamond, J. and Case, T., pp. 425–436. Harper and Row: New York, NY, USA.

Addicott, J. F., Aho, J. M., Antolin, M. F., Padilla, D. K., Richardson, J. S. and Soluk, D. A. (1987) Ecological neighborhoods: scaling environmental patterns. *Oikos*, **49**, 340–346.

Adolph, S. C. (1990) Influence of behavioral thermoregulation on microhabitat use by two *Sceloporus* lizards. *Ecology*, **71**, 315–327.

Agren, G. I. and Fagerstrom, T. (1984) Limiting dissimilarity in plants: randomness prevents exclusion of species with similar competitive abilities. *Oikos*, **43**, 369–375.

Alatalo, R. V. (1981) Habitat selection of forest birds in the seasonal environment of Finland. *Annales Zoologici Fennici*, **18**, 103–114.

Aldridge, H. D. J. N. and Rautenbach, I. L. (1987) Morphology, echolocation and resource partitioning in insectivorous bats. *Journal of Animal Ecology*, **56**, 763–778.

Alford, R. A. (1989) Variation in predator phenology affects predator performance and prey community composition. *Ecology*, **70**, 206–219.

Allen, T. F. N. and Starr, T. B. (1982) *Hierarchy: perspectives for ecological complexity*. Chicago University Press: Chicago, IL, USA.

Alscher, R. G. and Cumming, J. R. (ed.) (1990) *Stress responses in plants: adaptation and acclimation mechanisms*. Wiley: New York, NY, USA.

Anderson, S. H., Shugart, H. H., Jr and Smith, T. M. (1979) Vertical and temporal habitat utilization within a breeding bird community. In *The role of insectivorous birds in forest ecosystems*., ed. Dickson, J. G., Conner, R. N., Fleet, R. R., Kroll, J. C. and Jackson, J. A., pp. 261–268. Academic Press: New York, NY, USA.

Andow, D. A. (1991) Vegetational diversity and arthropod response. *Annual Review of Entomology*, **36**, 561–586.

Andrewartha, H. G. and Birch, L. C. (1954) *The distribution and abundance of animals*. Chicago University Press: Chicago, IL, USA.

Andrewartha, H. G. and Birch, L. C. (1984) *The ecological web. More on the distribution and abundance of animals*. Chicago University Press: Chicago, IL, USA.

Arditi, R. and Saïah, H. (1992) Empirical evidence of the role of heterogeneity in ratio-dependent consumption. *Ecology*, **73**, 1544–1551.

Armbruster, W. S. (1992) Phylogeny and the evolution of plant–animal interactions. *BioScience*, **42**, 12–20.

Armstrong, R. A. and McGehee, R. (1980) Competitive exclusion. *American Naturalist*, **115**, 151–170.

Askew, R. R. (1980) The diversity of insect communities in leaf-mines and plant galls. *Journal of Animal Ecology*, **49**, 817–829.

Asmussen, M. A. (1979) Density-dependent selection II. The Allee effect. *American Naturalist*, **114**, 796–809.

Atkinson, A. C. and Donev, A. N. (1992) *Optimum experimental designs*. Oxford University Press: Oxford, UK.

Atsatt, P. R. (1981) Lycaenid butterflies and ants: selection for enemy-free space. *American Naturalist*, **118**, 638–654.

Auger, P. (1985) Stability of interacting populations with age-class distributions. *Journal of Theoretical Biology*, **112**, 595–605.

Avise, J. C. (1983) Commentary: biochemical studies of microevolutionary processes. In *Perspectives in ornithology*, ed. Brush, A. H. and Clark, G. A., Jr, pp. 262–270. Cambridge University Press: Cambridge, UK.

Baer, J. G. (1951) *The ecology of animal parasites*. University of Illinois Press: Urbana, IL, USA.

Ball, J. P. (1990) Active diet selection or passive reflection of changing food availability: the underwater foraging of Canvasback Ducks. In *Behavioural mechanisms of food selection, 20*, ed. Hughes, R.N., pp. 95–107. Springer-Verlag: Berlin, Germany.

Barbosa, P. (1988) Some thoughts on 'the evolution of host range'. *Ecology*, **69**, 912–915.

Barbosa, P. and Krischik, V. A. (1987) Influence of alkaloids on feeding preference of deciduous trees by the gypsy moth *Lymantria dispar*. *American Naturalist*, **130**, 53–69.

Barbosa, P., Krischik, V. A. and Jones, C. G. (ed.) (1991) *Microbial mediation of plant–herbivore interactions*. Wiley: New York, NY, USA.

Barclay, R. M. R. and Brigham, R. M. (1991) Prey detection, dietary niche breadth, and body size in bats: why are aerial insectivorous bats so small? *American Naturalist*, **137**, 693–703.

Barker, K. M. and Chapman, A. R. O. (1990) Feeding preferences of periwinkles among four species of *Fucus*. *Marine Biology*, **106**, 113–118.

Barnard, C., Milinski, M., Peterson, C. H. and Godin, J-G. J. (1990) The role and importance of optimal foraging theory in ecology. In *Behavioural mechanisms of food selection, 20*, ed. Hughes, R.N., pp. 865–866. Springer-Verlag: Berlin, Germany.

Bazely, D. R. (1989) Carnivorous herbivores: mineral nutrition and the balanced diet. *Trends in Ecology and Evolution*, **4**, 155–156.

Bazzaz, F. A. (1991) Habitat selection in plants. *American Naturalist*, **137**, S116–S130.

Bazzaz, F. A. and Sultan, S. E. (1987) Ecological variation and the maintenance of plant diversity. In *Differentiation patterns in higher plants*, ed. Urbanska, K.M., pp. 69–93. Academic Press: London, UK.

Bell, H. L. and Ford, H. A. (1990) The influence of food shortage on interspecific niche overlap and foraging behavior of three species of Australian warblers (Acanthizidae). *Advances in Avian Biology*, **13**, 381–388.

Bell, G. P. (1980) Habitat use and response to patches of prey by desert insectivorous bats. *Canadian Journal of Zoology*, **58**, 1876–1883.

Belovsky, G. E. (1984) Herbivore optimal foraging: a comparative test of three models. *American Naturalist*, **124**, 97–115.

Belovsky, G. E. (1986) Generalist herbivore foraging and its role in competitive interactions. *American Zoologist*, **26**, 51–69.

Belovsky, G. E. (1990) How important are nutrient constraints in optimal foraging models or are spatial/temporal factors more important. In *Behavioural mechanisms of food selection, 20*, ed. Hughes, R.N., pp. 255–278. Springer-Verlag: Berlin, Germany.

Bence, J. R. (1986) Feeding rate and attack specialization: the roles of predator experience and energetic tradeoffs. *Environmental Biology of Fishes*, **16**, 113–121.

Benkman, C. W. (1988) Seed handling efficiency, bill structure, and the cost of specialization for crossbills. *Ibis*, **129**, 288–293.

Benkman, C. W. (1993) Adaptation to single resources and the evolution of crossbill (*Loxia*) diversity. *Ecological Monographs*, **63**, 305–325.

Bennett, B. A. and Branch, G. A. (1990) Relationships between production and consumption of prey species by resident fish in the Bot, a cool temperate South African estuary. *Estuarine, Coastal and Shelf Science*, **31**, 139–155.

Berenbaum, M. R. (1990) Evolution of specialization in insect–umbellifer associations. *Annual Review of Entomology*, **35**, 319–343.

Bergman, E. (1988) Foraging abilities and niche breadths of two percids, *Perca fluviatilis* and *Gymnocephalus cernua*, under different environmental conditions. *Journal of Animal Ecology*, **57**, 443–453.

Bernays, E. A. (1989) Host range in phytophagous insects: the potential role of generalist predators. *Evolutionary Ecology*, **3**, 299–311.

Bernays, E. A. and Graham, M. (1988) On the evolution of host specificity in phytophagous arthropods. *Ecology*, **69**, 886–892.

Bernstein, R. A. (1978) Slavery in the subfamily Dolichoderinae (F. Formicidae) and its ecological consequences. *Experientia*, **34**, 1281–1282.

Bernstein, R. A. (1979) Evolution of niche breadth in populations of ants. *American Naturalist*, **114**, 533–544.

Berryman, A. A. (1992) The origins and evolution of predator–prey theory. *Ecology*, **73**, 1530–1535.

Bierzychudek, P. and Eckhardt, V. (1988) Spatial segregation of the sexes in dioecious plants. *American Naturalist*, **132**, 34–43.

Bishir, J. and Namkoong, G. (1992) Density-dependent dynamics in size-structured populations. *Journal of Theoretical Biology*, **154**, 163–188.

Black, R. (1979) Competition between intertidal limpets: an intrusive niche on a steep resource gradient. *Journal of Animal Ecology*, **48**, 401–411.

Blake, J. G. and Hoppes, W. G. (1986) Influence of resource abundance on use of tree-fall gaps by bird in an isolated woodlot. *Auk*, **103**, 328–340.

Bloom, A. J., Chapin, F. S. I. and Mooney, H. A. (1985) Resource limitation in plants – an economic analogy. *Annual Review of Ecology and Systematics*, **16**, 363–392.

Bloom, S. A. (1981) Specialization and noncompetitive resource partitioning among sponge-eating dorid nudibranchs. *Oecologia (Berl.)*, **49**, 305–315.

Bock, C. E. (1987) Distribution–abundance relationships of some Arizona landbirds: a matter of scale? *Ecology*, **68**, 124–129.

Bock, C. E., Cruz, A., Jr, Grant, M. C., Aid, C. S. and Strong, T. R. (1992) Field experimental evidence for diffuse competition among southwestern riparian birds. *American Naturalist*, **140**, 815–828.

Bock, C. E. and Ricklefs, R. E. (1983) Range size and local abundance of some North American songbirds: a positive correlation. *American Naturalist*, **122**, 295–299.

Bolger, D. T., Alberts, A. C. and Soulé, M. E. (1991) Occurrence patterns of bird species in habitat fragments: sampling, extinction, and nested species subsets.

American Naturalist, **137**, 155–166.

Botsford, L. W. (1981) The effects of increased individual growth rates on depressed population size. *American Naturalist*, **117**, 38–63.

Boucher, D. H., James, S. and Keeler, K. H. (1983) The ecology of mutualism. *Annual Review of Ecology and Systematics*, **13**, 15–47.

Bouchon-Navaro, Y. (1986) Partitioning of food and space resources by chaetodontid fishes on coral reefs. *Journal of Experimental Marine Biology and Ecology*, **103**, 21–40.

Bourliere, F. (1956) *The natural history of mammals*. Alfred A. Knopf: New York, NY, USA.

Bowers, M. A. (1988) Relationships between local distribution and geographic range of desert heteromyid rodents. *Oikos*, **53**, 303–308.

Bowers, M. D., Collinge, S. K., Gamble, S. E. and Schmitt, J. (1992) Effects of genotype, habitat, and seasonal variation in iridoid content of *Plantago lanceolata* (Plantaginaceae) and the implications for insect herbivores. *Oecologia (Berl.)*, **91**, 210–207.

Boyce, M. S. and Daley, D. J. (1980) Population tracking of fluctuating environments and natural selection for tracking ability. *American Naturalist*, **115**, 480–491.

Braband, Å. (1985) Food of roach (*Rutilus rutilus*) and ide (*Leusiscus idus*): significance of diet shift for interspecific competition in omnivorous fishes. *Oecologia (Berl.)*, **66**, 461–467.

Braithwaite, R. W., Cockburn, A. and Lee, A. K. (1978) Resource partitioning by small mammals in lowland heath communities of south-eastern Australia. *Australian Journal of Ecology*, **3**, 423–445.

Brakefield, P. M. and Reitsma, N. (1991) Phenotypic plasticity, seasonal climate and the population biology of *Bicyclus* butterflies (Satyridae) in Malawi. *Ecological Entomology*, **16**, 291–303.

Bramwells, H. W. and Whiffin, T. (1984) Patterns of variation in *E. sideroxylon* A. Cunn. ex Woolls. 1. Variation in adult morphology. *Australian Journal of Botany*, **32**, 263–281.

Bratton, S. P. (1976) Resource division in an understory herb community. *American Naturalist*, **110**, 679–698.

Brew, J. S. (1982) Niche shift and the minimisation of competition. *Theoretical Population Biology*, **22**, 367–381.

Brew, J. S. (1984) An alternative to Lotka–Volterra competition in coarse-grained environments. *Theoretical Population Biology*, **25**, 265–288.

Brewin, N. J. (1991) Development of the legume root nodule. *Annual Review of Cell Biology*, **7**, 191–236.

Briand, F. (1983) Environmental control of food web structure. *Ecology*, **64**, 253–263.

Briand, F. and Cohen, J. E. (1984) Community food webs have scale-invariant structure. *Nature (Lond.)*, **307**, 264–267.

Briand, F. and Cohen, J. E. (1987) Environmental correlates of food chain length. *Science (Wash., DC)*, **238**, 956–960.

Briand, F. and Yodzis, P. (1982) The phylogenetic distribution of obligate mutualism: evidence for limiting similarity and global instability. *Oikos*, **39**, 273–275.

Briggs, J. C. (1974) The operation of zoogeographic barriers. *Systematic Zoology*, **23**, 248–256.

Brönmark, C. and Miner, J. G. (1992) Predator-induced phenotypical change in body morphology in crucian carp. *Science (Wash., DC)*, **258**, 1348–1350.

Bronstein, J. L. (1994) Conditional outcomes in mutualistic interactions. *Trends in Ecology and Evolution*, **9**, 214–217.

Brooks, D. R. (1988) Macroevolutionary comparisons of host and parasite phylogenies. *Annual Review of Ecology and Systematics*, **19**, 235–259.

Brosset, A. (1990) A long term study of the rain forest birds in M'Passa (Gabon). In *Biogeography and ecology of forest bird communities*, ed. Keast, A., pp. 259–274. SPB Academic Publishing: The Hague, Netherlands.

Brown, J. H. (1981) Two decades of homage to Santa Rosalia: toward a general theory of diversity. *American Zoologist*, **21**, 877–888.

Brown, J. H. (1984) On the relationship between abundance and distribution of species. *American Naturalist*, **124**, 255–279.

Brown, J. H. and Gibson, A. C. (1983) *Biogeography*. Mosby: St Louis, MO, USA.

Brown, J. H. and Kodric-Brown, A. (1979) Convergence, competition, and mimicry in a temperate community of hummingbird-pollinated plants. *Ecology*, **60**, 1022–1035.

Brown, J. H. and Maurer, B. A. (1989) Macroecology: the division of food and space among species on continents. *Science*, **243**, 1145–1150.

Brown, J. M. and Wilson, D. S. (1992) Local specialization of phoretic mites on sympatric carrion beetle hosts. *Ecology*, **73**, 463–478.

Brown, J. S. (1989) Coexistence on a seasonal resource. *American Naturalist*, **133**, 168–182.

Brown, J. S. and Pavlovic, N. B. (1992) Evolution in heterogeneous environments: effects of migration on habitat specialization. *Evolutionary Ecology*, **6**, 360–382.

Browning, T. O. (1963) *Animal populations*. Hutchinson: London, UK.

Bryan, J. E. and Larkin, P. A. (1972) Food specialization by individual trout. *Journal of the Fisheries Research Board of Canada*, **29**, 1615–1624.

Bryant, J. P., Kuropat, P. J., Cooper, S. M., Frisby, K. and Owen-Smith, N. (1989) Resource availability hypothesis and plant antiherbivore defence tested in a South African savanna ecosystem. *Nature (Lond.)*, **340**, 227–229.

Bull, C. M. (1991) Ecology of parapatric distributions. *Annual Review of Ecology and Systematics*, **22**, 19–36.

Burgman, M. A. (1989) The habitat volumes of scarce and ubiquitous plants: a test of the model of environmental control. *American Naturalist*, **133**, 228–239.

Bush, G. L. (1975) Modes of animal speciation. *Annual Review of Ecology and Systematics*, **6**, 339–364.

Butler, A. J. and Keough, M. J. (1990) A comment on short supply lines. *Trends in Ecology and Evolution*, **5**, 97.

Cain, S. A. (1944) *Foundations of plant geography*. Harper and Brothers: New York, NY, USA.

Calder, D. M. and Bernhardt, P. (ed.) (1983) *The biology of mistletoes*. Academic Press: Sydney, Australia.

Callow, J. A. (1977) Recognition, resistance and the role of plant lectins in host–parasite interactions. *Advances in Botanical Research*, **4**, 1–49.

Cameron, R. G. and Wyatt, R. (1990) Spatial patterns and sex ratios in dioecious and monoecious mosses of the genus *Splachnum*. *Bryologist*, **93**, 161–166.

Cameron, T. W. M. (1964) Host specificity and the evolution of helminthic parasites. *Advances in Parasitology*, **2**, 1–34.

Campbell, R. B. (1981) Some circumstances assuring monomorphism in subdivided populations. *Theoretical Population Biology*, **20**, 118–125.

Caraco, T. (1980) On foraging time allocation in stochastic environments. *Ecology*, **61**, 119–128.

Carefoot, T. H. (1987) *Aplysia*: its biology and ecology. *Oceanography and Marine Biology Annual Reviews*, **25**, 167–284.

Carnes, B. A. and Slade, N. A. (1982) Some comments on niche analysis in canonical space. *Ecology*, **63**, 888–893.

Carroll, C. R. (1979) A comparative study of two ant faunas: the stem-nesting ants of Liberia, West Africa and Costa Rica, Central America. *American Naturalist*, **113**, 551–561.

Carroll, G. C. and Wicklow, D. T. (ed.) (1992) *The fungal community. Its organization and role in the ecosystem*, 2nd edn. Marcel Decker: New York, NY, USA.

Carson, W. P. and Pickett, S. T. A. (1990) Role of resources and disturbance in the organization of an old-field plant community. *Ecology*, **71**, 226–238.

Case, T. J. (1980) MacArthur's minimization principle: a footnote. *American Naturalist*, **115**, 133–149.

Case, T. J. (1984) Niche overlap and resource weighting terms. *American Naturalist*, **124**, 604–608.

Case, T. J. (1990) Invasion resistance arises in strongly interacting species-rich model competition communities. *Proceedings of the National Academy of Science USA*, **87**, 9610–9614.

Case, T. J. and Bolger, D. T. (1991) The role of interspecific competition in the biogeography of island lizards. *Trends in Ecology and Evolution*, **6**, 135–139.

Case, T. J. and Casten, R. G. (1979) Global stability and multiple domains of attraction in ecological systems. *American Naturalist*, **113**, 705–714.

Caughley, G. and Lawton, J. H. (1981) Plant–herbivore systems. In *Theoretical Ecology*, ed. May, R. M., 2nd edn., pp. 132–166. 2nd edition. Blackwell Scientific: Oxford, UK.

Chandler, G. E. and Henderson, J. W. (1976) Studies on the nutrition and growth of *Drosera* species with reference to the carnivorous habit. *New Phytologist*, **76**, 129–141.

Chapin, F. S. and Bloom, A. (1976) Phosphate absorption: an adaptation of tundra graminoids to a low temperature, low phosphorus environment. *Oikos*, **26**, 111–121.

Chapman, A. and Rosenberg, K. V. (1991) Diets of four sympatric Amazonian woodcreepers (Dendrocolaptidae). *The Condor*, **93**, 904–915.

Charnov, E. L. (1976) Optimal foraging: the marginal value theorem. *Theoretical Population Biology*, **9**, 129–136.

Charnov, E. L., Orians, G. H. and Hyatt, K. (1976) Ecological implications of resource depression. *American Naturalist*, **110**, 247–259.

Cherrett, J. M. (1989) Key concepts: the results of a survey of our members. In *Ecological concepts. The contribution of ecology to an understanding of the natural world*, ed. Cherrett, J. M., Bradshaw, A. D., Goldsmith, F. B., Grubb, P. J. and Krebs, J. R., pp. 1–16. Blackwell Scientific Publications: Oxford, UK.

Chesson, P. (1990) MacArthur's consumer–resource model. *Theoretical Population Biology*, **37**, 26–38.

Chesson, P. (1991) A need for niches? *Trends in Ecology and Evolution*, **6**, 26–28.

Chesson, P. L. (1985) Coexistence of competitors in spatially and temporally varying environments: a look at the combined effects of different sorts of variability. *Theoretical Population Biology*, **28**, 263–287.

Chesson, P. L. (1988) Interactions between environment and competition: how fluctuations mediate coexistence and competitive exclusion. In *Lecture notes in biomathematics: Community ecology*, ed. Hastings, A., pp. 51–71. Springer-Verlag: Berlin, Germany.

Chew, F. S. (1981) Coexistence and local extinction in two pierid butterflies. *American Naturalist*, **118**, 655–672.

Christiansen, F. B. and Fenchel, T. M. (1977) *Theories of populations in biological communities*. Springer–Verlag: Berlin, Germany.

Clarke, R. D. (1977) Habitat distribution and species diversity of chaetodontid and pomacentrid fishes near Bimini, Bahamas. *Marine Biology*, **40**, 277–289.

Cloudsley-Thompson, J. L. (1991) *Ecophysiology of desert arthropods and reptiles*. Springer-Verlag: Berlin, Germany.

Cody, M. L. (1974) *Competition and the structure of bird communities*. Princeton University Press: Princeton, NJ, USA.

Cody, M. L. (1976) Habitat selection and interspecific competition among Mediterranean sylviid warblers. *Oikos*, **27**, 210–258.

Cody, M. L. (1978) Habitat selection and interspecific territoriality in sylviid warblers of England and Sweden. *Ecological Monographs*, **48**, 351–396.

Cody, M. L. (1983) Bird diversity and density in south African forests. *Oecologia (Berl.)*, **59**, 201–215.

Cohen, J. E. (1978) *Food webs and niche space*. Princeton University Press: Princeton, NJ, USA.

Cohen, J. E., Beaver, R. A., Cousins, S. H., DeAngelis, D. A., Goldwasser, L., Heong, K. L., Holt, R. D., Kohn, A. J., Lawton, J. H., Martinez, N., O'Malley, R., Page, L. M., Patten, B. C., Pimm, S. L., Polis, G. A., Rejm–nek, M., Schoener, T. W., Schoenly, K., Sprules, W. G., Teal, J. M., Ulanowicz, R. E., Warren, P. H., Wilbur, H. M. and Yodzis, P. (1993) Improving food webs. *Ecology*, **74**, 252–257.

Cohen, J. E., Briand, F. and Newman, C. M. (1990) *Community food webs*. Springer-Verlag: Berlin, Germany.

Coley, P. D. (1983) Herbivory and defensive characteristics of tree species in a lowland tropical forest. *Ecological Monographs*, **53**, 209–233.

Collins, S. L. and Glenn, S. M. (1990) A hierarchical analysis of species' abundance patterns in grassland vegetation. *American Naturalist*, **135**, 633–648.

Collins, S. L., James, F. C. and Risser, P. G. (1982) Habitat relationships of wood warblers (Parulidae) in northern central Minnesota. *Oikos*, **39**, 50–58.

Colwell, R. K. (1986) Population structure and sexual selection for host fidelity in the speciation of hummingbird flower mites. In *Evolutionary processes and theory*, ed. Karlin, S. and Nevo, E., pp. 475–495. Academic Press: New York, NY, USA.

Colwell, R. K. and Futuyma, D. (1971) On the measurement of niche breadth and overlap. *Ecology*, **52**, 567–576.

Comins, H. N. and Noble, I. R. (1985) Dispersal, variability, and transient niches: species coexistence in a uniformly variable environment. *American Naturalist*, **126**, 706–723.

Connell, J. H. (1978) Diversity of tropical rain forests and coral reefs. *Science (Wash., DC)*, **199**, 1302–1309.

Connell, J. H. (1980) Diversity and the coevolution of competitors, or the ghost of competition past. *Oikos*, **35**, 131–138.

Connell, J. H. (1983) On the prevalence and relative importance of interspecific competition: evidence from field experiments. *American Naturalist*, **122**, 661–696.

Connell, J. H. and Sousa, W. P. (1983) On the evidence needed to judge ecological stability or persistence. *American Naturalist*, **121**, 789–824.

Cooper, W. S. (1984) Expected time to extinction and the concept of fundamental fitness. *Journal of Theoretical Biology*, **107**, 603–629.

Cornell, H. V. and Lawton, J. H. (1992) Species interactions, local and regional processes, and limits to the richness of ecological communities: a theoretical perspective. *Journal of Animal Ecology*, **61**, 1–12.

Courtney, S. (1988) If it's not coevolution, it must be predation? *Ecology*, **69**, 910–911.

Cousins, S. (1985) The trophic continuum in marine ecosystems: structure and equations for a predictive model. *Canadian Journal of Fisheries and Aquatic Science*, **213**, 76–93.

Cousins, S. (1987) The decline of the trophic level concept. *Trends in Ecology and Evolution*, **2**, 312–316.

Cox, G. W. and Ricklefs, R. E. (1977) Species diversity and ecological release in Caribbean land bird faunas. *Oikos*, **28**, 113–122.

Cox, P. A. (1981) Niche partitioning between sexes of dioecious plants. *American Naturalist*, **117**, 295–307.

Crawford, R. M. M. (1992) Oxygen availability as an ecological limit to plant distribution. *Advances in Ecological Research*, **23**, 93–185.

Crawford, S. L. and Fall, T. C. (1990) Projection pursuit techniques for visualizing high-dimensional data sets. In *Visualization in scientific computing*, ed. Nielsen, G.M. and Shriver, B.D., pp. 94–108 IEEE Computer Society Press: Los Alamitos, CA, USA.

Crawley, M. J. (1983) *Herbivory. The dynamics of plant–animal interactions*. Blackwell Scientific: Oxford, UK.

Crist, T. O. and MacMahon, J. A. (1992) Harvester ant foraging and shrub–steppe seeds: interactions of seed resources and seed use. *Ecology*, **73**, 1768–1729.

Crompton, D. W. T. and Joyner, S. M. (1980) *Parasitic worms*. Wykeham Publications: London, UK.

Crowder, L. W., Rice, J. A., Miller, T. J. and Marschall, E. A. (1992) Empirical and theoretical approaches to size-based interactions and recruitment variability in fishes. In *Individual-based models and approaches in ecology. Populations, communities and ecosystems*, ed. DeAngelis, D.L. and Gross, L.J., pp. 237–255. Chapman and Hall: New York, NY, USA.

Crowley, P. H. and Johnson, D. M. (1982) Habitat and seasonality as niche axes in odonate communities. *Ecology*, **63**, 1064–1077.

Croy, M. I. and Hughes, R. N. (1990) The combined effects of learning and hunger in the feeding behaviour of the Fifteen-spined Stickleback (*Spinachia spinachia*). In *Behavioural mechanisms of food selection, 20*, ed. Hughes, R.N., pp. 215–233. Springer-Verlag: Berlin, Germany.

Cushing, J. M. (1992) A discrete model for competing stage-structured populations. *Theoretical Population Biology*, **41**, 372–387.

Cyr, H. and Pace, M. L. (1993) Magnitude and patterns of herbivory in aquatic and terrestrial ecosystems. *Nature (Lond.)*, **361**, 148–150.

Czárán, T. (1989) Coexistence of competing populations along an environmental gradient: a simulation study. *Coenoses*, **4**, 113–120.

Czárán, T. and Bartha, S. (1992) Spatiotemporal dynamic models of plant populations and communities. *Trends in Ecology and Evolution*, **7**, 38–42.

Czochor, R. J. and Leonard, K. J. (1982) Multiple-niche polymorphism in haploid microorganisms. *American Naturalist*, **119**, 293–296.

Dade, W. B., Jumars, P. A. and Penry, D. L. (1990) Supply-side optimization: maximizing absorptive rates. In *Behavioural mechanisms of food selection, 20*, ed. Hughes, R.N., pp. 531–555. Springer-Verlag: Berlin, Germany.

Damman, H. (1987) Leaf quality and enemy avoidance by the larvae of a pyralid moth. *Ecology*, **68**, 88–97.

Damuth, J. (1991) Of size and abundance. *Nature (Lond.)*, **351**, 268–269.

Danielson, B. J. (1991) Communities on a landscape: the influence of habitat heterogeneity on the interactions between species. *American Naturalist*, **138**, 1105–1120.

Danielson, B. J. (1992) Habitat selection, interspecific interactions and landscape composition. *Evolutionary Ecology*, **6**, 399–411.

Danielson, B. J. and Stenseth, N. C. (1992) The ecological and evolutionary implications of recruitment for competitively structured communities. *Oikos*, **65**, 34–44.

Darlington, P. J. (1957) *Zoogeography: the geographic distribution of animals*. Wiley: New York, NY, USA.

Davies, R. W., Wrona, F. J., Linton, L. and Wilkialis, J. (1981) Inter- and intra-specific analyses of the food niches of two sympatric species of Erpobdellidae (Hirudinoidea) in Alberta, Canada. *Oikos*, **37**, 105–111.

Dawson, T. E. and Bliss, L. C. (1989) Patterns of water use and the tissue water relations in the dioecious shrub, *Salix arctica*: the physiological basis for habitat partitioning between the sexes. *Oecologia (Berl.)*, **79**, 332–343.

Dawson, T. E. and Ehleringer, J. R. (1993) Gender-specific physiology, carbon isotope discrimination, and habitat distribution in boxelder, *Acer negundo*. *Ecology*, **74**, 798–815.

Day, P. R. (1974) *Genetics of host–parasite interaction*. Freeman: San Francisco, CA, USA.

Dayton, P. K. and Tegner, M. J. (1984) The importance of scale in community ecology: a kelp forest example with terrestrial analogs. In *A new ecology: novel approaches to interactive systems*, ed. Price, P.W., Slobodchikoff, C.N. and Gaud, W.S., pp. 457–481. Wiley: New York, NY, USA.

De Roos, A. M., Diekmann, O. and Metz, J. A. J. (1992) Studying dynamics of structured population models: a versatile technique and its application to *Daphnia*. *American Naturalist*, **139**, 123–147.

DeAngelis, D. L. and Gross, L. J. (ed.) (1992) *Individual-based models and approaches in ecology. Populations, communities and ecosystems.* Chapman and Hall: New York, NY, USA.

DeAngelis, D. L. and Waterhouse, J. C. (1987) Equilibrium and nonequilibrium concepts in ecological models. *Ecological Monographs*, **57**, 1–21.

Decho, A. W. and Fleeger, J. W. (1988) Ontogenetic feeding shifts in the meiobenthic harpacticoid copepod *Nitocra lacustris*. *Marine Biology*, **97**, 191–197.

Delbeek, J. C. and Williams, D. D. (1987) Food resource partitioning between sympatric populations of brackish water sticklebacks. *Journal of Animal Ecology*, **56**, 949–967.

Demment, H. W. and Van Soest, P. J. (1985) A nutritional explanation for body-size patterns in ruminant and nonruminant herbivores. *American Naturalist*, **125**, 641–672.

Dempster, E. R. (1955) Maintenance of genetic heterogeneity. *Cold Spring Harbor Symmposia on Quantitative Biology*, **20**, 25–32.

Denno, R. F. (1980) Ecotope differentiation in a guild of sap-feeding insects on the salt marsh grass, *Spartina patens*. *Ecology*, **61**, 702–714.

Denno, R. F., Larsson, S. and Olmstead, K. L. (1990) Role of enemy-free space and plant quality in host-plant selection by willow beetles. *Ecology*, **71**, 124–137.

Derrick, W. and Metzgar, L. (1991) Dynamics of Lotka–Volterra systems with exploitation. *Journal of Theoretical Biology*, **153**, 455–468.

Dial, K. P. and Marzluff, J. M. (1989) Nonrandom diversification within taxonomic assemblages. *Systematic Zoology*, **38**, 26–37.

Diamond, J. M. (1973) Distributional ecology of New Guinea birds. *Science (Wash., DC)*, **179**, 759–769.

Diamond, J. M. (1975) Assembly of species communities. In *Ecology and evolution of communities*, ed. Cody, M. L. and Diamond, J. M., pp. 342–444. Harvard University Press: Cambridge, MA, USA.

Diamond, J. M. (1978) Niche shifts and the rediscovery of interspecific competition. *American Scientist*, **66**, 322–331.

Diaz, N. F. and Valencia, J. (1985) Microhabitat utilization by two leptodactylid frogs in the Andes of central Chile. *Oecologia (Berl.)*, **66**, 353–357.

Diehl, S. (1992) Fish predation and benthic community structure: the role of omnivory and habitat complexity. *Ecology*, **73**, 1646–1661.

Dillon, W. R. and Goldstein, M. (1984) *Multivariate analysis – methods and applications*. Wiley: New York, NY, USA.

Dobkin, D. S. (1985) Heterogeneity of tropical floral microclimates and the response of hummingbird flower mites. *Ecology*, **66**, 536–543.

Dorgelo, J. (1976) Salt tolerance and the influence of temperature upon it. *Biological Reviews of the Cambridge Philosophical Society*, **51**, 255–290.

Doussard, D. E. (1993) Foraging with finesse: caterpillar adaptations for circumventing plant defences. In *Ecological and evolutionary constraints on caterpillars*, ed. Stamp, N. E. and Casey, T., pp. 92–131. Chapman and Hall: New York, NY, USA.

Doussard, D. E. and Denno, R. F. (1994) Host range of generalist caterpillars: trenching permits feeding on plants with secretory canals. *Ecology*, **77**, 69–78.

Downes, B. J. (1986) Guild structure in water mites (*Unionicola* spp.) inhabiting freshwater mussels: choice, competitive exclusion and sex. *Oecologia (Berl.)*, **70**, 457–465.

Downes, B. J. (1989) Host specificity, host location and dispersal: experimental conclusions from freshwater mites (*Unionicola* spp.) parasitizing unionid mussels. *Parisitology*, **98**, 189–196.

Downes, B. J. (1990) Host-induced morphology in mites: implications for host–parasite coevolution. *Systematic Zoology*, **39**, 162–168.

Downes, B. J. (1995) Spatial and temporal variation in recruitment and its effects on regulation of parasite populations. *Oecologia (Berl)*, in press.,

Drake, J. A. (1990) The mechanics of community assembly and succession. *Journal of Theoretical Biology*, **147**, 213–234.

du Toit, J. T. and Owen-Smith, N. (1989) Body size, population metabolism, and habitat specialization among large African herbivores. *American Naturalist*, **133**, 736–740.

DuBowy, P. J. (1988) Waterfowl communities and seasonal environments: temporal variability in interspecific competition. *Ecology*, **69**, 1439–1453.

Dueser, R. D. and Shugart, H. H., Jr (1979) Niche pattern in a forest-floor small-mammal fauna. *Ecology*, **60**, 108–118.

Dunham, A. E. (1983) Realized niche overlap, resource abundance, and intensity of interspecific competition. In *Lizard ecology. Studies of a model organism*, ed. Huey, R. B., Pianka, E. R. and Schoener, T. W., pp. 261–280. Harvard University Press: Cambridge, MA, USA.

Dunning, J. B., Danielson, B. J. and Pulliam, H. R. (1992) Ecological processes that affect populations in complex landscapes. *Oikos*, **65**, 169–174.

Ebenhöh, W. (1988) Coexistence of an unlimited number of algal species in a model system. *Theoretical Population Biology*, **34**, 130–144.

Ebenman, B. (1987) Niche differences between age classes and intraspecific competition in age-structured populations. *Journal of Theoretical Biology*, **124**, 25–33.

Ebenman, B. (1988) Competition between age classes and population dynamics. *Journal of Theoretical Biology*, **131**, 389–400.

Ebenman, B. (1992) Evolution of organisms that change their niches during the life cycle. *American Naturalist*, **139**, 990–1021.

Ebersole, J. P. (1985) Niche separation of two damselfish by aggression and differential microhabitat selection. *Ecology*, **66**, 14–20.

Eckhardt, V. M. (1992) Spatio-temporal variation in abundance and variation in foraging behavior of the pollinators of gynodioecious *Phacelia linearis* (Hydrophyllaceae). *Oikos*, **64**, 573–586.

Efron, B. (1982) *The jackknife, the bootstrap, and other resampling plans*. Society for Industrial Mathematics: Philadelphia, PA, USA.

Ehlinger, T. J. (1990) Habitat choice and phenotype-limited feeding efficiency in bluegill: individual differences and trophic polymorphism. *Ecology*, **71**, 886–896.

Ehrlich, P. R. and Murphy, D. D. (1988) Plant chemistry and host range in insect herbivores. *Ecology*, **69**, 908–909.

Ehrlich, P. R. and Raven, P. (1964) Butterflies and plants: a study in coevolution. *Evolution*, **18**, 586–608.

Eldredge, N. (1989) *Macroevolutionary dynamics: species, niches, and adaptive peaks.*

McGraw-Hill: New York, NY, USA.

Emlen, J. M. (1966) The role of time and energy in food preference. *American Naturalist*, **100**, 611–617.

Emlen, J. M. (1968) Optimal choice in animals. *American Naturalist*, **102**, 385–389.

Emlen, J. M. (1973) *Ecology: an evolutionary approach*. Addison-Wesley: Reading, MA, USA.

Emlen, J. M. (1981) Field estimation of competition intensity. *Theoretical Population Biology*, **19**, 275–287.

Emlen, J. T. (1981) Divergence in the foraging responses of birds on two Bahama islands. *Ecology*, **62**, 289–295.

Emlen, J. T. and DeJong, M. J. (1981) Intrinsic factors in the selection of foraging substrates by pine warblers: a test of an hypothesis. *The Auk*, **98**, 294–298.

Endler, J. A. (1977) *Geographic variation, speciation, and clines*. Princeton University Press: Princeton, NJ, USA.

Endler, J. A. (1986) *Natural selection in the wild*. Princeton University Press: Princeton, NJ, USA.

Ermentrout, G. B. and Edelstein-Keshet, L. (1993) Cellular automata approaches to biological modelling. *Journal of Theoretical Biology*, **160**, 97–133.

Esch, G. W. (ed.) (1977) *Regulation of parasite populations*. Academic Press: New York, NY, USA.

Esch, G. W. and Fernandez, J. C. (ed.) (1993) *A functional biology of parasitism: ecological and evolutionary implications*. Chapman and Hall: London, UK.

Estes, J. A. and Steinberg, P. D. (1988) Predation, evolution and kelp evolution. *Paleobiology*, **14**, 19–36.

Fahrig, L. and Merriam, G. (1985) Habitat patch connectivity and population survival. *Ecology*, **66**, 1762–1768.

Fairweather, P. G. and Underwood, A. J. (1983) The apparent diet of predators and biases due to different handling times of their prey. *Oecologia (Berl.)*, **56**, 169–179.

Farrell, B. D., Mitter, C. and Futuyma, D. J. (1992) Diversification at the insect–plant interface. Insights from phylogenetics. *BioScience*, **42**, 34–42.

Feeny, P. (1976) Plant apparency and chemical defense. *Recent Advances in Phytochemistry*, **10**, 1–40.

Feeny, P. P. (1991) Theories of plant chemical defense: a brief historical survey. In *Insects and plants*, ed. Jermy, T. and Szentesi, A., pp. 163–175. S P Bakker Academic Publishers: The Hague, Netherlands.

Feinsinger, P., Colwell, R. K., Terborgh, J. and Chaplin, S. B. (1979) Elevation and the morphology, flight energetics, and foraging ecology of tropical hummingbirds. *American Naturalist*, **113**, 481–497.

Feinsinger, P., Spears, E. E. and Poole, R. W. (1981) A simple measure of niche breadth. *Ecology*, **62**, 27–32.

Feinsinger, P., Swarm, L. A. and Wolfe, J. A. (1985) Nectar-feeding birds on Trinidad and Tobago: comparison of diverse and depauperate guilds. *Ecological Monographs*, **55**, 1–28.

Fenchel, T. (1987) *Ecology of Protozoa: the biology of free-living phagotrophic protists*. Springer-Verlag: New York, NY, USA.

Findley, J. S. and Black, H. (1983) Morphological and dietary structuring of a Zambian insectivorous bat community. *Ecology*, **64**, 625–630.

Fleming, T. H. (1985) Coexistence of five sympatric *Piper* (Piperaceae) species in a tropical dry forest. *Ecology*, **66**, 688–700.

Flint, R. W. and Kalke, R. D. (1986) Niche characterization of dominant benthic species. *Estuarine, Coastal and Shelf Science*, **22**, 657–674.

Folt, C. and Goldman, C. R. (1981) Allelepathy between zooplankton: a mechanism for interference competition. *Science (Wash., DC)*, **213**, 1133–1135.

Ford, H. A. (1990) Relationships between distribution, abundance and foraging specialization in Australian landbirds. *Ornis Scandinavica*, **21**, 133–138.

Ford, H. A. and Paton, D. C. (1977) The comparative ecology of ten species of honeyeaters in South Australia. *Australian Journal of Ecology*, **2**, 399–407.

Ford, H. A. and Paton, D. C. (1982) Partitioning of nectar sources in an Australian honeyeater community. *Australian Journal of Ecology*, **7**, 149–159.

Ford, H. A. and Paton, D. C. (1985) Habitat selection in Australian honeyeaters, with special reference to nectar productivity. In *Habitat selection in birds*, ed. Cody, M.L., pp. 367–388. Academic Press: Orlando, FL, USA.

Forman, R. T. T. and Godron, M. (1986) *Landscape ecology*. Wiley: New York, NY, USA.

Fox, C. W. and Lalonde, R. G. (1993) Host confusion and the evolution of insect diet breadths. *Oikos*, **67**, 577–581.

Fox, L. R. and Morrow, P. A. (1981) Specialization: species property or local phenomenon? *Science*, **211**, 887–893.

Fraser, D. F. (1976) Coexistence of salamanders in the genus *Plethodon*: a variation of the Santa Rosalia theme. *Ecology*, **57**, 238–251.

Freeman, D. C., Klikoff, L. G. and Harper, K. T. (1976) Differential resource utilization by the sexes of dioecious plants. *Science (Wash., D.C.)*, **193**, 597–599.

Freeman, P. W. (1979) Specialized insectivory: beetle-eating and moth-eating bats. *Journal of Mammalogy*, **60**, 467–479.

Fretwell, S. D. (1972) *Populations in a seasonal environment*. Princeton University Press: Princeton, NJ, USA.

Fretwell, S. D. (1978) Competition for discrete versus continuous resources: tests for predictions from the MacArthur–Levins model. *American Naturalist*, **112**, 73–81.

Fretwell, S. D. and Lucas, H. L., Jr (1970) On territorial behavior and other factors influencing habitat distribution in birds. *Acta Biotheoretica*, **19**, 16–36.

Futuyma, D. J. (1976) Food plant specialization and environmental predictability in Lepidoptera. *American Naturalist*, **110**, 286–292.

Futuyma, D. J. (1983) Selective factors in the evolution of host choice by phytophagous insects. In *Herbivorous insects: host seeking behavior and mechanisms*, ed. Ahman, S., pp. 227–244. Academic Press: New York, NY, USA.

Futuyma, D. J. and Moreno, G. (1988) The evolution of ecological specialization. *Annual Review of Ecology and Systematics*, **19**, 207–233.

Futuyma, D. J. and Peterson, S. C. (1985) Genetic variation in the use of resources by insects. *Annual Review of Entomology*, **30**, 217–238.

Futuyma, D. J. and Philippi, T. E. (1987) Genetic variation and variation in responses to host plants by *Alsophila pometaria* (Lepidoptera: Geometridae). *Evolution*, **41**, 269–279.

Gaines, S., Brown, S. and Roughgarden, J. (1985) Spatial variation in larval concentrations as a cause of spatial variation in settlement for the barnacle, *Balanus glandula. Oecologia (Berl.)*, **67**, 267–272.

Gaines, S. and Roughgarden, J. (1985) Larval settlement rate: a leading determinant of structure in an ecological community of the marine intertidal zone. *Proceedings of the National Academy of Science USA*, **82**, 3707–3711.

Gali-Muhtasib, H. U., Smith, C. C. and Higgins, J. J. (1992) The effect of silica in grasses on the feeding behavior of the prairie vole, *Microtus ochrogaster. Ecology*, **73**, 1724–1729.

Garnick, E. (1992) Niche breadth in parasites: an evolutionarily stable strategy model, with special reference to the protozoan parasite *Leishmania. Theoretical Population Biology*, **42**, 62–103.

Gascon, C. (1991) Population- and community-level analyses of species occurrences of central Amazonian rainforest tadpoles. *Ecology*, **72**, 1731–1746.

Gaston, K. J. (1993) Herbivory at the limits. *Trends in Ecology and Evolution*, **8**, 193–194.

Gaston, K. J. (1994) *Rarity*. Chapman and Hall: London, UK.

Gaston, K. J. and Lawton, J. H. (1989) Insect herbivores on bracken do not support the core–satellite hypothesis. *American Naturalist*, **134**, 761–777.

Gatto, M. (1982) Comments on 'MacArthur's minimization principle: a footnote'. *American Naturalist*, **119**, 140–144.

Gatto, M. (1990) A general minimum principle for competing populations: some ecological and evolutionary consequences. *Theoretical Population Biology*, **37**, 369–388.

Gerard, V. A. (1990) Ecotypic differentiation in the kelp *Laminaria saccharina*: phase-specific adaptation in a complex life cycle. *Marine Biology*, **107**, 519–528.

Geritz, S. A. H., Metz, J. A. J., Klinkhamer, P. G. L. and De Jong, T. J. (1988) Competition in safe-sites. *Theoretical Population Biology*, **33**, 161–180.

Getz, W. M. (1984) Population dynamics: a *per capita* resource approach. *Journal of Theoretical Biology*, **108**, 623–643.

Getz, W. M. and Kaitala, V. (1989) Ecogenetic models, competition, and heteropatry. *Theoretical Population Biology*, **36**, 34–58.

Gilbert, L. E. and Singer, M. C. (1975) Butterfly ecology. *Annual Review of Ecology and Systematics*, **6**, 365–397.

Gill, D. E. (1974) Intrinsic rate of increase, saturation density, and competitive ability. II. The evolution of competitive ability. *American Naturalist*, **108**, 103–116.

Giller, P. S. (1980) The control of handling time and its effects on foraging strategy of a heteropteran predator, *Notonecta. Journal of Animal Ecology*, **49**, 699–712.

Gilles, R. and Pequeux, A. (1983) Interactions of chemical and osmotic regulation with the environment. In *Biology of Crustacea, Vol. 8: Environmental adaptations*, ed. Vernberg, F.J. and Vernberg, W.B., pp. 109–177. Academic Press: New York, NY, USA.

Gillespie, J. (1974) Polymorphism in patchy environments. *American Naturalist*, **108**, 145–151.

Gilliam, J. F. (1990) Hunting by the hunted: optimal prey selection by foragers under predation hazard. In *Behavioural mechanisms of food selection*, ed. Hughes, R.N., pp. 797–818. Springer-Verlag: Berlin, Germany.

Gilpin, M. E. and Case, T. J. (1976) Multiple domains of attraction in competition communities. *Nature (Lond.)*, **261**, 40–42.

Gilpin, M. E. and Diamond, J. M. (1982) Factors contributing to non-randomness in species co-occurrences on islands. *Oecologia (Berl.)*, **52**, 75–84.

Gilpin, M. E. and Hanski, I. (1991) *Metapopulations*. Academic Press: London, UK.

Ginzburg, L. R. (1986) The theory of population dynamics: I. Back to first principles. *Journal of Theoretical Biology*, **122**, 385–399.

Ginzburg, L. R. and Akçakaya, H. R. (1992) Consequences of ratio-dependent predation for steady-state properties of ecosystems. *Ecology*, **73**, 1536–1543.

Ginzburg, L. R., Akçakaya, H. R. and Kim, J. (1988) Evolution of community structure: competition. *Journal of Theoretical Biology*, **133**, 513–523.

Givnish, T. J. (1989) Ecology and evolution of carnivorous plants. In *Plant–animal interactions*, ed. Abrahamson, W.G., pp. 243–290. McGraw-Hill: New York, NY, USA.

Gladfelter, W. B. and Johnson, W. S. (1983) Feeding niche separation in a guild of tropical reef fishes (Holocentridae). *Ecology*, **64**, 552–563.

Glasser, J. W. (1982) A theory of trophic strategies: the evolution of facultative specialists. *American Naturalist*, **119**, 250–262.

Glasser, J. W. (1983) Variation in niche breadth with trophic position: on the disparity between expected and observed species packing. *American Naturalist*, **122**, 542–548.

Glasser, J. W. (1984) Evolution of efficiencies and strategies of resource exploitation. *Ecology*, **65**, 1570–1578.

Glasser, J. W. and Price, H. J. (1982) Niche theory: new insights from an old paradigm. *Journal of Theoretical Biology*, **99**, 437–460.

Gleason, H. A. (1926) The individualistic concept of plant association. *Bulletin of the Torrey Botanical Club*, **53**, 7–26.

Gleeson, S. K. and Wilson, D. S. (1986) Equilibrium diet: optimal foraging and prey coexistence. *Oikos*, **46**, 139–144.

Godin, J-G. J. (1990) Diet selection under the risk of predation. In *Behavioural mechanisms of food selection*, ed. Hughes, R. N., pp. 739–769. Springer-Verlag: Berlin, Germany.

Good, R. A. (1974) *The geography of flowering plants*, 4th edn. Longmans: London, UK.

Gordon, D. R. and Rice, K. J. (1992) Partitioning of space and water between two California annual grassland species. *American Journal of Botany*, **79**, 967–976.

Gotelli, N. J. and Simberloff, D. (1987) The distribution and abundance of tallgrass prairie plants: a test of the core–satellite hypothesis. *American Naturalist*, **130**, 18–35.

Grace, J. B. and Tilman, D. (ed.) (1990) *Perspectives in plant competition*. Academic Press: New York, NY, USA.

Grant, P. R. (1975) The classical case of character displacment. *Evolutionary Biology*, **8**, 237–337.

Gray, L. (1979) The use of psychophysical unfolding theory to determine principal resource axes. *American Naturalist*, **114**, 695–706.

Gray, L. and King, J. A. (1986) The use of multidimensional scaling to determine principal resource axes. *American Naturalist*, **127**, 577–592.

Green, M. B., Schwarz, J. H. and Witten, E. (1987) *Superstring theory*. Cambridge University Press: Cambridge, UK.

Greenberg, R. (1983) The role of neophobia in determining the degree of foraging specialization in some migrant warblers. *American Naturalist*, **122**, 444–453.

Greenberg, R. (1990a) Ecological plasticity, neophobia, and resource use by birds. *Ornithological Monographs*, **13**, 431–437.

Greenberg, R. (1990b) Feeding neophobia and ecological plasticity: a test of the hypothesis with captive sparrows. *Animal Behaviour*, **39**, 375–379.

Greene, H. W. and Jaksić, F. M. (1983) Food–niche relationships among sympatric predators: effects of level of prey identification. *Oikos*, **40**, 151–154.

Greig-Smith, P. (1983) *Quantitative plant ecology*. Blackwell Scientific Publications: Oxford, UK.

Grenfell, B. T., Price, O. F., Albon, S. D. and Clutton-Brock, T. H. (1992) Overcompensation and population cycles in an ungulate. *Nature (Lond.)*, **355**, 823–826.

Grime, J. P. (1979) *Plant strategies and vegetation processes*. Wiley: Chichester.

Grime, J. P., Crick, J. C. and Rincon, J. E. (1986) The ecological significance of phenotypic plasticity. In *Plasticity in plants, 40*, ed. Jennings, D.H. and Trewavas, A.J., pp. 5–29. Company of Biologists: Cambridge, UK.

Grinnell, J. (1917) Field tests and theories concerning distributional control. *American Naturalist*, **51**, 115–128.

Grover, J. P. (1990) Resource competition in a variable environment: phytoplankton growing according to Monod's model. *American Naturalist*, **136**, 771–789.

Grubb, P. J. (1977) The maintenance of species-richness in plant communities: the importance of the regeneration niche. *Biological Reviews*, **52**, 107–145.

Gulve, P. S. (1994) Distribution and extinction patterns within a northern metapopulation of the pool frog, *Rana lessonae*. *Ecology*, **75**, 1357–1367.

Gurevitch, J., Morrow, L. L., Wallace, A. and Walsh, J. S. (1992) A meta-analysis of competition in field experiments. *American Naturalist*, **140**, 539–572.

Gurney, W. S. C. and Nisbet, R. M. (1980) Age- and density-dependent population dynamics in static and variable environments. *Theoretical Population Biology*, **17**, 321–344.

Gutierrez, A. P. (1992) Physiological basis of ratio-dependent predator–prey theory: the metabolic pool model as a paradigm. *Ecology*, **73**, 1552–1563.

Haefner, J. W. (1992) Parallel computers and individual-based models: an overview. In *Individual-based models and approaches in ecology. Populations, communities and ecosystems*, ed. DeAngelis, D.L. and Gross, L.J., pp. 126–164. Chapman and Hall: New York, NY, USA.

Haefner, J. W. and Edson, J. L. (1984) Community invasion by complex life cycles. *Journal of Theoretical Biology*, **108**, 377–404.

Haigh, J. and Maynard Smith, J. (1972) Can there be more predators than prey? *Theoretical Population Biology*, **3**, 290–299.

Hairston, N. G., Sr (1980) Species packing in the salamander genus *Desmognathus*: what are the interspecific interactions involved? *American Naturalist*, **115**, 354–366.

Hairston, N. G., Sr (1983) Alpha selection in competing salamanders: experimental verification of an a priori hypothesis. *American Naturalist*, **122**, 105–113.

Hairston, N. G., Sr (1985) The interpretation of experiments on interspecific competition. *American Naturalist*, **125**, 321–325.

Hairston, N. G., Jr and Hairston, N. G., Sr (1993) Cause–effect in energy flow, trophic structure, and interspecific interactions. *American Naturalist*, **142**, 379–411.

Hairston, N. G., Sr, Nishikawa, K. C. and Stenhouse, S. L. (1987) The evolution of competing species of terrestrial salamanders: niche partitioning or interference? *Evolutionary Ecology*, **1**, 247–262.

Hall, C. A. S. (1988) An assessment of several of the historically most influential theoretical models used in ecology and of the data provided in their support. *Ecological Modelling*, **43**, 5–31.

Hall, C. A. S. (1991) An idiosyncratic assessment of the role of mathematical models in environmental sciences. *Environmental International*, **17**, 507–517.

Hall, C. A. S. and DeAngelis, D. L. (1985) Models in ecology: paradigms found or paradigms lost? *Bulletin of the Ecological Society of America*, **66**, 339–346.

Hall, C. A. S., Stanford, J. A. and Hauer, F. A. (1992) The distribution and abundance of organisms as a consequence of energy balances along multiple environmental gradients. *Oikos*, **65**, 377–390.

Hall, S. J. and Raffaelli, D. G. (1991) Food web patterns: lessons from a species rich web. *Journal of Animal Ecology*, **60**, 823–841.

Hall, S. J. and Raffaelli, D. G. (1993) Food webs: theory and reality. *Advances in Ecological Research*, **24**, 187–239.

Hallett, J. G. (1980) Niche width and genetic variation in *Drosophila* reexamined. *American Naturalist*, **115**, 594–595.

Hallett, J. G. (1982) Habitat selection and the community matrix of a small-mammal fauna. *Ecology*, **63**, 1400–1410.

Hallett, J. G. and Pimm, S. L. (1979) Direct estimation of competition. *American Naturalist*, **113**, 593–600.

Hanski, I. (1978) Some comments on the measurement of niche metrics. *Ecology*, **59**, 168–174.

Hanski, I. (1982) Dynamics of regional distribution: the core and satellite species hypothesis. *Oikos*, **38**, 210–221.

Harder, L. D. (1985) Morphology as an predictor of flower choice by bumble bees. *Ecology*, **66**, 198–210.

Harley, J. L. and Harley, E. L. (1987) A check-list of mycorrhiza in the British flora. *New Phytologist*, **105 (supplement)**, 1–102.

Harley, J. L. and Smith, S. E. (1983) *Mycorrhizal symbiosis*. Academic Press: London, UK.

Harrison, G. W. (1979) Stability under environmental stress: resistance, resilience, persistence, and variability. *American Naturalist*, **113**, 659–669.

Hart, D. D. (1992) Community organization in streams: the importance of species interactions, physical factors, and chance. *Oecologia (Berl.)*, **91**, 220–228.

Hastings, A. (1980) Disturbance, coexistence, history, and competition for space. *Theoretical Population Biology*, **18**, 363–373.

Hastings, A. (1986) The invasion question. *Journal of Theoretical Biology*, **121**, 211–220.

Hastings, A. (1988) Food web theory and stability. *Ecology*, **69**, 1665–1668.

Hatfield, J. S. and Chesson, P. L. (1989) Diffusion analysis and stationary distribution of the two-species lottery competition model. *Theoretical Population Biology*, **36**, 251–266.

Hawkins, B. A. and Gross, P. (1992) Species richness and population limitation in insect parasitoid-host systems. *American Naturalist*, **139**, 417–423.

Hawkins, B. A., Shaw, M. R. and Askew, R. R. (1992) Relations among assemblage size, host specialization, and climatic variability in North American parasitoid communities. *American Naturalist*, **139**, 58–79.

Hawkins, C. P. and MacMahon, J. A. (1989) Guilds: the multiple meanings of a concept. *Annual Review of Entomology*, **34**, 423–451.

Hay, M. E. (1991) Marine–terrestrial contrasts in the ecology of plant chemical defenses against herbivores. *Trends in Ecology and Evolution*, **6**, 362–365.

Hay, M. E., Duffy, J. E. and Fenical, W. (1990) Host-plant specialization decreases predation on a marine amphipod: an herbivore in plant's clothing. *Ecology*, **71**, 733–743.

Hay, M. E. and Fenical, W. (1988) Marine plant–herbivore interactions: the ecology of chemical defense. *Annual Review of Ecology and Systematics*, **19**, 111–145.

Hay, M. E., Pawlick, J. R., Duffy, J. E. and Fenical, W. (1989) Seaweed–herbivore–predator interactions: host-plant specialization reduces predation on small herbivores. *Oecologia (Berl.)*, **81**, 418–427.

Hedrick, P. W. (1993) Sex-dependent habitat selection and genetic polymorphism. *American Naturalist*, **141**, 491–500.

Heithaus, E. R. (1979) Community structure of neotropical flower-visiting bees and wasps: diversity and phenology. *Ecology*, **60**, 190–202.

Henderson, R. W. (1986) The diet of the Hispaniolan colubrid snake, *Darlingtonia haetiana*. *Copeia*, **1986**, 529–531.

Hengeveld, R. (1990) *Dynamic biogeography*. Cambridge University Press: Cambridge, UK.

Hengeveld, R. and Haeck, J. H. (1982) The distribution of abundance. I. Measurements. *Journal of Biogeography*, **9**, 303–316.

Herbold, B. (1984) Structure of an Indiana stream fish community association: choosing an appropriate model. *American Naturalist*, **124**, 561–572.

Herrera, C. M. (1978a) Ecological correlates of residence and non-residence in a Mediterranean passerine bird community. *Journal of Animal Ecology*, **47**, 871–890.

Herrera, C. M. (1978b) Individual dietary differences associated with morphological variation in robins *Erithracus rubecula*. *Ibis*, **120**, 542–545.

Herrera, C. M. (1988) Variation in mutualisms: the spatio-temporal mosaic of a pollinator assemblage. *Biological Journal of the Linnean Society*, **35**, 95–125.

Hesse, R., Allee, W. C. and Schmidt, K. P. (1951) *Ecological animal geography*, 2nd edn. Wiley: New York, NY, USA.

Hildén, O. (1965) Habitat selection in birds: a review. *Annales Zoologica. Fennica*, **2**, 53–75.

Hill, K. D. and Johnson, L. A. S. (1991) Systematic studies in the eucalypts – 3. new taxa and combinations in *Eucalyptus* (Myrtaceae). *Telopea*, **4**, 223–267.

Hill, M. O. (1973) Diversity and evenness: a unifying notation and its consequences. *Ecology*, **54**, 427–432.

Hines, A. H. (1982) Coexistence in a kelp forest: size, population dynamics, and resource partitioning in a guild of spider crabs (Brachyura: Majidae). *Ecological Monographs*, **52**, 179–198.

Hockey, P. A. R. and Steele, W. K. (1990) Intraspecific kleptoparasitism and foraging efficiency as constraints on food selection by Kelp Gulls *Larus dominica-*

nus. In *Behavioural mechanisms of food selection, 20*, ed. Hughes, R.N., pp. 679–705. Springer-Verlag: Berlin, Germany.

Hofbauer, J. and Sigmund, K. (1988) *The theory of evolution and dynamical systems*. Cambridge University Press: Cambridge, UK.

Hoffmann, A. J., Fuentes, E. R., Cortes, I., Liberona, F. and Costa, V. (1986) *Tristerix tentrandrus* (Loranthaceae) and its host plants in the Chilean matorral: patterns and mechanisms. *Oecologia (Berl.)*, **69**, 202–206.

Hoffmaster, D. K. (1985) Resource breadth in orb-weaving spiders: a tropical–temperate comparison. *Ecology*, **66**, 626–629.

Holbrook, S. J. and Schmitt, R. J. (1988) The combined effects of predation risk and food reward on patch selection. *Ecology*, **69**, 125–134.

Holbrook, S. J. and Schmitt, R. J. (1992) Causes and consequences of dietary specialization in surfperches: patch choice and intraspecific competition. *Ecology*, **73**, 402–412.

Holling, C. S. (1959) The components of predation as revealed by a study of small-mammal predation of the European pine sawfly. *Canadian Entomologist*, **91**, 293–320.

Holling, C. S. (1965) The functional response of predators to prey density and its role in mimicry and population regulation. *Memoirs of the Entomological Society of Canada*, **45**, 5–60.

Holling, C. S. (1992) Cross-scale morphology, geometry, and dynamics of ecosystems. *Ecological Monographs*, **62**, 447–502.

Holmes, J. C. (1976) Host selection and its consequences. In *Ecological aspects of parasitism*, ed. Kennedy, C.R., pp. North-Holland: Amsterdam, Netherlands.

Holmes, J. C. (1979) Parasite populations and host community structure. In *Host-parasite interfaces*, ed. Nickol, B.B., pp. 27–46. Academic Press: New York, NY, USA.

Holmes, R. T., Bonney, R. E. J. and Pacala, S. W. (1979) Guild structure of the Hubbard Brook bird community: a multivariate approach. *Ecology*, **60**, 512–520.

Holmes, R. T. and Schultz, J. C. (1988) Food availability for forest birds: effects of prey distribution and abundance on bird foraging. *Canadian Journal of Zoology*, **66**, 720–728.

Holsinger, K. E. and Pacala, S. W. (1990) Multiple-niche polymorphisms in plant populations. *American Naturalist*, **135**, 301–309.

Holt, R. D. and Gaines, M. S. (1992) Analysis of adaptation in heterogeneous landscapes: implications for the evolution of fundamental niches. *Evolutionary Ecology*, **6**, 433–447.

Holt, R. D. and Kotler, B. P. (1987) Short-term apparent competition. *American Naturalist*, **130**, 412–430.

Hopf, F. A. and Hopf, F. W. (1985) The role of the Allee effect in species packing. *Theoretical Population Biology*, **27**, 27–50.

Horn, H. S. (1966) Measurement of 'overlap' in comparative ecological studies. *American Naturalist*, **100**, 419–424.

Horn, H. S. and May, R. M. (1977) Limits to similarity among co-existing competitors. *Nature (Lond.)*, **270**, 660–661.

Howe, H. F. (1984) Constraints on the evolution of mutualism. *American Naturalist*, **123**, 764–777.

Hubbell, S. P. and Foster, R. B. (1986) Biology, chance and history and the structure of tropical rain forest tree communities. In *Community ecology*, ed. Case, T.J. and Diamond, J., pp. 314–329. Harper and Row: New York, NY, USA.

Huey, R. B. and Hertz, P. E. (1984) Is a jack-of-all-temperatures a master of none? *Evolution*, **38**, 441–444.

Hughes, R. N. (1979) Optimal diets under the energy maximization premise: the effects of recognition time and learning. *American Naturalist*, **113**, 209–221.

Hughes, R. N. (ed.) (1990) *Behavioural mechanisms of food selection, 20*. Springer-Verlag: Berlin, Germany.

Humphrey, S. R., Bonaccorso, F. J. and Zinn, T. L. (1983) Guild structure of surface-gleaning bats in Panamá. *Ecology*, **64**, 284–294.

Hurlbert, S. H. (1978) The measurement of niche overlap and some relatives. *Ecology*, **59**, 67–77.

Hutchinson, G. E. (1957) Concluding remarks. *Cold Spring Harbor Symposia on Quantitative Ecology*, **22**, 415–427.

Hutchinson, G. E. (1959) Homage to Santa Rosalia, or why are there so many kinds of animals? *American Naturalist*, **93**, 145–159.

Hutchinson, G. E. (1978) *An introduction to population ecology*. Yale University Press: New Haven, CT, USA.

Hutto, R. L. (1985) Habitat selection by nonbreeding, migratory land birds. In *Habitat selection in birds*, ed. Cody, M.L., pp. 455–476. Academic Press: Orlando, FL, USA.

Ives, A. R. and May, R. M. (1985) Competition within and between species in a patchy environment: relations between microscopic and macroscopic models. *Journal of Theoretical Biology*, **115**, 65–92.

Ivlev, V. S. (1961) *Experimental ecology of the feeding of fishes*. Yale University Press: New Haven, CT, USA.

Jackson, J. B. C. (1973) The ecology of molluscs of *Thalassia* communities, Jamaica, West Indies. I. Distribution, environmental physiology, and ecology of common shallow-water species. *Bulletin of Marine Science*, **23**, 313–350.

Jackson, J. B. C. (1974) Biogeographic consequences of eurytopy and stenotopy among marine bivalves and their evolutionary significance. *American Naturalist*, **108**, 541–560.

Jackson, J. B. C. (1981) Interspecific competition and species' distributions: the ghosts of theories and data past. *American Zoologist*, **21**, 889–901.

Jaeger, R. G. and Lucas, J. (1990) On evaluation of foraging strategies through estimates of reproductive success. In *Behavioural mechanisms of food selection, 20*, ed. Hughes, R.N., pp. 83–94. Springer-Verlag: Berlin, Germany.

Jaenike, J. (1978) Resource predictability and niche breadth in the *Drosophila quinania* species group. *Evolution*, **32**, 676–678.

Jaenike, J. (1990) Host specialization in phytophagous insects. *Annual Review of Ecology and Systematics*, **21**, 243–273.

Jaenike, J. (1993) Rapid evolution of host specificity in a parasitic nematode. *Evolutionary Ecology*, **7**, 103–108.

Jaenike, J., Parker, E. D., Jr and Selander, R. K. (1980) Clonal niche structure in the parthogenetic earthworm *Octolasion tyrtaeum*. *American Naturalist*, **116**, 196–205.

Jaenike, J. and Selander, R. K. (1979) Ecological generalism in *Drosophila falleni*: genetic evidence. *Evolution*, **33**, 741–748.

Jaksić, F. M. and Braker, H. E. (1983) Food-niche relationships and birds of prey: competition versus opportunism. *Canadian Journal of Zoology*, **61**, 2230–2241.

James, C. D., Hoffmann, M. T., Lightfoot, D. C., Forbes, G. S. and Whitford, W. G. (1993) Pollination ecology of *Yucca elata*. *Oecologia (Berl.)*, **93**, 512–517.

James, F. C. (1971) Ordination of habitat relationships among breeding birds. *Wilson Bulletin*, **83**, 215–236.

James, F. C., Johnston, R. F., Wamer, N. O., Niemi, G. J. and Boecklen, W. J. (1984) The Grinnellian niche of the wood thrush. *American Naturalist*, **124**, 17–30.

James, S. D. (1991) Temporal variation in diets and trophic partitioning by coexisting lizards (*Ctenotus*: Scincidae) in central Australia. *Oecologia (Berl.)*, **85**, 553–561.

Janovy, J. J., Ferdig, M. T. and McDowell, M. A. (1990) A model of the dynamic behavior of a parasite species assemblage. *Journal of Theoretical Biology*, **142**, 517–529.

Jansen, W. (1987) A permanence theorem for replicator and Lotka–Volterra systems. *Journal of Mathematical Biology*, **25**, 411–422.

Janzen, D. H. (1981) The peak in North American ichneumonid species richness lies between 38° and 42°N. *Ecology*, **62**, 532–537.

Jennings, D. H. and Lee, D. L. (ed.) (1975) *Symbiosis*. Cambridge University Press: Cambridge, UK.

Jermy, T. (1984) Evolution of host/plant relationships. *American Naturalist*, **124**, 609–630.

Jermy, T. (1988) Can predation lead to narrow food specialization in phytophagous insects? *Ecology*, **69**, 902–904.

Joern, A. and Lawlor, L. R. (1980) Food and microhabitat utilization by grasshoppers from arid grasslands: comparisons with neutral models. *Ecology*, **61**, 591–599.

Joern, A. and Lawlor, L. R. (1981) Guild structure in grasshopper assemblages based on food and microhabitat resources. *Oikos*, **37**, 93–104.

Johanson, F. (1993) Intraguild predation and cannibalism in odonate larvae: effects of foraging behaviour and zooplankton availability. *Oikos*, **66**, 80–87.

Johnson, R. A. (1986) Intraspecific resource partitioning in the bumble bees *Bombus ternarius* and *B. pennsylvanicus*. *Ecology*, **67**, 133–138.

Jones, A. W. (1967) *Introduction to parasitology*. Addison-Wesley: Reading, MA, USA.

Jones, C. G., Hess, T. A., Whitman, D. W., Silk, P. J. and Blum, M. S. (1986) Idiosyncratic variation in chemical defenses among individual generalist grasshoppers. *Journal of Chemical Ecology*, **12**, 749–761.

Jordano, D., Rodríguez, J., Thomas, C. D. and Haeger, J. F. (1992) The distribution and density of a lycaenid butterfly in relation to Lasius ants. *Oecologia (Berl.)*, **91**, 439–446.

Jordano, D. and Thomas, C. D. (1992) Specificity of an ant–lycaenid butterfly interaction. *Oecologia (Berl.)*, **91**, 431–438.

Juanes, F. (1986) Population density and body size in birds. *American Naturalist*, **128**, 921–929.

Kadmon, R. (1993) Population dynamic consequences of habitat heterogeneity: an experimental study. *Ecology*, **74**, 816–825.

Kadmon, R. and Shmida, A. (1990) Quantifying spatiotemporal demographic processes: an approach and a case study. *American Naturalist*, **135**, 382–397.

Kareiva, P. (1990) Population dynamics in spatially complex environments: theory and data. *Philosophical Transactions of the Royal Society of London, B*, **330**, 175–190.

Karlsson, P. S., Nordell, K. O., Carlsson, B. Å. and Svensson, B. M. (1991) The effect of soil nutrient status on prey utilization in four carnivorous plants. *Oecologia (Berl.)*, **86**, 1–7.

Keast, A. (1977) Mechanisms expanding niche width and minimizing intraspecific competition in two centrarchid fishes. *Evolutionary Biology*, **10**, 333–395.

Keast, A. (1990) Biogeography of the North American broad-leafed deciduous forest avifauna. In *Biogeography and ecology of forest bird communities*, ed. Keast, A., pp. 109–120. SPB Academic Publishing: The Hague, Netherlands.

Kennedy, C. R. (1984) Host–parasite interrelationships: strategies of coexistence and coevolution. In *Producers and scroungers. Strategies of exploitation and parasitism*, ed. Barnard, C. J., pp. 34–60. Croom Helm: London, UK.

Kephart, S. R. (1983) The partitioning of pollinators among the species of *Asclepias*. *Ecology*, **64**, 120–133.

Kerfoot, W. C. (1987) Cascading effects and indirect pathways. In *Predation: direct and indirect impacts on aquatic commmunities*, ed. Kerfoot, W. C. and Sih, A., pp. University Press of New England: Hanover, NH, USA.

Kishimoto, K. (1990) Coexistence of any number of species in the Lotka–Volterra competitive system over two patches. *Theoretical Population Biology*, **38**, 149–158.

Kleijnen, J. P. C. (1992) Sensitivity analysis of simulation experiments: regression analysis and statistical design. *Mathematics and Computers in Simulation*, **34**, 297–315.

Klopfer, P. H. (1963) Behavioral aspects of habitat selection: the role of early experience. *Wilson Bulletin*, **75**, 15–22.

Klopfer, P. H. and Ganzhorn, J. U. (1985) Habitat selection: behavioral aspects. In *Habitat selection in birds*, ed. Cody, M. L., pp. 436–453. Academic Press: Orlando, FL, USA.

Kochmer, J. P. and Handel, S. N. (1986) Constraints and competition in the evolution of flowering phenology. *Ecology*, **56**, 303–325.

Kodric-Brown, A., Brown, J. H., Byers, G. S. and Gori, D. F. (1984) Organization of a tropical island community of hummingbirds and flowers. *Ecology*, **65**, 1358–1368.

Kohn, A. J. and Nybakken, J. W. (1975) Ecology of *Conus* on eastern Indian Ocean fringing reefs: diversity of species and resource utilization. *Marine Biology*, **29**, 211–234.

Kolasa, J. (1989) Ecological systems in hierarchical perspective: breaks in community structure and other consequences. *Ecology*, **70**, 36–47.

Kotliar, N. B. and Wiens, J. A. (1990) Multiple scales of patchiness and patch structure: a hierarchical framework for the study of heterogeneity. *Oikos*, **59**, 253–260.

Kratter, A. W. (1993) Geographic variation in the Yellow-billed Cacique, *Amblycercus holosericeus*, a partial bamboo specialist. *The Condor*, **95**, 641–651.

Krzysik, A. J. (1979) Resource allocation, coexistence, and the niche structure of a salamander community. *Ecological Monographs*, **49**, 173–194.

Kuijt, J. (1969) *The biology of flowering parasitic plants*. University of California Press: Berkeley, CA, USA.

Kuris, A. M. (1974) Trophic interactions: similarity of parasitic castrators to parasitoids. *Quarterly Review of Biology*, **49**, 129–148.

Kuris, A. M. and Norton, S. F. (1985) Evolutionary importance of overspecialization: insect parasitoids as an example. *American Naturalist*, **126**, 387–391.

Lack, D. (1937) The psychological factor in bird distribution. *British Birds*, **31**, 130–136.

Lack, D. (1954) *The natural regulation of animal numbers*. Oxford University Press: London, UK.

Lacy, R. C. (1984) Predictability, toxicity, and trophic niche breadth in fungus feeding Drosophiladae (Diptera). *Ecological Entomology*, **9**, 43–54.

Lambers, H. and Poorter, H. (1992) Inherent variation in growth rate between higher plants: a search for physiological causes and ecological consequences. *Advances in Ecological Research*, **23**, 188–261.

Lamont, B. (1985) Host distribution, potassium content, water relations and control of two co-occurring mistletoe species. *Journal of the Royal Society of Western Australia*, **68**, 21–25.

Lamont, B. B. and Bergl, S. M. (1991) Water relations, shoot and root architecture, and phenology of three co-occurring *Banksia* species: no evidence for niche differentiation in the pattern of water use. *Oikos*, **60**, 291–298.

Lance, G. N. and Williams, W. T. (1967) Mixed-data classificatory programs. I. Agglomerative systems. *Australian Computing Journal*, **1**, 15–20.

Lande, R. (1982) A quantitative genetic theory of life history evolution. *Ecology*, **63**, 607–615.

Lande, R. (1993) Risks of extinction for demographic and environmental stochasticity and random catastrophes. *American Naturalist*, **142**, 911–927.

Landres, P. B. and MacMahon, J. A. (1983) Community organization of arboreal birds in some oak woodlands of western North America. *Ecological Monographs*, **53**, 183–208.

Largent, D. L., Sugihara, N. and Brinitzer, A. (1980) *Ammanita gemmatum*, a non host-specific mycorrhizal fungus of *Arctostaphylos manzanita*. *Mycologia*, **72**, 435–459.

Law, R. and Blackford, J. C. (1992) Self-assembling food webs: a global viewpoint of coexistence of species in Lotka–Volterra communities. *Ecology*, **73**, 567–578.

Law, R. and Morton, R. D. (1993) Alternative permanent states of ecological communities. *Ecology*, **74**, 1347–1361.

Law, R. and Watkinson, A. R. (1989) Competition. In *Ecological concepts. The contribution of ecology to an understanding of the natural world*, ed. Cherrett, J. M., Bradshaw, A. D., Goldsmith, F. B., Grubb, P. J. and Krebs, J. R., pp. 243–284. Blackwell Scientific Publications: Oxford, UK.

Lawlor, L. R. (1980) Overlap, similarity, and competition coefficients. *Ecology*, **61**, 245–251.

Lawlor, L. R. and Maynard Smith, J. (1976) The coevolution and stability of competing species. *American Naturalist*, **110**, 79–99.

Lawton, J. H. (1993) Range, population abundance and conservation. *Trends in Ecology and Evolution*, **8**, 409–413.

Lawton, J. H. and Pimm, S. L. (1978) Reply to Saunders. *Nature (Lond.)*, **272**, 190.

Lawton, J. H. and Strong, D. R. (1981) Community patterns and competition in folivorous insects. *American Naturalist*, **118**, 317–338.

Le V. dit Durell, S. E. A., Goss-Custard, J. D. and Caldow, R. W. G. (1993) Sex-related differences in diet and feeding method in the oystercatcher *Haematopus ostralagus*. *Journal of Animal Ecology*, **62**, 205–215.

Lechowicz, M., Schoen, D. J. and Bell, G. (1988) Environmental correlates of habitat distribution and fitness components in *Impatiens capensis* and *Impatiens pallida*. *Journal of Ecology*, **76**, 1043–1054.

Lemen, C. A. (1978) Seed size selection in heteromyids. *Oecologia (Berl.)*, **35**, 13–19.

León, J. A. and Tumpson, D. B. (1975) Competition between two species for two complementary or substitutable resources. *Journal of Theoretical Biology*, **50**, 185–201.

Leong, R. T-S. (1975) Metazoan parasites of fishes of Cold Lake, Alberta. University of Alberta: Doctoral thesis

Levene, H. (1953) Genetic equilibrium when more than one niche is available. *American Naturalist*, **87**, 331–333.

Levey, D. J. (1988) Spatial and temporal variation in Costa Rica fruit and fruit-eating bird abundance. *Ecological Monographs*, **58**, 251–269.

Levin, S. A. (1992) The problem of pattern and scale in ecology. *Ecology*, **73**, 1943–1967.

Levins, R. (1968) *Evolution in changing environments*. Princeton University Press: Princeton, NJ, USA.

Levins, R. (1979) Coexistence in a variable environment. *American Naturalist*, **114**, 765–783.

Levins, R. and MacArthur, R. H. (1966) The maintenance of genetic polymorphism in a spatially heterogeneous environment: variations on a theme by Howard Levene. *American Naturalist*, **100**, 585–589.

Levins, R. and MacArthur, R. H. (1969) An hypothesis to explain the incidence of monophagy. *Ecology*, **50**, 910–911.

Leviten, P. J. and Kohn, A. J. (1980) Microhabitat resource use, activity patterns, and episodic catastrophes: *Conus* on tropical intertidal reef rock beaches. *Ecological Monographs*, **50**, 55–75.

Lewin, R. (1986) Supply-side ecology. *Science*, **234**, 25–27.

Lewis, W. M. J. (1978) Comparison of temporal and spatial variation in the zooplankton of a lake by means of variance components. *Ecology*, **59**, 666–671.

Li, K. T., Wetterer, J. K. and Hairston, N. G., Jr (1985) Fish size, visual resolution, and prey selectivity. *Ecology*, **66**, 1729–1735.

Lidicker, W. Z., Jr (1975) The role of dispersal in the demography of small mammals. In *Small mammals: their productivity and population dynamics*, ed. Golley, F.B., Petrusewicz, K. and Ryszkowski, L., pp. 103–128. Cambridge University Press: Cambridge, UK.

Lima, S. L. (1987) Vigilance while feeding and its relation to the risk of predation. *Journal of Theoretical Biology*, **124**, 303–316.

Linton, L. R., Davies, R. W. and Wrona, F. J. (1981) Resource utilization indices: an appraisal. *Journal of Animal Ecology*, **50**, 282–292.

Lishman, G. S. (1985) The food and feeding ecology of Adelie and Chinstrap Penguins at Signy Island, South Orkney Islands. *Journal of Zoology (Lond.)*, **205**, 245–263.

Lister, B. C. (1981) Seasonal niche relationships of rain forest anoles. *Ecology*, **62**, 1548–1560.

Llewellyn, J. B. and Jenkins, S. H. (1987) Patterns of niche shifts in mice: seasonal changes in microhabitat breadth and overlap. *American Naturalist*, **129**, 365–381.

LoBue, C. P. and Bell, M. A. (1993) Phenotypic manipulation by the cestode parasite *Schistocephalus solidus* of its intermediate host, *Gasterosteus aculeatus*, the three-spined stickleback. *American Naturalist*, **142**, 725–735.

Loeschcke, V. (1985) Coevolution and invasion in competitive guilds. *American Naturalist*, **126**, 505–520.

Loman, J. (1986) Use of overlap indices as competition coefficients: tests with field data. *Ecological Modelling*, **34**, 231–243.

Łomnicki, A. (1988) *Population ecology of individuals*. Princeton University Press: Princeton, NJ, USA.

Łomnicki, A. and Ombach, J. (1984) Resource partitioning within a single species population and population stability: a theoretical model. *Theoretical Population Biology*, **24**, 21–28.

Lowry, L. F., McElroy, A. J. and Pearse, J. S. (1974) The distribution of six species of gastropod molluscs in a California kelp forest. *Biological Bulletin*, **147**, 386–396.

Loyn, R. H., Rummalls, R. G., Forward, G. Y. and Tyers, J. (1983) Territorial bell miners and other birds affecting populations of insect prey. *Science (Wash., D.C.)*, **221**, 1411–1413.

Lucas, J. R. (1990) Time scale and diet choice decisions. In *Behavioural mechanisms of food selection, 20*, ed. Hughes, R.N., pp. 165–184. Springer-Verlag: Berlin, Germany.

Lynch, M. and Gabriel, W. (1987) Environmental tolerance. *American Naturalist*, **129**, 283–303.

M'Closkey, R. T. (1978) Niche separation and assembly in four species of Sonoran Desert rodents. *American Naturalist*, **112**, 683–694.

Mac Nally, R. C. (1981) On the reproductive energetics of chorusing males: energy depletion profiles, restoration, and growth in two sympatric species of *Ranidella* (Anura). *Oecologia*, **51**, 181–188.

Mac Nally, R. C. (1983a) On assessing the significance of interspecific competition to guild structure. *Ecology*, **64**, 1646–1652.

Mac Nally, R. C. (1983b) Trophic relationships of two sympatric species of *Ranidella* (Anura). *Herpetologica*, **39**, 130–140.

Mac Nally, R. C. (1985) Habitat and microhabitat distributions in relation to ecological overlap in two species of *Ranidella* (Anura). *Australian Journal of Zoology*, **33**, 329–338.

Mac Nally, R. C. (1987) Population energetics and food limitation in two sympatric species of *Ranidella* (Anura). *Proceedings of the Royal Society of Victoria*, **99**, 1–12.

Mac Nally, R. C. (1988a) On the statistical significance of the Hutchinsonian size-ratio parameter. *Ecology*, **69**, 1974–1982.

Mac Nally, R. C. (1988b) Simultaneous modelling of distributional patterns in a guild of eastern-Australian cicadas. *Oecologia (Berl.)*, **74**, 246–253.

Mac Nally, R. C. (1989) The relationship between habitat breadth, habitat position, and abundance in forest and woodland birds along a continental gradient. *Oikos*, **54**, 44–54.

Mac Nally, R. C. (1990a) Modelling distributional patterns in woodland birds along a continental gradient. *Ecology*, **71**, 360–374.

Mac Nally, R. C. (1990b) The roles of floristics and physionomy in avian community composition. *Australian Journal of Ecology*, **15**, 321–327.

Mac Nally, R. C. (1994a) On characterizing foraging versatility, illustrated by using birds. *Oikos*, **69**, 95–106.

Mac Nally, R. C. (1994b) Habitat-specific guild structure of forest birds in south-eastern Australia: a regional scale perspective. *Journal of Animal Ecology*, **63**, 988–1001.

Mac Nally, R. C. (1995) A protocol for classifying regional dynamics, exemplified by using of woodland birds in south-eastern Australia. *Australian Journal of Ecology*, **20**, in press.

Mac Nally, R. C. and Doolan, J. M. (1986) An empirical approach to guild structure: habitat relationships in a guild of cicadas. *Oikos*, **47**, 433–446.

MacArthur, R. H. (1957) On the relative abundance of bird species. *Proceedings of the National Academy of Science USA*, **45**, 293–295.

MacArthur, R. H. (1958) Population ecology of some warblers of north-eastern coniferous forests. *Ecology*, **39**, 599–619.

MacArthur, R. H. (1969) Species packing, and what competition minimizes. *Proceedings of the National Academy of Science USA*, **64**, 1369–1371.

MacArthur, R. H. (1970) Species packing and competitive equilibrium for many species. *Theoretical Population Biology*, **1**, 1–11.

MacArthur, R. H. (1972) *Geographical ecology*. Harper & Row: New York, NY, USA.

MacArthur, R. H. and Levins, R. (1964) Competition, habitat selection, and character displacement in a patchy environment. *Proceedings of the National Academy of Science USA*, **51**, 1207–1210.

MacArthur, R. H. and Levins, R. (1967) The limiting similarity, convergence, and divergence of coexisting species. *American Naturalist*, **101**, 377–385.

MacArthur, R. H. and Pianka, E. R. (1966) On the optimal use of a patchy environment. *American Naturalist*, **100**, 603–609.

MacArthur, R. H., Recher, H. and Cody, M. L. (1966) On the relation between habitat selection and species diversity. *American Naturalist*, **100**, 319–332.

MacArthur, R. H. and Wilson, E. O. (1967) *The theory of island biogeography*. Princeton University Press: Princeton, NJ, USA.

Mahdi, A., Law, R. and Willis, A. J. (1989) Large niche overlaps among coexisting plant species in a limestone grassland community. *Journal of Ecology*, **77**, 386–400.

Maiorana, V. C. (1978a) Difference in diet as an epiphenomenon: space regulates salamanders. *Canadian Journal of Zoology*, **56**, 1017–1025.

Maiorana, V. C. (1978b) What kinds of plants do herbivores really prefer? *American Naturalist*, **112**, 631–635.

Malanson, G. P. (1985) Spatial autocorrelation and distributions of plant species on environmental gradients. *Oikos*, **45**, 278–280.

Malmquist, H. J. (1992) Phenotype-specific feeding behaviour of two arctic charr *Salvelinus alpinus* morphs. *Oecologia (Berl.)*, **92**, 354–361.

Mangel, M. and Clark, C. W. (1988) *Dynamic modeling in behavioral ecology*. Princeton University Press: Princeton, NJ, USA.

Mann, K. H. (1965) Energy transformations by a population of fish in the River Thames. *Journal of Animal Ecology*, **34**, 253–273.

Martinez, N. D. (1991) Artifacts or attributes? Effects of resolution on the Little Rock Lake food web. *Ecological Monographs*, **61**, 367–392.

Matessi, C. and Gatto, M. (1984) Does *K*-selection imply prudent predation? *Theoretical Population Biology*, **25**, 347–363.

Matessi, C. and Jayakar, S. D. (1981) Coevolution of species in competition: a theoretical study. *Proceedings of the National Academy of Science USA*, **78**, 1081–1084.

Maurer, B. A. (1984) Interference and exploitation in bird communities. *Wilson Bulletin*, **96**, 380–395.

Maurer, B. A. (1985) Avian community dynamics in desert grasslands: observational scale and hierarchical structure. *Ecological Monographs*, **55**, 295–312.

May, R. M. (1973) *Stability and complexity in model ecosystems.* Princeton University Press: Princeton, NJ, USA.

May, R. M. (1974a) Ecosystem patterns in randomly fluctuating environments. *Progress in Theoretical Biology*, **3**, 1–50.

May, R. M. (1974b) On the theory of niche overlap. *Theoretical Population Biology*, **5**, 297–332.

May, R. M. (1975) Some notes on estimating the competition matrix, α. *Ecology*, **56**, 737–741.

May, R. M. (1978) The dynamics and diversity of insect faunas. In *Diversity of insect faunas*, ed. Mound, L. A. and Waloff, N., pp. Blackwell: Oxford, UK.

May, R. M. (ed.) (1981) *Theoretical ecology*. 2nd edition. Blackwell Scientific: Oxford, UK.

May, R. M. and MacArthur, R. H. (1972) Niche overlap as a function of environmental variability. *Proceedings of the National Academy of Science USA*, **69**, 1109–1113.

May, R. M. and Southwood, T. R. E. (1990) Introduction. In *Living in a patchy environment*, ed. Shorrocks, B. and Swingland, I. R., pp. 219–235. Oxford University Press: Oxford, UK.

Mayr, E. (1976) *Evolution and the diversity of life*. Harvard University Press: Cambridge, MA, USA.

McCauley, E., Wilson, W. G. and de Roos, A. M. (1993) Dynamics of age-structured and spatially structured predator-prey interactions: individual-based models and population-level formulations. *American Naturalist*, **142**, 412–442.

McEvoy, P. B. (1986) Niche partitioning in spittlebugs (Homoptera: Cercopidae) sharing shelters on host plants. *Ecology*, **67**, 465–478.

McIvor, C. C. and Odum, W. E. (1988) Food, predation risk, and microhabitat selection in a marsh fish assemblage. *Ecology*, **69**, 1341–1351.

McLaughlin, J. F. and Roughgarden, J. (1992) Predation across spatial scales in heterogeneous environments. *Theoretical Population Biology*, **41**, 277–299.

McLean, R. C. and Ivimey-Cook, W. R. (1973) *Textbook of theoretical botany*, Vol. 4. Longman: London, UK.

McMurtrie, R. (1976) On the limit to niche overlap for nonuniform niches. *Theoretical Population Biology*, **10**, 96–107.

McNamara, J. M. and Houston, A. I. (1986) The common currency for behavioral decisions. *American Naturalist*, **127**, 358–378.

McNaughton, S. J. and Wolf, L. L. (1970) Dominance and the niche in ecological systems. *Science (Wash., D.C.)*, **167**, 131–139.

Mercurio, K. S., Palmer, A. R. and Lowell, R. B. (1985) Predator-mediated microhabitat partitioning by two species of visually cryptic, intertidal limpets.

Ecology, **66**, 1417–1425.

Meserve, P. L. (1976) Habitat and resource utilization by rodents of a California coastal scrub community. *Journal of Animal Ecology*, **45**, 647–666.

Meserve, P. L. (1981a) Resource partitioning in a Chilean semi-arid small mammal community. *Journal of Animal Ecology*, **50**, 745–757.

Meserve, P. L. (1981b) Trophic relationships among small mammals in a Chilean thorn scrub community. *Journal of Mammalogy*, **62**, 304–314.

Meyer, A. (1987) Phenotypic plasticity and heterochrony in *Cichlasoma managuense* and their implications for speciation in cichlid fishes. *Evolution*, **41**, 1357–1369.

Michaud, J. P. (1992) Further considerations on host range evolution in herbivorous insects. *Oikos*, **64**, 587–590.

Milbrath, L. R., Tauber, M. J. and Tauber, C. A. (1993) Prey specificity in *Chrysopa*: an interspecific comparison of larval feeding and defensive behavior. *Ecology*, **74**, 1384–1393.

Miller, J. C. (1980) Niche relationships among parasitic insects occurring in a temporary habitat. *Ecology*, **61**, 270–275.

Miller, J. S. (1992) Host-plant associations among prominent moths. *BioScience*, **42**, 50–55.

Miller, R. S. (1967) Pattern and process in competition. *Advances in Ecological Research*, **4**, 1–74.

Milligan, B. G. (1986) Invasion and coexistence of two phenotypically variable species. *Theoretical Population Biology*, **30**, 245–270.

Mills, N. J. (1992) Parasitoid guilds, lifestyles and host ranges in the parasitoid complexes of tortricoid hosts (Lepidoptera: Tortricoidea). *Environmental Entomology*, **21**, 230–239.

Mills, N. J. (1993) Species richness and structure in the parasitoid complexes of tortricoid hosts. *Journal of Animal Ecology*, **62**, 45–58.

Milne, B. T., Turner, M. G., Wiens, J. A. and Johnson, A. R. (1992) Interactions between the fractal geometry of landscapes and allometric herbivory. *Theoretical Population Biology*, **41**, 337–353.

Mitchell, J. C. (1979) Ecology of southeastern Arizona whiptail lizards (*Cnemidiphorus*: Teiidae): population densities. *Canadian Journal of Zoology*, **57**, 1487–1499.

Mitchell, M. J. (1978) Vertical and horizontal distributions of orbatid mites (Acari: Cryptostigmata) in an aspen woodland soil. *Ecology*, **59**, 516–525.

Mitter, C., Farrell, B. and Wiegmann, B. (1988) The phylogenetic study of adaptive zones: has phytophagy promoted insect diversification? *American Naturalist*, **132**, 107–128.

Mooney, H. A., Winner, W. E. and Pell, E. J. (ed.) (1991) *Response of plants to multiple stresses*. Academic Press: New York, NY, USA.

Moran, N. A. (1988) The evolution of host-plant alternation in aphids: evidence for specialization as a dead end. *American Naturalist*, **132**, 681–706.

Moran, N. A. (1992) The evolutionary maintenance of alternative phenotypes. *American Naturalist*, **139**, 971–989.

Morisita, M. (1959) Measuring of interspecific association and similarity between communities. *Memoirs of the Faculty of Science, Kyushu Univ., Series E (Biology)*, **3**, 65–80.

Morris, D. W. (1987a) Ecological scale and habitat use. *Ecology*, **68**, 362–369.

Morris, D. W. (1987b) Spatial scale and the cost of density-dependent habitat selection. *Evolutionary Ecology*, **1**, 379–388.

Morris, D. W. (1989) Density-dependent habitat selection: testing the theory with fitness data. *Evolutionary Ecology*, **3**, 80–94.

Morris, D. W. (1992) Scales and costs of habitat selection in heterogeneous landscapes. *Evolutionary Ecology*, **6**, 412–432.

Morse, D. H. (1974) Niche breadth as a function of social dominance. *American Naturalist*, **108**, 818–830.

Mosse, B. (1973) Advances in the study of vesicular-arbuscular mycorrhiza. *Annual Review of Phytopathology*, **11**, 171–195.

Mountainspring, S. and Scott, J. M. (1985) Interspecific competition among Hawaiian forest birds. *Ecological Monographs*, **55**, 219–239.

Moyle, P. B. and Vondracek, B. (1985) Persistence and structure of the fish assemblage in a small California stream. *Ecology*, **66**, 1–13.

Mueller, L. D. and Altenberg, L. (1985) Statistical inference on measures of niche overlap. *Ecology*, **66**, 1204–1210.

Mueller, L. D. and Ayala, F. J. (1981) Dynamics of single-species population growth: experimental and statistical analysis. *Theoretical Population Biology*, **20**, 101–117.

Muotka, T. (1990) Coexistence in a guild of filter-feeding caddis larvae: do different instars act as different species? *Oecologia (Berl.)*, **85**, 281–292.

Murphy, G. I. (1968) Patterns in life history and the environment. *American Naturalist*, **102**, 390–404.

Murray, B. G., Jr (1981) The origins of adaptive interspecific territorialism. *Biological Reviews (Cambridge)*, **56**, 1–22.

Murray, B. G., Jr (1986) The structure of theory, and the role of competition in community dynamics. *Oikos*, **46**, 145–158.

Murray, M. G. and Gerrard, R. (1985) Putting the challenge into resource exploitation: a model of contest competition. *Journal of Theoretical Biology*, **115**, 367–389.

Mushinsky, H. R. and Hebrard, J. J. (1977) The use of time by sympatric water snakes. *Canadian Journal of Zoology*, **55**, 1545–1550.

Mushinsky, H. R., Hebrard, J. J. and Vodopich, D. S. (1982) Ontogeny of water snake foraging ecology, *Ecology*, **63**, 1624–1629.

Myers, A. A. and Giller, P. S. (ed.) (1988) *Analytical biogeography: an integrated approach to the study of animal and plant distributions*. Chapman and Hall: London, UK.

Nee, S., Read, A. F., Greenwood, J. J. D. and Harvey, P. H. (1991) The relationship between abundance and body size in British birds. *Nature (Lond.)*, **351**, 312–313.

Neuweiter, G. (1989) Foraging ecology and audition in echolocating bats. *Trends in Ecology and Evolution* **4**, 160–166.

Newman, E. I. (ed.) (1982) *The plant community as a working mechanism*. Blackwell Scientific Publications: Oxford, UK.

Niemelä, J. (1993) Interspecific competition in ground-beetle assemblages (Carabidae): what have we learned? *Oikos*, **66**, 325–335.

Nisbet, R. M. and Gurney, W. S. C. (1982) *Modeling fluctuating populations*. Wiley Interscience: New York, NY, USA.

Nisbet, R. M., Gurney, W. S. C. and Pettipher, M. A. (1978) Environmental fluctuations and the theory of the ecological niche. *Journal of Theoretical Biology*, **75**, 223–237.

Nitao, J. K., Ayres, M. P., Lederhouse, R. C. and Scriber, J. M. (1991) Larval adaptation to lauraceous hosts: geographic divergence of the spicebush swallowtail butterfly. *Ecology*, **72**, 1428–1435.

Noble, I. R. (1989) Ecological traits of the *Eucalyptus* L'Hérit. subgenera *Monocalyptus* and *Symphyomyrtus*. *Australian Journal of Botany*, **37**, 207–224.

Nold, A. (1979) Competitive overlap and coexistence. *Theoretical Population Biology*, **15**, 232–245.

Norbury, G. L. and Sanson, G. D. (1992) Problems with measuring diet selection of terrestrial, mammalian herbivores. *Australian Journal of Ecology*, **17**, 1–7.

Nudds, T. D. (1983) Niche dynamics and organization of waterfowl guilds in variable environments. *Ecology*, **64**, 319–330.

Nunney, L. (1980) Density compensation, isocline shape and single-level competition models. *Journal of Theoretical Biology*, **86**, 323–349.

Nunney, L. (1981) Interactive competition models and isocline shape. *Mathematical Biosciences*, **56**, 77–110.

O'Brien, W. J., Slade, N. A. and Vinyard, G. L. (1976) Apparent size as the determinant of prey selection by bluegill sunfish (*Lepomis macrochirus*). *Ecology*, **57**, 1304–1310.

O'Connor, R. J. (1987) Organization of avian assemblages – the influence of intraspecific habitat dynamics. *British Ecological Society Symposium*, **27**, 163–183.

O'Dowd, D. J. and Williamson, G. B. (1979) Stability conditions in plant defence guilds. *American Naturalist*, **114**, 379–382.

Oaten, A. and Murdoch, W. W. (1975) Switching, functional response, and stability in predator–prey systems. *American Naturalist*, **109**, 299–318.

Oksanen, L. (1991a) A century of community ecology: how much progress? *Trends in Ecology and Evolution*, **6**, 294–296.

Oksanen, L. (1991b) Trophic levels and trophic dynamics: a concensus emerging? *Trends in Ecology and Evolution*, **6**, 58–60.

Opdam, P. (1991) Metapopulation theory and habitat fragmentation: a review of holarctic breeding bird studies. *Landscape Ecology*, **5**, 93–106.

Orians, G. H. and Wittenberger, J. F. (1991) Spatial and temporal scales in habitat selection. *American Naturalist*, **137**, S29–S49.

Osborne, B. A. and Whittington, W. J. (1981) Eco-physiological aspects of interspecific and seasonal variation in nitrate utilization in the genus *Agrostris*. *New Phytologist*, **87**, 595–614.

Owen-Smith, N. (1988) *Megaherbivores: the influence of very large body size on ecology*. Cambridge University Press: Cambridge, UK.

Owen-Smith, N. and Novellie, P. (1982) What should a clever ungulate eat? *American Naturalist*, **119**, 151–178.

Pacala, S. W. and Roughgarden, J. (1982) The evolution of resource partitioning in a multidimensional resource space. *Theoretical Population Biology*, **22**, 127–145.

Pagel, M. D., May, R. M. and Collie, A. R. (1991) Ecological aspects of the geographical distribution and diversity of mammalian species. *American Naturalist*, **137**, 791–815.

Pahl-Wostl, C. (1993) Food webs and ecological networks across temporal and spatial scales. *Oikos*, **66**, 415–432.
Paine, R. T. (1984) Ecological determinism in the competition for space. *Ecology*, **65**, 1339–1348.
Paine, R. T. (1988) Food webs: road maps of interactions or grist for theoretical development? *Ecology*, **69**, 1648–1654.
Paine, R. T. (1992) Food-web analysis through field measurement of per capita interaction strength. *Nature (Lond.)*, **355**, 73–75.
Palmer, M. W. (1992) The coexistence of species in fractal landscapes. *American Naturalist*, **139**, 375–397.
Parrish, J. A. and Bazzaz, F. A. (1976) Underground niche separation in successional plants. *Ecology*, **57**, 1281–1288.
Parrish, J. A. D. and Bazzaz, F. A. (1979) Differences in pollination niche relationships in early and late successional plant communities. *Ecology*, **60**, 597–610.
Parrish, J. A. D. and Bazzaz, F. A. (1985) Ontogenetic niche shifts in old-field annuals. *Ecology*, **66**, 1296–1302.
Parsons, P. A. (1978) Boundary conditions for *Drosophila* resource utilization in temperate regions, especially at low temperatures. *American Naturalist*, **112**, 1063–1074.
Paul, V. J. and Hay, M. E. (1986) Seaweed susceptibility to herbivory: chemical and morphological correlates. *Marine Ecology Progress Series*, **33**, 255–264.
Pearson, D. L. and Mury, E. J. (1979) Character divergence and convergence among tiger beetles (Coleoptera: Cincindelidae). *Ecology*, **60**, 557–566.
Pease, C. M., Lande, R. and Bull, J. J. (1989) A model of population growth, dispersal, and evolution in a changing environment. *Ecology*, **70**, 1657–1664.
Peckarsky, B. L., Cowan, C. A. and Anderson, C. R. (1994) Consequences and plasticity of the specialized predatory behavior of stream-dwelling stonefly larvae. *Ecology*, **75**, 166–181.
Pellmyr, O. (1989) The cost of mutualism: interactions between *Trollis europaeus* and its pollinating parasites. *Oecologia (Berl.)*, **78**, 53–59.
Pennings, S. C., Nadeau, M. T. and Paul, V. J. (1993) Selectivity and growth of the generalist herbivore *Dolabella auricularia* feeding upon complementary resources. *Ecology*, **74**, 879–890.
Pennings, S. C. and Paul, V. J. (1992) Effect of plant toughness, calcification, and chemistry on herbivory by *Dolabella auricularia*. *Ecology*, **73**, 1606–1619.
Persson, L. (1983) Food consumption and the significance of detritus and algae to intraspecific competition in roach *Rutilus rutilus* in a shallow eutrophic lake. *Oikos*, **41**, 118–125.
Persson, L. (1985) Asymmetrical competition: are larger animals competitively superior? *American Naturalist*, **126**, 261–266.
Persson, L. and Greenberg, L. A. (1990) Juvenile competitive bottlenecks: the perch (*Perca fluviatilis*)–roach (*Rutilus rutilus*) interaction. *Ecology*, **71**, 44–56.
Peters, R. H. (1976) Tautology in evolution and ecology. *American Naturalist*, **110**, 1–12.
Peters, R. H. (1983) *The ecological implications of body size*. Cambridge University Press: Cambridge, UK.

Peters, R. H. (1988) Some general problems for ecology illustrated by food web theory. *Ecology*, **69**, 1673–1676.

Peters, R. H. (1991) *A critique for ecology*. Cambridge University Press: Cambridge, UK.

Petraitis, P. S. (1979) Likelihood measures of niche breadth and overlap. *Ecology*, **60**, 703–710.

Petraitis, P. S. (1981) Algebraic and graphical relationships among niche breadth measures. *Ecology*, **62**, 545–548.

Petraitis, P. S. (1983) Presentation of niche measure relationships when more than three resource classes are involved. *Ecology*, **64**, 1318–1320.

Petraitis, P. S. (1985) The relationship between likelihood niche measures and replicated tests for goodness-of-fit. *Ecology*, **66**, 1983–1985.

Pfennig, D. W. (1992) Proximate and functional causes of polyphenism in an anuran tadpole. *Functional Ecology*, **6**, 167–174.

Phipps, M. J. (1992) From local to global: the lesson of cellular automata. In *Individual-based models and approaches in ecology. Populations, communities and ecosystems*, ed. DeAngelis, D. L. and Gross, L. J., pp. 165–187. Chapman and Hall: New York, NY, USA.

Pianka, E. R. (1973) The structure of lizard communities. *Annual Review of Ecology and Systematics*, **4**, 53–74.

Pianka, E. R. (1974) Niche overlap and diffuse competition. *Proceedings of the National Academy of Science U.S.A.*, **71**, 2141–2145.

Pianka, E. R. (1975) Niche relations of desert lizards. In *Ecology and evolution of communities*, ed. Diamond, J. and Cody, M. L., pp. 292–314. Belknap Press of Harvard University: Cambridge, MA, USA.

Pielou, E. C. (1975) *Ecological diversity*. Wiley: New York, NY, USA.

Pierce, N. E. (1989) Butterfly–ant mutualisms. In *Towards a more exact ecology*, ed. Grubb, P. J. and Whittaker, J. B., pp. 299–324. Blackwell Scientific Publications: Oxford, UK.

Pierce, N. E. and Elgar, M. A. (1985) The influence of ants on host plant selection by *Jalmenus evagoras*, a myrmecophilous lycaenid butterfly. *Behavioral Ecology and Sociobiology*, **16**, 209–222.

Pimm, S. L. (1982) *Food webs*. Chapman and Hall: London, UK.

Pimm, S. L. (1984) The complexity and stability of ecosystems. *Nature (Lond.)*, **307**, 321–326.

Pimm, S. L., Jones, H. L. and Diamond, J. (1988) On the risk of extinction. *American Naturalist*, **132**, 757–785.

Pimm, S. L. and Kitching, R. L. (1988) Food web patterns: trivial flaws or the basis of an active research program? *Ecology*, **69**, 1669–1672.

Pimm, S. L. and Lawton, J. H. (1978) On feeding on more than one trophic level. *Nature (Lond.)*, **275**, 542–544.

Pimm, S. L. and Pimm, J. W. (1982) Resource use, competition, and resource availability in Hawaiian honeycreepers. *Ecology*, **63**, 1468–1480.

Pimm, S. L. and Redfearn, A. (1988) The variability of population densities. *Nature*, **334**, 613–614.

Pimm, S. L. and Rosenzweig, M. L. (1981) Competitors and habitat use. *Oikos*, **37**, 1–6.

Pindyck, R. S. and Rubinfeld, D. L. (1981) *Econometric models and economic forecasts*,

2nd edn. McGraw-Hill Book Company: Singapore.

Platt, T. and Denman, K. L. (1975) Spectral analysis in ecology. *Annual Review of Ecology and Systematics*, **6**, 189–210.

Platt, W. J. and Weis, I. M. (1977) Resource partitioning and competition within a guild of fugitive prairie plants. *American Naturalist*, **111**, 479–513.

Polansky, P. (1979) Invariant distributions for multi-population models in random environments. *Theoretical Population Biology*, **16**, 25–34.

Polis, G. A. (1984) Age structure component of niche width and intraspecific resource partitioning: can age groups function as ecological species? *American Naturalist*, **123**, 541–564.

Polis, G. A. (1991) Complex trophic interactions in deserts: an empirical critique of food web theory. *American Naturalist*, **138**, 123–155.

Polis, G. A. and McCormick, S. J. (1987) Intraguild predation and competition among desert scorpions. *Ecology*, **68**, 332–343.

Polis, G. A., Myers, C. A. and Holt, R. D. (1989) The ecology and evolution of intraguild predation: potential competitors that eat each other. *Annual Review of Ecology and Systematics*, **20**, 297–330.

Pomerantz, M. J. and Gilpin, M. E. (1979) Community covariance and coexistence. *Journal of Theoretical Biology*, **79**, 67–81.

Possingham, H. P., Davies, I., Noble, I. R. and Norton, T. W. (1992) A metapopulation simulation model for assessing the likelihood of plant and animal extinctions. *Mathematics and Computers in Simulation*, **33**, 367–372.

Poulin, B., Lefebvre, G. and McNeil, R. (1992) Tropical avian phenology in relation to abundance and exploitation of food resources. *Ecology*, **73**, 2295–2309.

Powell, J. A. (1980) Evolution of larval food preferences in microlepidoptera. *Annual Review of Entomology*, **25**, 133–159.

Powell, J. R. (1971) Genetic polymorphism in varied environments. *Science (Wash., D.C.)*, **174**, 1035–1036.

Powell, T. and Richerson, P. J. (1985) Temporal variation, spatial heterogeneity, and competition for resources in plankton systems: a theoretical model. *American Naturalist*, **125**, 431–464.

Preston, C. R. (1990) Distribution of raptor foraging in relation to prey biomass and habitat structure. *The Condor*, **92**, 107–112.

Preston, F. W. (1962) The canonical distribution of commonness and rarity. *Ecology*, **43**, 185–215.

Price, P. W. (1980) *The evolutionary biology of parasites*. Princeton University Press: Princeton, NJ, USA.

Price, P. W. (1983) Hypotheses on organization and evolution in herbivorous insect communities. In *Variable plants and herbivores in natural and managed systems*, ed. Denno, R.F. and McClure, M.S., pp. 559–598. Academic Press: New York, NY, USA.

Prout, T. (1986) The delayed effect on fertility of preadult competition: two-species population dynamics. *American Naturalist*, **127**, 809–818.

Provenza, F. D. and Balph, D. F. (1990) Applicability of five diet-selection models to various foraging challenges ruminants encounter. In *Behavioural mechanisms of food selection, 20*, ed. Hughes, R.N., pp. 423–459. Springer-Verlag: Berlin, Germany.

Pulliam, H. R. (1985) Foraging efficiency, resource partitioning, and the coexistence of sparrow species. *Ecology*, **60**, 1829–1836.

Pulliam, H. R. (1986) Niche expansion and contraction in a variable environment. *American Zoologist*, **26**, 71–79.

Pulliam, H. R. (1988) Sources, sinks, and population regulation. *American Naturalist*, **132**, 652–661.

Pulliam, H. R. and Danielson, B. J. (1991) Sources, sinks and habitat selection: a landscape perspective on population dynamics. *American Naturalist*, **137**, S50–S66.

Pyke, G. H. (1984) Optimal foraging theory: a critical review. *Annual Review of Ecology and Systematics*, **15**, 523–576.

Pyke, G. H., Pulliam, H. R. and Charnov, E. L. (1977) Optimal foraging: a selective review of theory and tests. *Quarterly Review of Biology*, **52**, 137–154.

Quinn, G. P. and Keough, M. J. (1993) Potential effect of enclosure size on field experiments with herbivorous intertidal gastropods. *Marine Ecology Progress Series*, **98**, 199–201.

Rabinowitz, D. (1981) Seven forms of rarity. In *The biological aspects of rare plant conservation*, ed. Synge, H., pp. 205–217. Wiley: Chichester, UK.

Rabinowitz, D., Cairns, S. and Dillon, T. (1986) Seven forms of rarity and their frequency in the flora of the British Isles. In *Conservation biology: the science of scarcity and diversity*, ed. Soul–, M.E., pp. 182–204. Sinauer: Sunderland, MA, USA.

Rahel, F. J. (1990) The hierarchical nature of community persistence: a problem of scale. *American Naturalist*, **136**, 328–344.

Raley, C. M. and Anderson, S. H. (1990) Availability and use of arthropod food resources by Wilson's Warblers and Lincoln's Sparrows in southeastern Wyoming. *The Condor*, **92**, 141–150.

Rapoport, E. H. (1982) *Areography*. Pergamon Press: Oxford, UK.

Rausher, M. D. (1984) The evolution of habitat preference in subdivided populations. *Evolution*, **38**, 596–608.

Rausher, M. D. (1988) Is coevolution dead? *Ecology*, **69**, 898–901.

Rayner, A. D. M. and Todd, N. K. (1979) Population and community structure and dynamics of fungi in decaying wood. *Advances in Botanical Research*, **7**, 334–420.

Real, L. A. (1975) A general analysis of resource allocation by competing individuals. *Theoretical Population Biology*, **8**, 1–11.

Recher, H. F. (1989) Foraging segregation of Australian warblers (Acanthizidae) in open forests near Sydney, New South Wales. *Emu*, **89**, 204–215.

Rees, M. J. (1987) The emergence of structure in the universe: galaxy formation and dark matter. In *Three hundred years of gravitation*, ed. Hawking, S. and Israel, W., pp. 459–498. Cambridge University Press: Cambridge, UK.

Remane, A. and Schlieper, C. (1971) *Biology of brackish water*. Wiley Interscience: New York, NY, USA.

Remsen, J. V., Jr, Stiles, F. G. and Scott, P. E. (1986) Frequency of arthropods in stomachs of tropical hummingbirds. *Auk*, **103**, 436–441.

Rescigno, A. and Richardson, I. W. (1965) On the competitive exclusion principle. *Bulletin of Mathematical Biophysics Supplement*, **27**, 85–89.

Rhoades, D. F. and Cates, R. G. (1976) Toward a general theory of plant antiherbivore chemistry. In *Biochemical interactions between plants and insects, Vol.*

10, ed. Wallace, J.W. and Mansell, R.L., pp. 168–213. Plenum: New York, NY, USA.

Rice, J., Anderson, B. W. and Ohmart, R. D. (1980) Seasonal habitat selection by birds in the lower Colorado valley. *Ecology*, **61**, 1402–1411.

Rice, K. J. and Menke, J. W. (1985) Competitive reversals and environment-dependent resource partitioning in *Erodium*. *Oecologia (Berl.)*, **67**, 430–434.

Ricklefs, R. E. (1987) Community diversity: relative roles of local and regional processes. *Science (Wash., DC)*, **235**, 167–171.

Ricklefs, R. E. and Lau, M. (1980) Bias and dispersion of overlap indices: results of some Monte Carlo simulations. *Ecology*, **61**, 1019–1024.

Riechert, S. E. (1991) Prey abundance vs diet breadth in a spider test system. *Evolutionary Ecology*, **5**, 327–338.

Riechert, S. E. and Cady, A. B. (1983) Patterns of resource use and tests for competitive release in a spider community. *Ecology*, **64**, 899–913.

Robinson, S. K., Terborgh, J. and Munn, C. A. (1990) Lowland tropical forest bird communities of a site in Western Amazonia. In *Biogeography and ecology of forest bird communities*, ed. Keast, A., pp. 229–258. SPB Academic Publishing: The Hague, Netherlands.

Rogers, A. R. (1986) Population dynamics under exploitation competition. *Journal of Theoretical Biology*, **119**, 363–368.

Rohde, K. (1979) A critical evaluation of intrinsic and extrinsic factors responsible for niche restriction in parasites. *American Naturalist*, **114**, 648–671.

Rosenthal, G. A. and Berenbaum, M. R. (ed.) (1991) *Herbivores: their interactions with secondary plant metabolites. 1. The chemical participants*. 2nd edn. Academic Press: New York, NY, USA.

Rosenthal, G. A. and Janzen, D. H. (1979) *Herbivores: their interaction with secondary plant metabolites*. Academic Press: New York, NY, USA.

Rosenzweig, M. L. (1966) Community structure in sympatric carnivora. *Journal of Mammalogy*, **47**, 602–612.

Rosenzweig, M. L. (1981) A theory of habitat selection. *Ecology*, **62**, 327–335.

Rosenzweig, M. L. (1985) Some theoretical aspects of habitat selection. In *Habitat selection in birds*, ed. Cody, M.L., pp. 517–540. Academic Press: Orlando, FL, USA.

Rosenzweig, M. L. (1991) Habitat selection and population interactions: the search for mechanism. *American Naturalist*, **137**, S5–S28.

Rosenzweig, M. L. and MacArthur, R. H. (1963) Graphical representation and stability conditions of predator–prey interactions. *American Naturalist*, **97**, 209–223.

Rotenberry, J. T. (1985) The role of habitat in avian community composition: physiognomy or floristics? *Oecologia*, **67**, 213–217.

Roughgarden, J. (1972) The evolution of niche width. *American Naturalist*, **106**, 683–718.

Roughgarden, J. (1974a) Niche width: biogeographic patterns among *Anolis* lizard populations. *American Naturalist*, **108**, 429–442.

Roughgarden, J. (1974b) Species packing and the competition function with illustrations from coral reef fish. *Theoretical Population Biology*, **5**, 163–186.

Roughgarden, J. (1979) *Theory of population genetics and evolutionary ecology: an introduction*. Macmillan: New York, NY, USA.

Roughgarden, J. (1983) Competition and theory in community ecology. *American Naturalist*, **122**, 538–601.

Roughgarden, J. (1989) The structure and assembly of communities. In *Perspectives in ecological theory*, ed. Roughgarden, J., May, R.M. and Levin, S.A., pp. 203–226. Princeton University Press: Princeton, NJ, USA.

Ruiz, A. and Heed, W. B. (1988) Host-plant specificity in the cactophilic *Drosophila mulleri* species complex. *Journal of Animal Ecology*, **57**, 237–249.

Rummel, J. D. and Roughgarden, J. (1985) A theory of faunal buildup for competition communities. *Evolution*, **39**, 1009–1033.

Rusterholtz, K. A. (1981a) Niche overlap among foliage-gleaning birds: support to Pianka's niche overlap hypothesis. *American Naturalist*, **117**, 395–399.

Rusterholtz, K. A. (1981b) Competition and the structure of an avian foraging guild. *American Naturalist*, **118**, 173–190.

Ryti, R. T. (1987) Allocation of energy to reproduction: effects of competition and population structure. *Journal of Theoretical Biology*, **128**, 499–512.

Sabath, M. D. (1974) Niche breadth and genetic variability in sympatric natural populations of drosophilid flies. *American Naturalist*, **108**, 533–540.

Sale, P. F. (1990) Recruitment of marine species: is the bandwagon rolling in the right direction? *Trends in Ecology and Evolution*, **5**, 25–26.

Sanders, I. R. (1993) Temporal infectivity and specificity of vesicular-arbuscular mycorrhizas in co-existing grassland species. *Oecologia (Berl.)*, **93**, 349–355.

Sano, M. (1990) Patterns of habitat and food utilization in two coral-reef sandperches (Mugiloididae): competitive or noncompetitive coexistence. *Journal of Experimental Marine Biology and Ecology*, **140**, 209–223.

Scheiner, S. M. (1993) Genetics and the evolution of phenotypic plasticity. *Annual Review of Ecology and Systematics*, **24**, 35–68.

Schemske, D. W. (1983) Limits to specialization and coevolution in plant–animal mutualisms. In *Coevolution*, ed. Nitecki, M.H., pp. 67–109. Chicago University Press: Chicago, IL, USA.

Schemske, D. W. and Brokaw, N. (1981) Treefalls and the distribution of understory birds in a tropical forest. *Ecology*, **62**, 938–945.

Schimel, D. S., Kittel, T. G. F., Knapp, A. K., Seastedt, T. R., Parton, W. J. and Brown, V. B. (1991) Physiological interactions along resource gradients in a tallgrass prairie. *Ecology*, **72**, 672–684.

Schlichting, C. D. (1986) The evolution of phenotypic plasticity in plants. *Annual Review of Ecology and Systematics*, **17**, 667–693.

Schluter, D. (1988) The evolution of finch communities on islands and continents: Kenya vs Galapágos. *Ecological Monographs*, **58**, 229–249.

Schoener, T. W. (1968) Sizes of feeding territories among birds. *Ecology*, **49**, 123–140.

Schoener, T. W. (1970) Nonsynchronous spatial overlap of lizards in patchy habitats. *Ecology*, **51**, 408–418.

Schoener, T. W. (1971) Theory of feeding strategies. *Annual Review of Ecology and Systematics*, **2**, 369–403.

Schoener, T. W. (1974a) Resource partitioning in ecological communities. *Science*, **185**, 27–39.

Schoener, T. W. (1974b) Some methods for calculating competition coefficients from resource-utilization spectra. *American Naturalist*, **108**, 332–340.

Schoener, T. W. (1976) Alternatives to Lotka–Volterra competition: models of intermediate complexity. *Theoretical Population Biology*, **10**, 309–333.

Schoener, T. W. (1983) Field experiments on interspecific competition. *American Naturalist*, **122**, 240–285.

Schoener, T. W. (1989a) The ecological niche. In *Ecological concepts: the contribution of ecology to an understanding of the natural world*, ed. Cherrett, J.M., & Bradshaw, A.D., pp. 79–114. Blackwell Scientific Publications: Oxford, UK.

Schoener, T. W. (1989b) Food webs from the small to the large. *Ecology*, **70**, 1559–1589.

Schoener, T. W. and Adler, G. H. (1991) Greater resolution of distributional complementarities by controlling for habitat affinities: a study with Bahamian lizards and birds. *American Naturalist*, **137**, 669–692.

Schoener, T. W. and Gorman, G. C. (1968) Some niche differences in three lesser Antillean lizards of the genus *Anolis*. *Ecology*, **49**, 819–830.

Schoener, T. W., Huey, R. B. and Pianka, E. R. (1979) A biogeographic extension of the compression hypothesis: competitors in narrow sympatry. *American Naturalist*, **113**, 295–317.

Schoener, T. W. and Spiller, D. A. (1987) High population persistence in a system with high turnover. *Nature (Lond.)*, **330**, 474–477.

Scott, J. A. and McClelland, G. A. H. (1977) A model of polymorphism with several seasons and several habitats, and its application to the mosquito *Aedes aegypti*. *Theoretical Population Biology*, **11**, 342–355.

Seagle, S. W. and McCracken, G. F. (1986) Species abundance, niche position, and niche breadth for five terrestrial animal assemblages. *Ecology*, **67**, 816–828.

Seagle, S. W. and Shugart, H. H. (1985) Faunal richness and turnover on dynamic landscapes: a simulation study. *Journal of Biogeography*, **12**, 499–508.

Seastadt, T. R. and Knapp, A. K. (1993) Consequences of nonequilibrium resource availability across multiple time scales: the transient maxima hypothesis. *American Naturalist*, **141**, 621–633.

Selander, R. K. (1966) Sexual dimorphism and differential niche utilization in birds. *The Condor*, **68**, 113–151.

Shaffer, M. L. (1981) Minimum population sizes for species conservation. *BioScience*, **31**, 131–134.

Shaffer, M. L. and Samson, F. B. (1985) Population size and extinction: a note on determining critical population sizes. *American Naturalist*, **125**, 144–152.

Shannon, C. E. and Weaver, W. (1949) *The mathematical theory of communication*. University of Illinois Press: Urbana, IL, USA.

Shapiro, A. M. (1976) Seasonal polyphenism. *Evolutionary Biology*, **9**, 259–333.

Sheehan, W. (1991) Host range patterns of hymenopteran parasitoids of exophytic lepidopteran foliovores. In *Insect–plant interactions*, Vol. III, ed. Bernays, E., pp. 209–248. CRC Press: Boca Raton FL, USA.

Shelly, T. E. (1985) Ecological comparisons of robber fly species (Diptera: Asilidae) coexisting in a neotopical rainforest. *Oecologia (Berl.)*, **67**, 57–70.

Sherry, T. W. (1984) Comparative dietary ecology of sympatric, insectivorous neotropical flycatchers (Tyrannidae). *Ecological Monographs*, **54**, 313–338.

Sherry, T. W. (1990) When are birds dietarily specialized? Distinguishing ecological from evolutionary approaches. *Advances in Avian Biology*, **13**, 337–352.

Shine, R. (1977) Habitats, diets, and symaptry in snakes: a study from Australia. *Canadian Journal of Zoology*, **55**, 1118–1128.

Shipley, B. (1987) The relationship between dynamic game theory and the Lotka–Volterra competition equations. *Journal of Theoretical Biology*, **125**, 121–123.

Shmida, A. and Wilson, M. V. (1985) Biological determinants of species diversity. *Journal of Biogeography*, **12**, 1–21.

Shorrocks, B. and Rosewell, J. (1986) Guild size in drosophilids: a simulation model. *Journal of Animal Ecology*, **55**, 527–541.

Shugart, H. H., Jr and Patten, B. C. (1972) Niche quantification and the concept of niche pattern. In *Systems analysis and simulation in ecology*, Vol II, ed. Patten, B.C., pp. 284–327. Academic Press: New York, NY, USA.

Siegismund, H. R., Loeschke, V. and Jacobs, J. (1990) Interspecific competition and components of niche width in age structured populations. *Theoretical Population Biology*, **37**, 291–319.

Sih, A. (1987) Prey refuges and predator–prey stability. Theoretical Population Biology, **31**, 1–12.

Silvertown, J. (1987) *Introduction to plant population ecology*, 2nd edn. Longman: London, UK.

Silvertown, J. and Law, R. (1987) Do plants need niches? Some recent developments in plant community ecology. *Trends in Ecology and Evolution*, **2**, 24–26.

Simberloff, D. and Abele, L. G. (1982) Refuge design and island biogeographic theory: effects of fragmentation. *American Naturalist*, **120**, 41–50.

Simberloff, D. and Boeklen, W. (1981) Santa Rosalia reconsidered: size ratios and competition. *Evolution*, **35**, 1206–1228.

Simberloff, D. and Dayan, T. (1991) The guild concept and the structure of ecological communities. *Annual Review of Ecology and Systematics*, **22**, 115–143.

Simpson, G. G. (1965) *The geography of evolution*. Chilton: Philadelphia, PA, USA.

Singer, M. C. and Parmesan, C. (1993) Sources of variations in patterns of plant–insect association. *Nature (Lond.)*, **361**, 251–253.

Sitaram, B. R. and Varma, V. S. (1984) Statistical mechanics of the Gompertz model of interacting species. *Journal of Theoretical Biology*, **110**, 253–256.

Skeate, S. T. (1987) Interactions between birds and fruits in a northern Florida hammock community. *Ecology*, **68**, 297–309.

Slagsvold, T. (1975) Habitat selection in birds: on the presence of other bird species with special reference to *Turdus pilaris*. *Journal of Animal Ecology*, **49**, 523–536.

Slatkin, M. and Lande, R. (1976) Niche width in a fluctuating environment–density independent model. *American Naturalist*, **110**, 31–55.

Slobodchickoff, C. N. and Shulz, W. C. (1980) Measures of niche overlap. *Ecology*, **61**, 1051–1055.

Smiley, J. T. (1978) Plant chemistry and the evolution of host specificity: new evidence from *Heliconius* and *Passiflora*. *Science (Wash., DC)*, **201**, 745–747.

Smith, D. C. and Douglas, A. E. (1987) *The biology of symbiosis*. Edward Arnold: London, UK.

Smith, J. N. M., Grant, P. R., Grant, B. R., Abbott, I. J. and Abbott, L. K. (1978) Seasonal variation in feeding habits of Darwin's ground finches. *Ecology*, **59**, 1137–1150.

Smith, S. M. (1978) The "underworld" in a territorial sparrow: adaptive strategy for floaters. *American Naturalist*, **112**, 571–582.

Smith, T. J., III, Chan, H. T., McIvor, C. C. and Robblee, M. B. (1989) Comparisons of seed predation in tropical, tidal forests from three continents. *Ecology*, **70**, 146–151.

Snow, B. K. and Snow, D. W. (1972) Feeding niches of hummingbirds in a Trinidad valley. *Journal of Animal Ecology*, **41**, 471–485.

Snyder, M. A. (1992) Selective herbivory by Abert's squirrel mediated by chemical variability in ponderosa pine. *Ecology*, **73**, 1730–1741.

Soberón, J. (1986) The relationship between use and suitability of resources and its consequences to insect population size. *American Naturalist*, **127**, 338–357.

Sogard, S. M., Powell, G. V. N. and Holmquist, J. G. (1989) Spatial distribution and trends in abundance of fishes residing in seagrass meadows on Florida Bay mudbanks. *Bulletin of Marine Science*, **44**, 179–199.

Solé, R. V., Bascompte, J. and Valls, J. (1992) Stability and complexity of spatially extended two-species competition. *Journal of Theoretical Biology*, **159**, 469–480.

Solomon, M. E. (1949) The natural control of animal populations. *Journal of Animal Ecology*, **18**, 1–35.

Soulé, M. and Stewart, B. R. (1970) The "niche" variation hypothesis: a test of alternatives. *American Naturalist*, **104**, 85–97.

Southwood, T. R. E., May, R. M., Hassell, M. P. and Conway, G. R. (1974) Ecological strategies and population parameters. *American Naturalist*, **108**, 791–804.

Speith, P. T. (1979) Environmental heterogeneity: a problem of contradictory selection pressures, gene flow, and local polymorphism. *American Naturalist*, **113**, 247–260.

Spence, J. A. (1983) Pattern and process in co-existence of water-striders (Heteroptera: Gerridae). *Journal of Animal Ecology*, **52**, 497–511.

Spiller, D. A. (1986) Consumptive–competition coefficients: an experimental analysis with spiders. *American Naturalist*, **127**, 604–614.

Sprules, W. G. and Bowerman, J. E. (1988) Omnivory and food chain length in zooplankton food webs. *Ecology*, **69**, 418–426.

Stamp, N. E. and Ohmart, R. D. (1978) Resource utilization by desert rodents in the lower Sonoran Desert. *Ecology*, **59**, 700–707.

Stamps, J. A. (1977) The relationship between resource competition, risk, and aggression in a tropical territorial lizard. *Ecology*, **58**, 349–358.

Starr, M. P. (1975) A generalized scheme for classifying organismic associations. In *Symbiosis*, *XXIX*, ed. Jennings, D. H. and Lee, D. L., pp. 1–20. Cambridge University Press: Cambridge, UK.

Stearns, S. C. (1989) The evolutionary significance of phenotypic plasticity. *BioScience*, **3**, 436–445.

Steenhof, K. and Kochert, M. N. (1988) Dietary responses of three raptor species to changing prey densities in a natural environment. *Journal of Animal Ecology*, **57**, 37–48.

Steinberg, P. D. and van Altena, I. (1992) Tolerance of marine invertebrate herbivores to brown algal phlorotannins in temperate Australia. *Ecological Monographs*, **62**, 189–222.

Steiner, W. W. M. (1977) Niche width and genetic variation in Hawaiian

Drosophila. *American Naturalist*, **111**, 1037–1045.
Stephens, D. W. (1987) On economically tracking a variable environment. *Theoretical Population Biology*, **32**, 15–25.
Stephens, D. W. and Krebs, J. R. (1986) *Foraging theory*. Princeton University Press: Princeton, NJ, USA.
Stevens, G. C. (1989) The latitudinal gradient in geographical range: how so many species coexist in the tropics. *American Naturalist*, **133**, 240–256.
Stiling, P. D. (1980) Competition and coexistence among *Eupteryx* leafhoppers. *Journal of Animal Ecology*, **49**, 793–805.
Stiling, P. D. and Strong, D. R. (1983) Weak competition among *Spartina* stem borers by means of murder. *Ecology*, **64**, 770–778.
Stockoff, B. A. (1993) Diet heterogeneity: implications for growth of a generalist herbivore, the Gypsy Moth. *Ecology*, **74**, 1939–1949.
Storey, K. B. (1990) Life in a frozen state: adaptive strategies for natural freeze tolerance in amphibians and reptiles. *American Journal of Physiology*, **258**, R559–R568.
Strebel, D. E. (1985) Environmental fluctuations and extinction – single species. *Theoretical Population Biology*, **27**, 1–26.
Strobeck, C. (1973) *N* species competition. *Ecology*, **54**, 650–654.
Strong, D. R., Lawton, J. H. and Southwood, T. R. E. (1984) *Insects on plants: community patterns and mechanisms*. Blackwell Scientific: Oxford, UK.
Sugihara, G. (1980) Minimal community structure: an explanation of species abundance patterns. *American Naturalist*, **116**, 770–787.
Svärdson, G. (1949) Competition and habitat selection in birds. *Oikos*, **1**, 157–174.
Sweatman, H. P. A. (1985) The influence of adults of some coral reef fishes on larval recruitment. *Ecological Monographs*, **55**, 469–485.
Swift, S. M., Racey, P. A. and Avery, M. I. (1985) Feeding ecology of *Pipistrellus pipistrellus* (Chiroptera: Vespertilionidae) during pregnancy and lactation. II. Diet. *Journal of Animal Ecology*, **54**, 217–225.
Taghon, G., Houston, A. I., Belovsky, G. E. and McNamara, J. M. (1990) Can there be a general theory of diet selection? In *Behavioural mechanisms of food selection, 20*, ed. Hughes, R. N., pp. 863–864. Springer-Verlag: Berlin, Germany.
Talbot, J. J. (1979) Time budget, niche overlap, inter- and intraspecific aggression in *Anolis humilis* and *A. limifrons* from Costa Rica. *Copeia*, **1979**, 472–481.
Tanaka, K. (1991) Food consumption and diet composition in the web-building spider *Agelena limbata* in two habitats. *Oecologia (Berl.)*, **86**, 8–15.
Taper, M. L. and Case, T. J. (1985) Quantitative genetic models for the coevolution of character displacement. *Ecology*, **66**, 355–371.
Taylor, A. A., Se-Felice, J. and Havill, D. C. (1982) Seasonal variation in nitrogen availability and utilization in an acidic and calcareous soil. *New Phytologist*, **92**, 141–152.
Taylor, C. E., Jefferson, D. R., Turner, S. R. and Goldman, S. R. (1989) RAM: artificial life for the exploration of complex biological systems. In *Artificial life. The proceedings of an interdisciplinary workshop on the synthesis and simulation of living systems*, ed. Langton, C., pp. 275–295. Addison-Wesley: Redwood City, CA, USA.
Taylor, D. R. and Aarssen, L. W. (1990) Complex competitive relationships among

genotypes of three perennial grasses: implications for species coexistence. *American Naturalist*, **136**, 305–327.

Taylor, P. D. and Jonker, L. B. (1978) Evolutionary stable strategies and game dynamics. *Mathematical Biosciences*, **40**, 145.

Taylor, P. J. (1988a) Consistent scaling and parameter choice for linear and generalized Lotka–Volterra models used in community ecology. *Journal of Theoretical Biology*, **135**, 543–568.

Taylor, P. J. (1988b) The construction and turnover of complex community models having generalized Lotka–Volterra dynamics. *Journal of Theoretical Biology*, **135**, 569–588.

Taylor, R. A. J. and Taylor, L. R. (1979) A behavioural model for the evolution of spatial dynamics. In *Population dynamics*, ed. Anderson, R.M., Turner, B.D. and Taylor, L.R., pp. 1–27. Blackwell Scientific Publications: Oxford, UK.

Templeton, A. R. and Rothman, E. D. (1974) Evolution in heterogeneous environments. *American Naturalist*, **108**, 409–428.

Templeton, A. R. and Rothman, E. D. (1981) Evolution in fine-grained environments. II. Habitat selection as a homeostatic mechanism. *Theoretical Population Biology*, **19**, 326–340.

Terborgh, J. (1985) The role of ecotones in the distribution of Andean birds. *Ecology*, **66**, 1237–1246.

Terborgh, J. and Faaborg, J. (1973) Turnover and ecological release in the avifaua of Monas Island, Puerto Rico. *Auk*, **90**, 759–779.

Terborgh, J., Faaborg, J. and Brockmann, H. J. (1978) Island colonization by Lesser Antillean birds. *Auk*, **95**, 59–72.

Thomas, C. D., Vasco, D., Singer, M. C., Ng, D., White, R. R. and Hinkley, D. (1990) Diet divergence in two sympatric congeneric butterflies: community or species level phenomenon? *Evolutionary Ecology*, **4**, 67–74.

Thompson, J. N. (1982) *Interaction and coevolution*. Wiley-Interscience: New York, NY, USA.

Thompson, J. N. (1988a) Coevolution and alternative hypotheses on insect/plant interactions. *Ecology*, **69**, 893–895.

Thompson, J. N. (1988b) Variation in interspecific interactions. *Annual Review of Ecology and Systematics*, **19**, 65–87.

Thompson, J. N. (1989) Concepts of coevolution. *Trends in Ecology and Evolution*, **4**, 179–183.

Thompson, J. N. and Burdon, J. J. (1992) Gene-for-gene coevolution between plants and parasites. *Nature (Lond.)*, **360**, 121–125.

Thompson, J. N. and Pellmyr, O. (1991) Evolution of oviposition behavior and host preference in Lepidoptera. *Annual Review of Entomology*, **36**, 65–89.

Thompson, J. N. and Pellmyr, O. (1992) Mutualism with pollinating seed parasites amid co-pollinators: constraints on specialization. *Ecology*, **73**, 1780–1791.

Thompson, J. N. and Willson, M. F. (1979) Evolution of temperate fruit/bird interactions: phenological strategies. *Evolution*, **33**, 973–982.

Thorman, S. and Wiederholm, A-M. (1986) Food, habitat and time niches in a coastal fish species assemblage in a brackish water bay in the Bothnian Sea, Sweden. *Journal of Experimental Marine Biology and Ecology*, **95**, 67–86.

Tilman, D. (1982) *Resource competition and community structure*. Princeton University

Press: Princeton, NJ, USA.

Tilman, D. (1985) The resource-ratio hypothesis of plant succession. *American Naturalist*, **125**, 827–852.

Tilman, D. (1986) A consumer–resource approach to community structure. *American Zoologist*, **26**, 5–22.

Tilman, D. (1988) *Plant strategies and the dynamics and structure of plant communities.* Princeton University Press: Princeton, NJ, USA.

Tilman, D. (1994) Competition and biodiversity in spatially structured habitats. *Ecology*, **75**, 2–16.

Toft, C. A. (1985) Resource partitioning in amphibians and reptiles. *Copeia*, **1985**, 1–21.

Tokeshi, M. (1993) Species abundance patterns and community structure. *Advances in Ecological Research*, **24**, 111–186.

Tracy, C. R. and George, T. L. (1992) On the determinants of extinction. *American Naturalist*, **139**, 102–122.

Travis, C. C., Post, W. M., DeAngelis, D. L. and Perkowski, J. (1980) Analysis of compensatory Leslie matrix models for competing species. *Theoretical Population Biology*, **18**, 16–30.

Trivelpiece, W. Z., Trivelpiece, S. G. and Volkman, N. J. (1987) Ecological segregation of Adelie, Gentoo, and Chinstrap Penguins at King George Island, Antarctica. *Ecology*, **68**, 351–361.

Trowbridge, C. D. (1991) Diet specialization limits herbivorous sea slug's capacity to switch among food species. *Ecology*, **72**, 1880–1888.

Tschumy, W. O. (1982) Competition between juveniles and adults in age-structured populations. *Theoretical Population Biology*, **21**, 255–268.

Tugwell, S. and Branch, G. M. (1992) Effects of herbivore gut surfactants on kelp polyphenol defenses. *Ecology*, **73**, 205–215.

Tuljapurkar, S. D. and Orzack, S. H. (1980) Population dynamics in variable environments I. Long-run growth rates and extinction. *Theoretical Population Biology*, **18**, 314–342.

Turelli, M. (1977) Random environments and stochastic calculus. *Theoretical Population Biology*, **12**, 140–178.

Turelli, M. (1981) Niche overlap and invasion of competitors in random environments I. Models without demographic stochasticity. *Theoretical Population Biology*, **20**, 1–56.

Turelli, M. and Gillespie, J. H. (1980) Conditions for the existence of stationary densities for some two-dimensional diffusion processes with applications to population biology. *Theoretical Population Biology*, **17**, 167–189.

Underwood, A. J. and Denley, E. J. (1984) Paradigms, explanations and generalizations in models for the structure of intertidal communities on rocky shores. In *Ecological communities: conceptual issues and the evidence*, ed. Strong, D. R. Jr., Simberloff, D., Abele, L.G. and Thistle, A.B., pp. 151–180. Princeton University Press: Princeton, NJ, USA.

Underwood, A. J. and Fairweather, P. G. (1989) Supply-side ecology and benthic marine assemblages. *Trends in Ecology and Evolution*, **4**, 16–19.

Underwood, T. (1986) The analysis of competition by field experiments. In *Community ecology: pattern and process*, ed. Kikkawa, J. and Anderson, D.J., pp. 240–268. Blackwell Scientific Publications: Melbourne, Australia.

Vadas, R. L. (1990) The importance of omnivory and predator regulation of prey in freshwater fish assemblages of North America. *Environmental Biology of Fishes*, **27**, 285–302.

Van Buskirk, J. (1992) Competition, cannibalism, and size-class dominance in a dragonfly. *Oikos*, **65**, 455–464.

Van Horne, B. (1983) Density as a misleading indicator of habitat quality. *Journal of Wildlife Management*, **47**, 893–901.

Van Horne, B. and Ford, R. G. (1982) Niche breadth analysis based on discriminant analysis. *Ecology*, **63**, 1172–1174.

Van Valen, L. and Grant, P. R. (1970) Variation and niche width reexamined. *American Naturalist*, **104**, 587–590.

Vance, R. R. (1978) Predation and resource partitioning in one predator–two prey model communities. *American Naturalist*, **112**, 797–813.

Vance, R. R. (1985) The stable coexistence of two competitors for one resource. *American Naturalist*, **126**, 72–86.

Vandermeer, J. H. (1972) Niche theory. *Annual Review of Ecology and Systematics*, **3**, 107–132.

Vanderplank, J. E. (1978) *Genetic and molecular basis of plant pathogenesis*. Springer-Verlag: Berlin, Germany.

Vanderploeg, H. A., Paffenhöfer, G-A. and Liebig, J. R. (1990) Concentration-variable interactions between calanoid copepods and particles of different food quality: observations and hypotheses. In *Behavioural mechanisms of food selection, 20*, ed. Hughes, R.N., pp. 595–613. Springer-Verlag: Berlin, Germany.

Verlinden, C. and Wiley, R. H. (1989) The constraints on digestive rate: an alternative model of diet selection. *Evolutionary Ecology*, **3**, 264–273.

Via, S. (1987) Genetic constraints on the evolution of phenotypic plasticity. In *Genetic constraints on adaptive evolution*, ed. Loeschcke, V., pp. 47–71. Springer-Verlag: Berlin, Germany.

Via, S. (1991) Specialized host plant performance of pea aphid clones is not altered by experience. *Ecology*, **72**, 1420–1427.

Villa, F. (1992) New computer architectures as tools for ecological thought. *Trends in Ecology and Evolution*, **7**, 179–183.

Vitt, L. J., van Loben Sels, R. C. and Ohmart, R. D. (1981) Ecological relationships among arboreal desert lizards. *Ecology*, **62**, 398–410.

Wagner, J. L. (1981) Seasonal change in guild structure of oak woodland insectivorous birds. *Ecology*, **62**, 973–981.

Wainwright, P. C. (1988) Morphology and ecology: functional basis of feeding constraints in Caribbean labrid fishes. *Ecology*, **69**, 635–645.

Wainwright, S. J. (1980) Plants in relation to salinity. *Advances in Botanical Research*, **8**, 221–261.

Waldorf, E. S. (1976) Spider size, microhabitat selection, and use of food. *American Midland Naturalist*, **96**, 76–87.

Walter, D. E. and O'Dowd, D. J. (1992a) Leaf morphology and predators: effect of domatia on the distribution of phytoseiid mites (Acari: Phytoseiidae). *Environmental Entomology*, **21**, 478–484.

Walter, D. E. and O'Dowd, D. J. (1992b) Leaves with domatia have more mites. *Ecology*, **73**, 1514–1518.

Walton, W. E., Hairston, N. G., Jr and Wetterer, J. K. (1992) Growth-related constraints on diet selection by sunfish. *Ecology*, **73**, 429–437.

Ward, M. A. and Thorpe, J. P. (1989) Assessment of space utilization in a subtidal temperate bryozoan community. *Marine Biology*, **103**, 215–224.

Waser, N. M. and Real, L. A. (1979) Effective mutualism between sequentially flowering plant species. *Nature (Lond.)*, **281**, 670–672.

Wcislo, W. T. (1989) Behavior as a pacemaker of evolution. *Annual Review of Ecology and Systematics*, **20**, 137–169.

Wecker, S. C. (1963) The role of early experience in habitat selection in the Prairie Deermouse *Peromyscus maniculatus bairdi*. *Ecological Monographs*, **33**, 307–325.

Weider, L. J. and Hebert, P. D. N. (1987) Ecological and physiological differentiation among low-Arctic clones of *Daphnia pulex*. *Ecology*, **68**, 188–198.

Werner, D. (1992) *Symbiosis of plants and microbes*. Chapman and Hall: London, UK.

Werner, E. E. (1988) Size, scaling, and the evolution of complex life cycles. In *Size-structured populations: ecology and evolution*, ed. Ebenman, B. and Persson, L., pp. 60–81. Springer: Berlin, Germany.

Werner, E. E. (1992) Individual behavior and higher-order species interactions. *American Naturalist*, **140**, S5–S32.

Werner, E. E. and Gilliam, J. F. (1984) The ontogenetic niche and species interactions in size structured populations. *Annual Review of Ecology and Systematics*, **15**, 393–425.

Werner, E. E., Gilliam, J. F., Hall, D. J. and Mittelbach, G. G. (1983) An experimental test of the effects of predation risk on habitat use in fish. *Ecology*, **64**, 1525–1539.

Werner, E. E. and Hall, D. J. (1974) Optimal foraging and size-selection of prey by the blue sunfish (*Lepomis macrochirus*). *Annual Review of Ecology and Systematics*, **15**, 393–425.

Werner, P. A. and Platt, W. J. (1976) Ecological relationships of co-occurring goldenrods (*Solidago*: Compositae). *American Naturalist*, **110**, 959–971.

Werner, T. K. and Sherry, T. W. (1987) Behavioral feeding specialization in *Pinaroloxias inornata*, the "Darwin's Finch" of Cocos Island, Costa Rica. *Proceedings of the National Academy of Science USA*, **84**, 5506–5510.

West, L. (1988) Prey selection by the tropical snail *Thais melones*: a study of interindividual variation. *Ecology*, **69**, 1839–1854.

West-Eberhard, M. J. (1989) Phenotypic plasticity and the origins of diversity. *Annual Review of Ecology and Systematics*, **20**, 249–278.

Westoby, M. (1974) An analysis of diet selection by large generalist herbivores. *American Naturalist*, **108**, 290–304.

Westoby, M. (1978) What are the biological bases of varied diets? *American Naturalist*, **112**, 627–631.

Wheeler, D. E. (1991) The developmental basis of worker caste polymorphism in ants. *American Naturalist*, **138**, 1218–1238.

Wheelwright, N. T. (1985) Fruit size, gape width, and the diets of fuit-eating birds. *Ecology*, **66**, 808–818.

Whiteside, M. C., Williams, J. B. and White, C. P. (1978) Seasonal abundance and pattern of Chydorid, Cladocera in mud and vegetative habitats. *Ecology*, **59**, 1177–1188.

Whitfield, P. (1990) Individual feeding specializations of wintering turnstone *Arenaria interpres*. *Journal of Animal Ecology*, **59**, 193–211.

Whittaker, R. H. (1975) *Communities and ecosystems*. Macmillan: New York, NY, USA.

Whittam, T. S. and Siegel-Causey, D. (1981) Species interactions and community structure in Alaskan seabird colonies. *Ecology*, **62**, 1515–1524.

Wiegmann, B. M., Mitter, C. and Farrell, B. (1993) Diversification of carnivorous parasitic insects: extraordinary radiation or a specialized dead end? *American Naturalist*, **142**, 737–754.

Wiens, J. A. (1977) On competition and variable environments. *American Scientist*, **65**, 590–597.

Wiens, J. A. (1984) Resource systems, populations and communities. In *A new ecology: novel approaches to interactive systems*, ed. Price, P. W., Slobodchikoff, C. N. and Gaud, W. S., pp. 397–436. Wiley: New York, NY, USA.

Wiens, J. A. (1989a) *The ecology of bird communities. 1. Foundations and patterns*. Cambridge University Press: Cambridge, UK.

Wiens, J. A. (1989b) *The ecology of bird communities. 2. Processes and variations*. Cambridge University Press: Cambridge, UK.

Wiens, J. A., Rotenberry, J. T. and Van Horne, B. (1987) Habitat occupancy patterns of North American shrubsteppe birds: the effects of spatial scale. *Oikos*, **48**, 132–147.

Wilbur, H. M. (1972) Competition, predation and the structure of the *Amblystoma–Rana sylvatica* community. *Ecology*, **53**, 3–21.

Wilbur, H. M. (1980) Complex life cycles. *Annual Review of Ecology and Systematics*, **11**, 67–93.

Wilbur, H. M. and Collins, J. P. (1973) Ecological aspects of amphibian metamorphosis. *Science (Wash., DC)*, **182**, 1305–1314.

Williams, P. H. (1988) Habitat use by bumble bees. *Ecological Entomology*, **13**, 223–237.

Wilson, D. S. (1975) The adequacy of body size as a niche difference. *American Naturalist*, **109**, 769–784.

Wilson, D. S. (1992) Complex interactions in metacommunities, with implications for biodiversity and higher levels of selection. *Ecology*, **73**, 1984–2000.

Wilson, D. S. and Turelli, M. (1986) Stable underdominance and the evolutionary invasion of empty niches. *American Naturalist*, **127**, 835–850.

Wilson, J. B. and Agnew, A. D. Q. (1992) Positive-feedback switches in plant communities. *Advances in Ecological Research*, **23**, 263–336.

Winemiller, K. O. (1990) Spatial and temporal variation in tropical fish trophic networks. *Ecological Monographs*, **60**, 331–367.

Wissel, C. and Stöcker, S. (1991) Extinction of populations by random influences. *Theoretical Population Biology*, **39**, 315–328.

Wissinger, S. A. (1992) Niche overlap and the potential for competition and intraguild predation between size-structured populations. *Ecology*, **73**, 1431–1444.

Woinarski, J. C. Z., Tidemann, S. C. and Kerin, S. (1988) Birds in a tropical mosaic: the distribution of bird species in relation to vegetation patterns. *Australian Wildlife Research*, **15**, 171–196.

Wolda, H. and Roubik, D. W. (1986) Nocturnal bee abundance and seasonal bee activity in a Panamanian forest. *Ecology*, **67**, 426–433.

Wootton, J. T. (1993) Indirect effects and habitat use in an intertidal community:

interaction chains and interaction modifications. *American Naturalist*, **141**, 71–89.
Wootton, J. T. (1994) Predicting direct and indirect effects: an integrated approach using experiments and path analysis. *Ecology*, **75**, 151–165.
Yan, A. (1989) Host specificity of *Lysiana exocarpi* subsp. *exocarpi* and other mistletoes in southern South Australia. *Australian Journal of Botany*, **38**, 475–486.
Yeaton, R. I. and Cody, M. L. (1974) Competitive release in island song sparrow populations. *Theoretical Population Biology*, **5**, 42–58.
Yodzis, P. (1984) How rare is omnivory? *Ecology*, **65**, 321–323.
Yodzis, P. (1986) Competition, mortality, and community structure. In *Community Ecology*, ed. Diamond, J. and Case, T. J., pp. 480–491. Harper and Row: New York, NY, USA.
Yodzis, P. (1988) The indeterminacy of ecological interactions as perceived through perturbation experiments. *Ecology*, **69**, 508–515.
Yokoyama, S. and Schaal, B. A. (1985) A note on multiple-niche polymorphisms in plant populations. *American Naturalist*, **125**, 158–163.
Yoshiyama, R. M. and Roughgarden, J. (1977) Species packing in two dimensions. *American Naturalist*, **111**, 107–121.
Young, J. P. W. and Johnston, A. W. B. (1989) The evolution of specificity in the legume–Rhizobium symbiosis. *Trends in Ecology and Evolution*, **4**, 341–349.
Yu, D. S., Luck, R. F. and Murdoch, W. W. (1990) Competition, resource partitioning and coexistence of an endoparasite *Encarsia perniciosi* and an ectoparasite *Aphytis melinus* of the California red scale. *Ecological Entomology*, **15**, 469–480.
Zicarelli, J. (1975) Mathematical analysis of a population model with several predators on a single prey. University of Minnesota: PhD thesis.
Zobel, M. (1992) Plant species coexistence – the role of historical, evolutionary and ecological factors. *Oikos*, **65**, 314–320.
Zwarts, L. and Esselink, P. (1989) Versatility of male Curlews *Numenius arquata* upon *Nereis diversicolor*: deploying contrasting capture modes dependent on prey availability. *Marine Ecology Progress Series*, **56**, 255–269.

Index